Advances in Science, Technology & Innovation

IEREK Interdisciplinary Series for Sustainable Development

Advances in Science, Technology & Innovation (ASTI) is a series of peer-reviewed books based on the best studies on emerging research that redefines existing disciplinary boundaries in science, technology and innovation (STI) in order to develop integrated concepts for sustainable development. The series is mainly based on the best research papers from various IEREK and other international conferences, and is intended to promote the creation and development of viable solutions for a sustainable future and a positive societal transformation with the help of integrated and innovative science-based approaches. Offering interdisciplinary coverage, the series presents innovative approaches and highlights how they can best support both the economic and sustainable development for the welfare of all societies. In particular, the series includes conceptual and empirical contributions from different interrelated fields of science, technology and innovation that focus on providing practical solutions to ensure food, water and energy security. It also presents new case studies offering concrete examples of how to resolve sustainable urbanization and environmental issues. The series is addressed to professionals in research and teaching, consultancies and industry, and government and international organizations. Published in collaboration with IEREK, the ASTI series will acquaint readers with essential new studies in STI for sustainable development.

More information about this series at http://www.springer.com/series/15883

Chaham Alalouch · Hassan Abdalla ·
Emmanuel Bozonnet · George Elvin ·
Oscar Carracedo
Editors

Advanced Studies in Energy Efficiency and Built Environment for Developing Countries

Proceedings of IEREK Conferences: Improving Sustainability Concept in Developing Countries (ISCDC-2), Egypt 2017 and Alternative and Renewable Energy Quest in Architecture and Urbanism (AREQ-2), Spain 2017

Springer

Editors
Chaham Alalouch
College of Engineering
Sultan Qaboos University
Muscat, Oman

Hassan Abdalla
School of Architecture, Computing and Engineering
University of East London
London, UK

Emmanuel Bozonnet
Faculty of Science
University of La Rochelle
La Rochelle, France

George Elvin
School of Architecture
North Carolina State University
Raleigh, NC, USA

Oscar Carracedo
School of Design and Environment
National University of Singapore
Singapore, Singapore

Series Editor
Mourad Amer
International Experts for Research Enrichment and Knowledge Exchange (IEREK)
Cairo, Egypt

ISSN 2522-8714 ISSN 2522-8722 (electronic)
Advances in Science, Technology & Innovation
IEREK Interdisciplinary Series for Sustainable Development
ISBN 978-3-030-10855-7 ISBN 978-3-030-10856-4 (eBook)
https://doi.org/10.1007/978-3-030-10856-4

Library of Congress Control Number: 2019931959

Cover image by IEREK (M. Amer)

This Springer imprint is published by the registered company Springer Nature Switzerland AG
The registered company address is: Gewerbestrasse 11, 6330 Cham, Switzerland

Conference Scientific Committee

Agis Papdopoulos, Professor, Director of the Process Equipment Design Laboratory, Energy Department, Laboratory of Process Equipment Design, Aristotle University of Thessaloniki, Greece

Ferdinando Trapani, Professor of Architecture and Urban Planning, School of Architecture, Palermo University, Italy

Dinko Vukadinović, Full Professor, University of Split, Faculty of Electrical Engineering, Mechanical Engineering and Naval Architecture, Boškovića, Croatia

Chaham Alalouch, Assistant Professor in Architectural Engineering, College of Engineering, Sultan Qaboos University, Muscat, Oman

Randa A. Mahmoud, Associate Professor at the French University of Egypt UFE, Ain Shams University, Urban Planning and Design Department

Nabil Mohareb, Associate Professor, Head of Faculty of Architectural Engineering, Arts and Design Program Branch, Faculty of Architectural Engineering, and Arts and Design Program, Beirut Arab University, Tripoli Campus

Antonio Frattari, Professor of Building Construction and Wooden Architecture, Faculty of Engineering, University of Trento, Italy, and the Chief of the Laboratory of Building Design (LBD)—Chief of CUnEdI

Iman M. A. Amad, Director, Unit of Architectural Conservation, Urban Planning and Risk Reduction Centre Associate Professor of Architecture, An-Najah National University

Mohamed Mandour, Professor of Urban Design and Planning at Faculty of Architecture at Helwan University, Egypt

Ahmed Shehata, Professor of Urban and Environmental Design at Effat University

Veronica Lucia Castaldo, Master Degree in Building Engineering Professional Engineer Ph.D. candidate in Energy Engineering, Columbia University

Wael Mohamed Youssef, Professor of Urban Planning, Faculty of Engineering at Cairo, Al-Azhar University, Egypt

Yasser Mahgoub, Chairman of the Department of Architecture and Urban Planning, College of Engineering, Qatar University

Ahmed Hoseny Radwan, Professor of Architecture, College of Fine Arts Helwan University

Daniele Ronsivalle, Ph.D. Architect, Urban Planning Researcher at the Department of Architecture of the University of Palermo

Lotfy Azaz, Associate Professor of GIS and Remote Sensing Applications in Urban Planning

Ashraf Kamal, Professor of Architecture and Urban Planning Housing and Building National Research Center, Egypt

Preface

The world's population is continuously growing causing a multitude of environmental challenges such as global climate change, excessive fossil fuel dependency, significant increase in demand for energy, unprecedented high consumption rates, pollution, and serious environmental degradation. These challenges are major challenges in this current century and are likely to be among the greatest and most serious issues that humanity will encounter in the coming centuries. In fact, the negative impact of these environmental problems is already noticeable. Humanity has to immediately act to mitigate the impact of these challenges. In this regard, creating sustainable buildings and cities and using renewable energy resources are fundamental principles that should guide the design of our future built environment. This book addresses some issues related to sustainability in the developing countries. It is a necessary resource for researchers, planners, urban designers, architects, policy makers, and those who have a role in shaping our built environment. The aim of this book is to contribute to the creation of a more sustainable built environment in the developing countries by careful consideration from the early design stage.

Being a culmination of highly regarded research papers that aim to educate readers about the importance of improving sustainability in developing countries, this book attempts to emphasize, enhance, and spread the role of urban designers and architects in improving the quality of the built environment in countries where the sustainability concept is still to be a common stream. The book discusses important issues related to sustainability in the developing countries such as energy-saving measures in buildings and cities, the potential of renewable energy resources, and the use of efficient and sustainable building materials along with other design issues and technical solutions that could help to enhance the environmental efficiency. Nonetheless, the book discusses the role of digital tools and simulation in the design process of a sustainable built environment.

The book is a compilation of carefully selected papers participated in the conferences organized by IEREK in different parts of the World. All chapters of this book were subjected to rigorous and careful blinded peer-review process as well as a review process by the book editors and book series editors.

This book is a valuable resource for scholars looking for cutting-edge research and inspiration for future research agenda. It is also a guide for academics who are looking to update their knowledge in the fields of sustainability, urban planning, and renewable energy. The research and materials in this book are also directed to those who are actively engaged in the decision-making processes and at a heterogeneous audience who has an interest to critically examine the new literature available in the field.

A special mention should be made to the editors of this book and to all the authors and co-authors of the chapters who collectively provided the academic community with unique and increasingly valuable literature.

Muscat, Oman — Chaham Alalouch
London, UK — Hassan Abdalla
La Rochelle, France — Emmanuel Bozonnet
North Carolina, USA — George Elvin
Singapore — Oscar Carracedo

Acknowledgements

We would like to thank the authors of the research papers that were chosen to be added to this book. We would also like to thank the scientific committee of reviewers who helped us select these papers and the editors of this book. Lastly, a special thanks go to the IEREK team for supporting the publication of the best research papers submitted to the conference.

Muscat, Oman	Chaham Alalouch
London, UK	Hassan Abdalla
La Rochelle, France	Emmanuel Bozonnet
North Carolina, USA	George Elvin
Singapore	Oscar Carracedo

Contents

About the Editors

Chaham Alalouch is architect, researcher, and educator with a background in urbanism. He is concerned about how to improve people's quality of life through sustainable and smart built environment. He believes that the built environment (building and cities) is made for people and it should be designed and constructed as such. His research was recognized for excellence. His Ph.D. was awarded the MacFarlane Prize by Heriot-Watt University in the UK, and one of his peer-reviewed journal papers was awarded the Highly Commended Award at the Emerald Literati Network Awards for Excellence, UK. He has authored, co-authored and edited a large number of articles, technical reports, and books covering a wide range of topics related to his broad area of interest including healthcare building design, privacy, spatial experience of space users, space syntax, restorative environments, stakeholder's involvement in the design process, and passive and sustainable design and construction. He is also a member of Oman Eco-House Project and currently undertaking research in architectural pedagogy and "livability" of eco-architecture. He held invited talks and keynote speeches at several international conferences. He also acts as a reviewer for a range of reputable journals and as a scientific responsible for a number of international design schools, demonstrating academic leadership. He is currently teaching and researching at Sultan Qaboos University in Oman, having previously researched and taught at universities in the UK and Syria. There, he also worked in several architectural consultancy firms and in two European Commission-funded projects. He is Series Editor responsible for book proposals submitted in the field of sustainable architecture and urbanism and architectural heritage.

Hassan Abdalla is International Authority in the field of sustainability, engineering design, factories of the future, artificial intelligence in design, and future smart cities. He has over 20 years of experience in both academia and industry. He is Fellow of the Royal Society for the Encouragement of Arts, Manufacture and Commerce, Fellow of the IMechE, and Principal Fellow of the Higher Education Academy. He was awarded a Medal of Appreciation by the IMS Centre in Tokyo, which is run under the auspices of the Japanese Government, in recognition of his innovation and substantial contribution to the international collaboration with Kajima Corporation, Hazama Ando Corporation, The University of Tokyo, and Waseda University. He is Author and Co-Author of over 100 papers published in international journals and conferences worldwide. He is currently Dean of the School of Architecture, Computing and Engineering (ACE) at the University of East London and Visiting Professor at Cranfield University. He is the Series Editor responsible for evaluating book proposals submitted in the field of sustainability and artificial intelligence in design.

Emmanuel Bozonnet is Associate Professor at the University of La Rochelle in France. He has more than 10 years of academic and professional experience in building engineering physics and civil engineering. He specialized in building energy in his studies on building interactions with urban microclimate and on urban street canyons on a district scale. This modeling work has been corroborated by various scale experiments from the building envelope component to the development of a reduced-scale mock-up for urban canyons and in situ buildings. Moreover, he has more than 50 publications in the field, including articles, book chapters, and academic papers. His publications and studies focus on the development of cooling strategies to mitigate building energy demand and urban heat islands. He is also involved in various scientific conference committees and in project management. He contributes to the board of IBPSA France (International Building Performance Simulation Association) and to the board of the European Cool Roof Council (ECRC). He is also Co-Leader of the group on urban microclimate and energy in the research federation IRSTV (Institute for Research on Urban Sciences and Techniques, France). He is Series Editor responsible for evaluating book proposals submitted in the field of urban planning, sustainable development and energy management.

George Elvin's research and teaching focus on emerging design strategies and technologies for green building. His current emphasis is on Architecture for Extreme Environments. This project takes students to some of the world's most extreme environments to develop design strategies that can help us adapt to a rapidly changing climate. In January 2018, the Web site won a Flux Student Design Award from AIGA Blue Ridge. He is also Founder of Gone Studio, the company pioneering post-petroleum design and zero-plastic, zero-waste, zero-electricity manufacturing. His designs have been featured by the Discovery Channel, Macworld, Treehugger, and over fifty other green design venues. He is Author of over forty books and articles, and his works on green design and technology have been published by Routledge, Wiley, and Princeton Architectural Press, among others. His most recent book, Post-Petroleum Design, has been called "a must read for anyone wondering how we built our plastics economy—and how we might move beyond it." His talks and workshops have inspired audiences in Europe, Asia, the Middle East, and the Americas for decades. He spent his childhood losing himself in the 600 acres of forest that stretched from his back door to the Potomac River. During high school, he worked for the CIA, although it was in the cafeteria at their Langley headquarters. After high school, as his friends went off to college, Dr. Elvin went to train at a Zen Buddhist monastery. In 1981, he founded his own design-build firm in Washington DC, and in 1998, he received his Ph.D. in architecture from the University of California at Berkeley. He has been Assistant Professor of Architecture at the University of Illinois at Urbana-Champaign and Visiting Fellow at the University of Edinburgh's Institute for Advanced Studies in the Humanities.

Oscar Carracedo is Architect, Urbanist, and Educator, and currently Assistant Professor at the Department of Architecture, National University of Singapore. He is Director of the DRIA-Designing Resilience in Asia International Research Programme and Director of the InnerHoods Lab, where he develops his research on resilience, sustainability, integrated urban planning, and informal urbanism. His expertise also includes health cities, and the relationship between urban development and well-being, urban regeneration and development, with a special focus on developing countries. He is also Co-Founder Director of CSArchitects, an urban planning, urban design, and architecture firm based in Barcelona, Spain. Spanning over 17 years of international professional experience, he has been responsible for more than 30 masterplans, an extensive number of projects and consultancies in urban design, site, physical and spatial planning, architecture and public spaces, as well as many projects with underprivileged communities. He has won two national urban planning and design prizes, more than 40 national and international competitions, and his work and research have been awarded and published nationally and internationally. He is Author of numerous books and articles, and drawn on his research, he has recently published "Ibid. In the same place", where he explores on-site resilient revitalization strategies for low-income neighborhoods, "Indushoods. From industries to Neighbourhoods" a study on how to renew and reposition industrial areas, and "Naturban" a reflection about the relationship and integration between urban and natural milieu in cities.

Part I
Resilience in the Built Environment

The first part of this book heavily and critically focuses on urban environments. It starts with chapters focusing on Egypt's urban environment which discusses ways to mitigate various deficiencies in urban processes that have caused a catastrophic decline in the Egyptian urban situation. Given Egypt's status as a developing country, the authors have narrowed down possible factors that could have been contributing to the decline in the urban environment and have put forth a variety of solutions ranging from the use of geographic information systems to analyzing institutional frameworks within the country. The part precedes with chapters addressing similar issues in other developing countries such as Ecuador.

In the first chapter titled "Governing the Transition of RE Electrification in Egypt", the author provided an analysis of the current institutional framework of the Supreme Council of Energy that governs the transition to renewable energy sources in Egypt. The analysis relies on the Network Governance Theory and the Transition Management Theory to understand the institutional dynamics of actors in charge of renewable energy transition in Egypt. In another chapter titled "Contemporary Urban Centres as a Mean Towards Sustainability", the author also focuses her research on ways to improve sustainability in Egypt. The chapter analyzes urban construction in the country and provides an analytical narrative for the concept of mixed-use development to highlight ways to use urban centers to improve sustainability.

In the following chapter titled "Geographic Information System as a Support Tool for Planners to Choose the Best Suitable Location for the Development of New Economic Cities", the authors also analyze the Egypt's current urban situation and attempts to provide solutions using geographic information systems (GIS). The authors highlight that only 7.6% of Egypt's land is cultivated and inhabited causing overcrowding and inefficiency in those areas and leaving 92.4% of the land unused. They propose GIS-based model that can be used support planners and local policy-makers in choosing locations for new city. The chapter tests the feasibility of the model in Sinai Peninsula.

In chapter four titled "The Energy Poverty Situation: A Review of Developing Countries", the author reviews inadequacies in energy provision in South Asia, sub-Saharan Africa, and East Asia. The author highlights the weaknesses in energy systems contributing to energy poverty at a national level and discusses overlooked issues that translate to energy poverty in households and redefines energy poverty while providing various solutions to this issue.

In the chapters titled "The Recent Legislation for Eco-efficient Design in Quito, Ecuador: A Review and Case Study" and "R.I.S.Q: RIsk aSsessment Tool for Quito", the author heavily focuses on Quito, Ecuador, in providing solutions to improve sustainability. In a quest to map out eco-efficient design in Quito, the author analyzes how the new resolution in eco-efficient design for Quito helps out this task and becomes a versatile instrument, part of a broader tool, in contributing to the goals of eco-efficient design. The author also attempts to identify vulnerable settlements and map out disaster potential using risk index for decision making.

Aiming to optimize the use of urban space and making cities more resilient, the author of the chapters titled "Resilient Thinking: An Important Tool to Contribute to Sustainable Development" and "Car Space Recycling Algorithm, a Powerful Urban Planning and Decision Making Tool" identifies and reviews the literature on resilience and links it to recent debits on sustainability. The authors also propose a better use for the urban space allocated for private automobiles to allow for a more sustainable, livable city.

Governing the Transition of RE Electrification in Egypt

Farah Shoukry

Abstract

"Governing the Transition Toward RE Electrification in Egypt" investigates the institutional challenges that hinder the achievement of the RE 2020 targets, announced by the Supreme Council of Energy (SCE) back in 2008. The objective of this research is to explain how the institutional framework is arranged that govern the renewable energy transition and hence improve the governance operational efficiencies of the RE power sector. The analytical methodology relies on the network governance theory and transition management theory to understand the institutional dynamics of actors in charge of the renewable energy transition in Egypt. The background data of the analyzed governance networks was supported by a set of primary interviews with representatives of the identified critical actors and networks. The results of the research pinpointed some of the core reasons of why such institutional barriers persist and in what way they hinder the achievement of the RE targets, leading to the assessment of the overall status of RE transition in Egypt as a work in progress. In the absence of an independent monitoring institution for achieving a non-binding target of RE, the weight of institutional barriers amplifies, especially since the macroeconomic conditions, as well as the market environment, are not a driving force either. The study argues that well-defined indicators for political commitment are a key ingredient to managing the RE transition process successfully, less so are the aspects of institutional culture and participation. The research hints on a set of recommendations to ensure a more effective institutional structure to promote renewable energy: removal of subsidies, one stops shop to overcome administrative bureaucracy and the focus on building stronger institutions rather than projects.

Keywords

Egypt • RE transition • Electricity • Governance • Institutional barriers

Abbreviations

CBE	Central Bank of Egypt
D	Distribution
EEAA	Egyptian Environmental Affairs Agency
EEHC	Egyptian Electricity Holding Company
EETC	Egyptian Electricity Transmission Company
EgyptERA	Egyptian Electric Utility and Consumer Protection Agency
FiT	Feed-in-Tariff Program
G	Generation
GAFI	General Authority for Investment and Free Zones
GoE	Government of Egypt
JCEE	Joint Committee for Energy and Energy Efficiency
MoERE	Ministry of Electricity and Renewable Energy
MoF	Ministry of Finance
MoP	Ministry of Planning
MoPMW	Ministry of Petroleum and Mineral Wealth
NREA	New and Renewable Energy Agency
PPA	Power Purchase Agreements
RCREEE	Regional Center for Renewable Energy and Energy Efficiency
RE	Renewable Energy
SCE	Supreme Council of Energy
T	Transmission
WB	The World Bank Group

F. Shoukry (✉)
Chemonics Egypt, Ad Doqi, Egypt
e-mail: farahshoukry@aucegypt.edu

C. Alalouch et al. (eds.), *Advanced Studies in Energy Efficiency and Built Environment for Developing Countries*, Advances in Science, Technology & Innovation, https://doi.org/10.1007/978-3-030-10856-4_1

1 Introduction

The governance challenges to RE electrification are numerous and other than the long-term learning curve for the RE policies; institutions (public, private, financial, and academic) in charge of the RE process can either catalyze or hinder the process. According to the network governance theory (NGT) and Transition theory, RE institutions function within a macroeconomic energy scene and micro-level experimentation dynamics where the behavior of actors within such network of institutions is difficult to predict (Loorbach 2007; Klijn and Koppenjan 2016). This setting translates to a diverse network of actors and changing relationships between players, in which each institution functions according to a set of rules and regulations. In that respect, understanding the institutional challenges within the transition scene is highly important to facilitate the realization of an inevitable process.

Within the changing dynamics of the energy scene in Egypt, a specific set of institutional barriers is the focus of this research: the politicization of RE policies, the lack of stakeholder participation in the decision-making process, and the bureaucratic institutional culture which is reflected on both transparency issues, and reluctance to experiment. Those institutional barriers are analyzed as an entry to point to study the competencies of the current institutional structure, and whether it is capable of achieving targets set by the government—20% by 2020. To that reason, the nature of the RE targets needs to be clarified, and to understand the degree by which it is legally, and politically binding. Concisely, the aim of this research is to analyze the institutional barriers within the Egyptian governance system and investigate the extent of which it is capable of achieving the proposed RE targets.

2 Case of Egypt

Egypt is following a global movement toward RE electrification. Indeed, the RE for power generation in Egypt is a complex process that is hampered by challenges and changes. The drive toward renewable energy in Egypt stems from the need to cover up for the increasing electricity demands of an exponentially growing population of 90 million inhabitants. The energy crisis is becoming a pressing issue for the Egyptian government and citizens as the rate of summer blackouts during peak hours is increasing, and major industries—relying on electricity—are losing on profitability (El-Refai 2015). As a result, the Egyptian electricity sector had initiated a reform program in order to cope with the shortages crisis and to enable the use of renewable. The reforms' program was made available due to the existence of other positive driving forces such as the geographic suitability of wind and solar energy, the land availability to execute the projects, and the decreasing costs of solar and wind technology worldwide, which encourages the integration of RE resources within the energy mix.

Egypt has set a 20% of the electric energy demand from renewable energy resources to be achieved by 2020. This ambitious vision was mandated in 2008 in line with the National Energy Strategy in which it is updated regularly by the Ministry of Renewable Energy and Electricity (MOEE). Implying a 12 GW installed capacity, the strategic energy mix seeks to include 12% from wind power, 8% from hydropower and solar energy (New and Renewable Energy Authority (NREA) 2016). It is important to note that the targets were later postponed to 2022 (RES4MED 2015). This decision was undertaken due to the political instability in the aftermath of the 2011 revolution (Bottoms 2016). Moreover, an EU Advisory Board consulted the MoERE to prepare the issued reforms as well as the FiT program (Bottoms 2016). This also included setting medium- and long-term targets (30% by 2035 and 65% by 2050), respectively, reaffirming the government's long-term commitment to integrate renewables within the energy mix (Bottoms 2016). Currently, the RE generation capacity falls short of almost 70% where 7.5 GW need to be installed within the coming 4 years, indicating that fast-forwarding steps need to be made (Davies et al. 2015).

Two major due reforms made electrification through renewable resources a feasible alternative: the plan to lift up the subsidies on the electricity bill and the issuance of a new electricity law. The first reform dealt with the removal of subsidies. For the fiscal year 2014/2015, the government had to increase tariffs by 30% in the residential, commercial and industrial sectors. Egypt was considered the sixth highest ranking country on energy subsidies prior to the reform (Kieffer and Couture 2015). The Egyptian government foresaw numerous benefits of reforming the energy subsidies other than draining national budget resources. For one, the removal of the subsidies would help remove the distorted inequalities of the benefits of the energy subsidies, having the one-fifth of the population enjoying those benefits. Most importantly, this reform was among the biggest drivers for the RE electricity market to emerge. By removing the subsidies, the prices of RE electricity generation—price per kWh—become economically competitive when compared to the traditional methods of electricity generation (Freris and Infield 2008). In other words, the project financing for the private sector to build, own, and operate solar and wind power plants would be a valid option.

As a matter of fact, the importance of the second reform —the New Electricity Law—lies in the facilitation of the private sector to generate electricity from RE sources. The introduction of the New Electricity Law which entails a feed-in-tariff program sets the main guidelines to allow for the gradual process of liberating the electricity market through. The government has introduced a feed-in tariff to incentivize private sector investment in solar and wind energy, with 4.3 MW expected to be commissioned within the coming two years (Economic Ministerial Committee 2015). The new electricity law which was issued in June 2015 permitted the private sector to be an active party in electricity generation with a profit margin set within the legal framework of feed-in Tariff (FiT) and power purchase agreements (PPAs) (Egyptian Electricity Holding Company 2014). The new electricity law allowed the use of the transmission grids by the renewable energy generators—solar and wind—to reach the end users. The electricity tariff restructuring and the inclusion of the private sector in the RE electricity generation are the primary steps that would enable the process of renewable energy transition to take place.

The reform program sets Egypt on an early path of energy transition, both on a physical level and a structural one. The physical level translates to moving away from the reliance on natural gas power production, with the aim of securing energy supply and the diversifying the energy mix. The structural transition, on the other hand, refers to moving away from a highly centralized energy production entity to a liberalized market model. To clarify, the mandated strategic plan by the MoERE does not intend a full market liberation reform on the short run; however, the partial deregulation plan—emphasized by the New Electricity Law—would also require restructuring of the existing centralized hierarchy (El-Markabi 2015). The organizational unbundling of the centralized structure is perceived as one of the major challenges on the prospects of RE transition in Egypt. The institutional structure is highly centralized and is almost 90% publically owned. The Egyptian Electricity Holding Company (EEHC), which is supervised by the Ministry of Energy and Renewable Energy (MoERE), owns the six generating energy company, the sole transmission company (EETC), and the nine distributor companies across the country (Egyptian Electricity Holding Company 2014). Besides the public distribution companies, there are several BOOT power plants that were commissioned in the 1990s, and their generation capacity is transferred to the transmission company (Osman 2015). The centrality of the governance structure is further emphasized by the Supreme Council of Energy (SCE) which—theoretically—promulgates the national strategies concerning energy.

Within the changing dynamics of the energy scene in Egypt—the introduction of the new electricity law, the governments' scheme of FiT and PPA projects, the rise of renewable energy companies—it is evident that the kick-off for the RE transition is a complex process. Between the governmental ministerial bodies, the private companies that are eager for participation, and the supporting financial and regulatory institutions, it becomes necessary to decipher the roles and responsibilities of RE networks and actors, and to investigate the viability of the current institutional arrangement whether it is capable of achieving the promised target.

3 Research Questions

The main research question is: To what extent do the current institutional barriers* of the RE electricity sector in Egypt influence the achievement of the proposed 2020 Targets? The studied barriers include lack of political commitment. The research sub-questions include:

- To what extent does the level of political commitment of the Government of Egypt (GoE) influence the RE transition process?
- To what extent is the local stakeholders participating in the RE transition process?
- How does the institutional culture (degree of transparency and experimentation) impact the implementation of the RE targets?
- To what extent is the 2020 RE targets currently met? ment, lack of stakeholder participation, lack of transparency and experimentation.

4 Governance Theories

The main theories reviewed for the purpose of this research are the network governance theory (NGT) and the transition management (TM) theory to help analyze the institutional governance structure. The network governance theory (NGT) stresses the complexities involved in the decision-making process and emphasizes the interdependencies and interlinkages between actors (Klijn and Koppenjan 2016). Erik Hans and Joop Koppenjan further categorize the complexity of governance networks in the public sector according to three overlapping themes: substantive complexity, strategic complexity, institutional complexity, and the overarching theme of network management complexity. On the other hand, transition management (TM) theory was reviewed in order to understand the implications of the energy transition process based on the Ph.D. work of (Loorbach 2007), and the article of Voß et al. (2009). The TM theory helps to explain how institutions function—or a projected to function—within such a setting. Transition can be defined as a set of "transformation processes in which existing structures, institutions, culture, and practices are broken down, and new ones are established (Loorbach 2007)." This definition fits the purpose of this research, and it hints that institutional restructuring is inevitable in the prospects of energy transition.

Having the Egyptian case in mind—which has been earlier analyzed as an early transition phase—the NGT becomes a more viable source for analysis. While NGT and TM can be viewed as complementary theories, NGT remains a subset of TM. Both theories have their strengths and limitations. Though the NGT provides an almost comprehensive framework to analyze the institutional setup, it is not without disadvantages. Ellen van Beuren points out four major pitfalls of adopting the network process management (Voß et al. 2009). First of all, the process emphasizes the communication patterns and interaction between the actors which is not an end in itself. Secondly, the process becomes unnecessarily complicated and time-consuming due to increased communication and negotiations which will discourage some actors to participate. Thirdly, the quality of the decisions made will not be based purely on knowledge and the end result may lead to the implementation of sub-optimal solutions. The fourth downfall is that it allows actors to enforce their decisions upon others or slow the decision-making process. I agree that the network process management does over-emphasize the communication patterns, but that is part of the proposed solution for complex problems. In other words, if the problem is simple enough then the problem-solving process would neither be time-consuming nor encounter power struggles. Though NGT has cleverly put a framework to analyze the internal institutional structures, it has not succeeded in analyzing the external factors which influence the functional strategies of those institutions. In that sense, transition management theory comes in handy to provide an analytical framework for the external conditions.

5 Institutional Analysis and Development Framework

This section aims to provide an overview of the literature on frameworks to analyze institutions within the RE transition context and meanwhile highlighting barriers within institutions. Koster and Anderies (2013) adopt an analytical framework to assess institutions—institutional analysis and development framework (IADF)—in the context of energy transition. Key features were identified by the authors to ensure successful governance of energy transition (Koster and Anderies 2013): government commitment, polycentric structure, stakeholder participation, institutional transparency and willingness to experiment, binding international agreements, grid connectivity, and monitoring mechanisms. However, in the absence of the above-mentioned drivers, they act as institutional barriers to energy transition. Painuly (2001) put forward an analytical framework to analyze the different barriers pertaining to the penetration of RE, which includes the institutional barriers, and they happen to be almost an exact antonym of the institutional drivers. Though the research is focused on studying the barriers to political commitment, stakeholder participation, and institutional culture, the whole set of institutional analysis factors is overviewed to provide a comprehensive understanding. This extensive discussion refers back to the IADF model—Fig. 1 explained below—which suggests an overlapping nature between the strategic outcomes and the decision-making process. The IADF model suggests an interdependency of the studied institutional factors.

The IADF summarizes the institutional drivers to successfully facilitate the process of RE transition and is represented in Fig. 1 (Koster and Anderies 2013). As shown in Fig. 1, elements are interrelated where the strategic outcomes are a result of a dynamic equation: external variables and the decision-making arena. The biophysical conditions refer to the macroeconomic context and external factors shaping the surrounding. Biophysical conditions in that sense are the macro-level or the energy landscape identified by the transition theory. The attributes of the community are the social conditions of the stakeholders, i.e., willingness to participate and their perception and awareness of RE sources. The rules of the house refer to the policies, rules, and regulations that shape the transition process. The second set of factors refers to the decision-making process where key decision-makers are the main actors, and their action is influenced by the external factors, societal characteristics, and governing regulations, giving rise to institutional outcomes. One key aspect of the IAD framework is the emphasis on the feedback loop, which is in line with previous governance theories of NGT and TM. In the context of this study, it highlights the dynamism of decisions regarding energy transition.

5.1 Conceptual Framework

The reviewed governance literature, institutional characteristics, and RE targets culminate into the following conceptual framework. Inspired by the institutional analysis and development framework on analyzing the drivers by which institutions lead the process of energy transition (Koster and Anderies 2013), the research—in reverse—investigates the impact of barriers to the achievement of targets. Specifically, the conceptual framework is derived from the reviewed literature of network governance theory (NGT) and transition management (TM) theory, which shows the conceptual framework in relation to theory. The independent variable—institutional barriers—is analyzed within a larger sphere of governance theories. The dependent variable is the RE targets and is used as a benchmark to monitor the transition process. Though the RE targets act as a benchmark to evaluate the RE transition process, they are understood as an

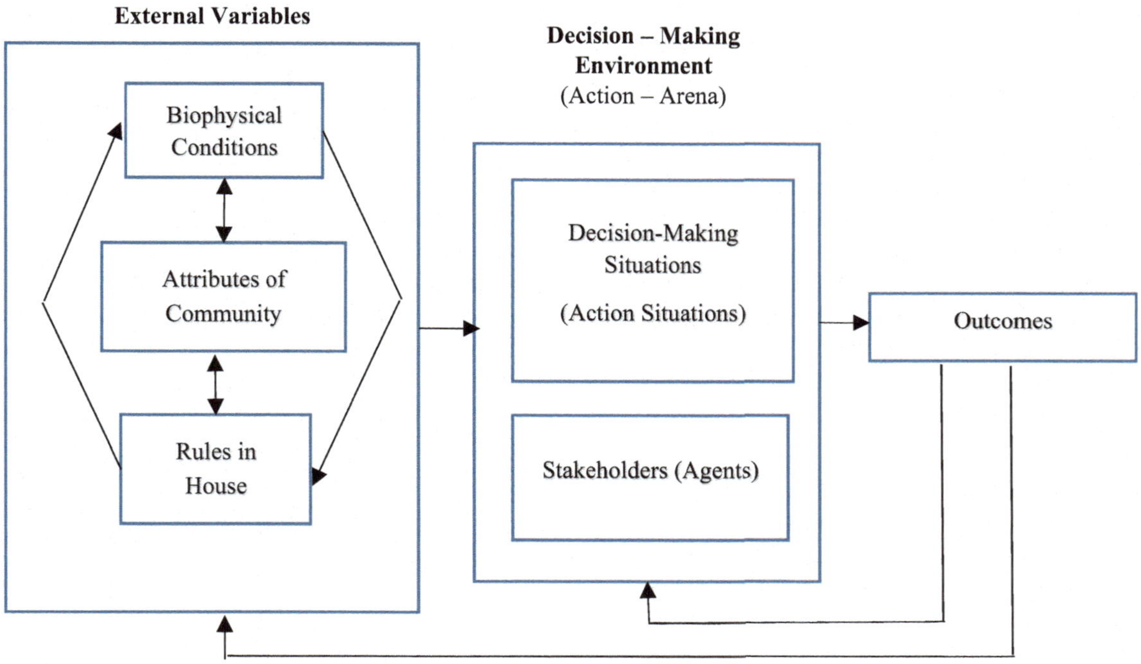

Fig. 1 Institutional analysis and development framework (Koster and Anderies 2013; Ostrom 2010)

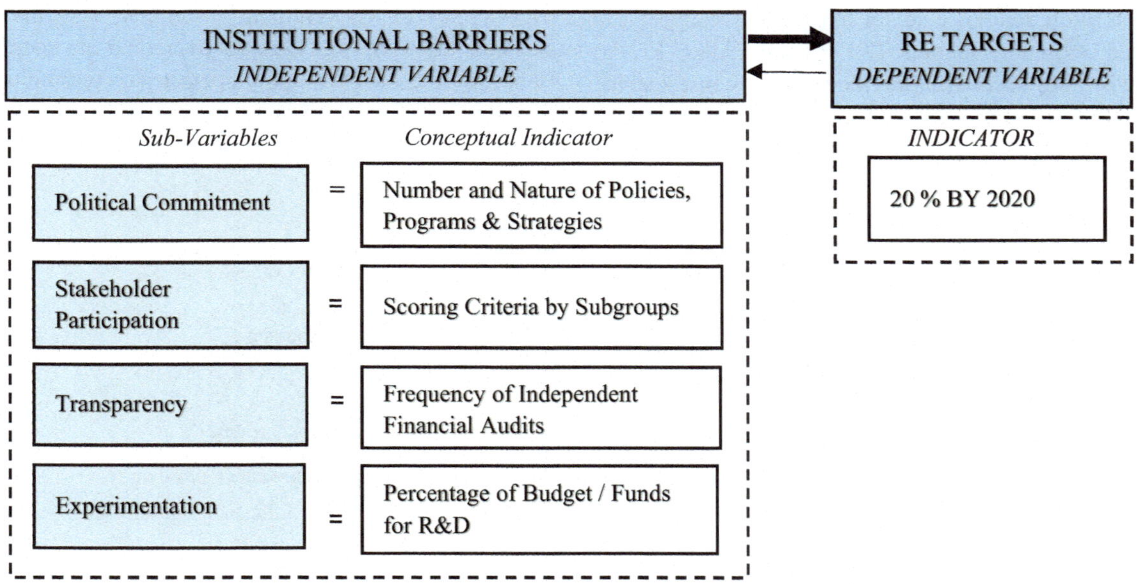

Fig. 2 Conceptual framework indicating variables

institutional outcome—refer to the IADF—and thus, imply a feedback loop. The studied barriers of lack of government commitment, lack of stakeholder participation, lack of transparency, and intolerance to experimentation are all characteristics of unfit institutions to lead the RE transition process. The selected barriers act as (independent) sub-variables in which the extent of their existence hinders the transition process—shown in Fig. 2.

- Independent Variable: Institutional Barriers
- Independent Sub-Variables: Government Commitment, Stakeholder Participation, Transparency, Experimentation
- Dependent Variable: RE Targets
- → Strong Link
- ← Weak Link

6 Methodology

The conceptual foundations of this research is based on both theoretical and empirical grounds. The analyzed governance theories include: the Network Governance Theory (NGT), and Transition Management theory (TM), as well as the IAD framework. As for the empirical grounds: the research carried out a number of in-depth interviews with Egyptian RE stakeholders in order to complement and validate the literature findings following both a purposive and snowballing sample.

The main reason for applying the purposive sample was that the main target is key stakeholders within the RE electrification process with a profile of expertise and are key decision-makers within their institutions. On the other hand, the snowballing sampling—also known as network chain referral or reputational sampling—relied on extending the network of interviewees based on the starting point of contacts, i.e., a snowballing effect of interrelationships. This strategy stemmed out from the networking methods required to schedule interviews. However, preliminary contact was established within the researcher's scope, and a minimum number of interviews was attained at the point of saturation. Saturation translates to the repetition of answers by respondents with reference to the research questions. The interviewed experts include: (7) Private Solar Energy Companies—both service providers and consultancies; (4) Academics; (4) Public Institutions; (1) Finance. Originally, 20 interviews were targeted; however, after transcribing 16 interviews a sufficient degree of saturation was achieved (Table 1).

7 Research Findings

7.1 Political Commitment

Fundamentally, the nature of the Egyptian RE targets is non-binding and that is a fact that topples down the commitment discussion. What is then the purpose of having non-binding targets? Reading in between the lines, the answer could relate to the initiation of an RE transition scenario on the short term, that is, duty-free of legal and hence economic obligations. It also relates to being influenced by international policy lessons for the Egyptian targets which were announced shortly after Germany's initiative and the Kyoto Protocol; however, it missed out on the whole of the context.

The macroeconomic challenges, combined with the electricity sector's heavy fiscal debt burdens, are the actual reason behind the decision to flood the market with conventional electricity. In the years of political instability 2011–2014, the over-supply of conventional electricity was the cheaper decision taken by the GoE to overcome an energy crisis. Hence, even if the oversupply of electricity was pinpointed as a challenge to RE transition by the private sector, in actuality, it was the GoE's solution to an immediate power shortage

Table 1 List of respondents

#	Position	Entity	Institution
R1	Alternative Energy—General Manager	Private Equity Company	Financial
R2	Sustainable Tourism Consultant	Ministry of Tourism	Public
R3	Renewable Energy Projects—Engineer	EETC	Academic
R4	Professor and Board Member	Cairo University—EEHC	Academic
R5	General Manager	Private Environmental Consultancy	Private
R6	General Manager	Private Energy Efficiency Consultancy & Store	Private
R7	Professor	The American University in Cairo	Academic
R8	Managerial Position	JCEE	Public
R9	Managerial Position	RCREEE	Public
R10	Renewable Energy Projects—Engineer	EgyptERA	Public
R11	Senior Associates	Private Law Firm	Private
R12	Professor	The American University in Cairo	Academic
R13	Managerial Position	Private Management Consultancy	Private
R14	Chief Technical Officer	Private Solar Energy Company	Private
R15	Business Developer	Private Solar Energy Company	Private
R16	Solar Energy Expert—Managerial Position	Private Renewable Energy Company	Private

emergency. The "oversupply" is an example of the substantive complexity where the problem at hand is viewed differently. While the private sector saw the crisis as a legit motive to foster RE transition, the GoE desperately opted to "efficiently" solve the problem by embarking on different energy scenarios—both conventional and renewable. Undeniably, the energy portfolio is highly politicized, and the geopolitics is an influential factor on de-prioritizing the RE agenda. The "emergency plan" to tackle the energy crisis was financed by Gulf countries—oil-rich countries—who have paid a token of allegiance to the presidential cabinet post the revolution. That may be one reason why the conventional route ousted the RE transition off-take, and in return, limited the RE market to grow. In addition, the market de-growth is coupled by the project-financing challenges that are a result of a struggling economic context and most importantly ill-thought-out regulations: The FiT over-pricing and the onshore arbitration issues stand as a witness. That again can be analyzed on the substantive complexity line of thought.

Next to the project-financing obstacles is the absence of an official entity responsible for monitoring the targets achievement. The closest to this description is both EgyptERA and NREA. However, they are concerned with the projects following up rather than strategizing on the RE energy mix. The good news is that the 2030 strategy suggests a supra-governmental authority for the monitoring and evaluation responsibilities. As we speak, though, such an entity does not exist. Conversely, what supports the GoE's position in the political commitment discussion is that most of the challenges above are recognized, and noted down in the form of a to-do list to be resolved in the future—and that again refers to the 2030 document.

To sum up, though there are no shortcuts in evaluating the degree of the GoE's political commitment, three factors stand out as the most influential on the targets achievement: the non-binding nature of the targets, the project-financing obstacles encountered by the private sector, and thirdly the lack of a supra-monitoring entity to supervise an overall strategic energy vision. With that in mind, we move forward to the next section discussing "Stakeholder Participation."

7.2 Stakeholder Participation

The participation of the private sector and end users is almost nil in the policy-making of RE generation given that there are no formal operating mechanisms to do so. However, there is room to discuss the private sector participation in RE generation put forward by the recent legislation and the FiT program.

Given that the private sector is participating in energy generation but not in policy-making, this marks a gap in the conceptual framework of this research. Participation was highlighted as a main driver of the energy transition process, within a fragmented polycentric governance structure where the end users represent a major stakeholder, or at least this had worked for the success case studies as in Germany, Netherlands, and Denmark—where community-driven RE generation is a noticeable pattern. Theoretically, the participation of both the private sector and the end users in the decision-making process is a principle driver for RE generation as the awareness of the benefits of RE integration and environmental concerns exhibit their motivation toward being active prosumers. This is, however, a far cry from the case in a developing country like Egypt where only 40% of the collection costs are actually recovered of subsidized (conventional) electricity bill. Thus, the responsibility of driving RE transition falls primarily on the GoE—where its non-binding level of commitment had been discussed earlier—and secondly on the private sector.

For the private sector to work within an emerging market, a whole set of obstacles arise. The policies are still being tested where there are mismatches between the conditions of the legal framework and available financing mechanisms for the RE projects: firstly, the absence of sovereign guarantor for the power purchase agreements (PPAs) and, secondly, the objection of international financiers to backup RE projects in Egypt with an onshore arbitration clause. Adding to that, there are no substantial incentives given by the GoE for the private sector to participate—with the exception of the high FiT prices which was proved earlier a non-sustainable financial arrangement. The direct form of incentives was meant to be a tax reduction on RE equipment, and even that is not fully offered.

This discussion is not meant to take sides with the private sector, but rather to explain the conditions in which they are meant to participate. The GoE's decision-making on the RE agenda is a result of overlapping variables, external, institutional, and less so of a micro-level. Thus, the challenges encountered by the private sector are not meant to blame the GoE, but rather to find middle grounds to mediate solutions. But then again, why is it important for the private sector to participate in RE generation? To put the argument simply, it does not make sense for a burdened governmental sector to continue overtaking the responsibility of power provision, especially since the central gain of RE electrification is decentralized generation.

Accounting for the private sector's role in RE generation and the degree of political commitment discussed in the earlier section, the following section builds on the discussion of institutional barriers discussing the degree of transparency and experimentation.

7.3 Institutional Culture

Expectedly, the Egyptian public governmental institutions are frequently reported to behold low transparency levels. However, a clarification must be made to what does the "low level" of transparency entails. Clarifying the various interrelated meanings with the term transparency is necessary to analyze the respondents' answers objectively. According to the reviewed literature, transparency is often presented as an antonym of corruption, or it is associated with the availability of data and ease of access, and thus, it is related to the flow of communication between the different actors. More importantly, the terms transparency and accountability are often intermingled where a high level of institutional transparency translates to direct proportionality to the degree of accountability. The WB report on transparency argues that both transparency and accountability in Egypt's electricity sector are limited due to the fact that "EEHC subsidiaries are not held responsible—as specified in the companies' Article of Association and their licensing agreements," as well as the issues of reporting within EEHC evaluating it as highly underdeveloped, especially those connected with customer satisfaction toward the companies' performance (Hassan et al. 2015).

Taking into consideration the above conversation, how does the "low level" of transparency within public institutions influence the achievement of the RE targets? Well, the keyword here is operational efficiency. Slow-paced communication in a learning-policy-curve environment would just not work. The re-sale market of RE generation licenses resultant of the lenient FiT qualification procedures stands as a witness on how administrative inefficiencies can result in a parallel system of unregulated operations of solar energy companies. Consequently, the market competitiveness—from the private sector's perspective—is at stake, and this, in totality, repels investors from backing-up a growing RE market and hence decreases the probability of achieving the targets.

7.4 Willingness to Experiment

Again, a gap arises between the reviewed literature and the practical reality in Egypt with reference to the "experimentation" concept. According to the IADF, the degree of experimentation served as a driver to promote innovation and thus allowed the RE technology more penetration within the market as well as industries. The experimentation concept as a driver is founded on the logic that technological innovation increases the efficiency of RE generation, and in that sense, R&D is a core activity for the success of the RE transition process. However, the developments in Egypt are in a completely different phase to the extent that the term "experimentation" is not even the right word to use; simply, it is "RE technology." Not that the technology is new, but rather the context in which it works.

Back to the question of how the lack of experimentation influences the achievement of the RE targets? To that, the answer will have to split into a theoretical argument and a contextual one. Theoretically speaking, non-experimentation means that a whole lot of stakeholders are not active—or active enough—to carry on the transition process. Primarily, R&D institutions that are backed-up by commercial applications should be in the lead. Those R&D institutions include both corporations—governmentally or privately funded—as well as the academic network of universities, non-profit organizations, and individual experts. Again the R&D institutions are connected to the industrial sector where actual manufacturing of equipment and tools takes place. All those stakeholders work on making the RE technology more cost-competitive by nationalizing the technology where local materials are labors found the RE industry. Cost-competitiveness and a promise of an emerging market increase the viability of adopting RE electrification and hence increase the probability of achieving the targets. However, such a discussion remains a far-fetched scenario for Egypt. To a great extent, the technology is imported—e.g., solar and wind—and the market dominance—after the public utilities—remains in the hands of international consultants and corporations, and thus, it is nowhere near localization. That is not to undermine the role of Egyptian private sector, or small industries contributing to the development of the RE construction schemes, but when we talk of large mega-projects of RE targets, the foreign dominance outweighs the local contributions.

So is a lack of experimentation really a barrier? I would say it is in a developed context of transition. However, in the case of an early phase of transition as in Egypt, the non-experimentation is not an initial priority. The first step is actually to localize the knowledge and expertise, and then the next step would be to localize the technology manufacturing.

7.5 RE 20 Targets

As it had been hypothesized in the conceptual framework of this research, the barriers of RE transition influence the achievability of the targets, and the targets themselves impact the severity of the barriers—i.e., feedback loop. Taking the lack of political commitment as an example, the non-binding nature of the targets serves as a core reason on why commitment is on the low side. However, on the other hand, the targets themselves propel the GoE to work on the RE agenda. Another conceptual visualization is that the barriers are all inter-linked, and they do influence one another and in return influence the RE targets—in line with the transition and network theories. That has been the case for institutional misdesign which is a common theme that affects the political commitment in terms of the absence of a monitoring entity and also results in accountability issues in reporting between the public sector's electricity institutions. It also affects the market competitiveness in which the private sector participates. Furthermore, the circle of influence of each barrier on the targets achievement is not at all equal. While the lack of political commitment comes about as the most influential, the institutional culture is less so. Meaning, the institutional characteristics of low levels of transparency and experimentation are indirectly related to the RE achievement as they act as a catalyst to better operationalize the RE transition. However, their existence—as barriers—does not exactly pause the transition process. In fact, the obstacles faced by the private sector in project funding and development are the most direct barrier to achieving the RE target. To put the argument more clearly, there is no RE transition without the participation of the private sector, and currently, the concern levels to seal the deals—for the FiT program as an example—are on the high rise. The FiT program constitutes a 4.3 GW installed capacity, and the concern issue is the 2 GW of solar projects which represent an almost 20% of the 20% of RE electrification to reach the targets. However, that 20% of the FiT program are the actual representation of the participation of the Egyptian private sector's dire need of liberalizing the energy market; and its rate of success would fast-forward the attempt to liberalize the market and optimally integrate RE electrification.

Knowing that the RE electrification target is almost halfway, with the greater percentage still representing hydropower, it becomes necessary to understand the significance of assigning the targets. In other words, the numbers are useless unless the motive behind them is unveiled. Theoretically, RE target setting should be reflective of the priorities of the energy Agenda. The priorities may include: diversification of energy sources, securitization in a politically unstable region, national sovereignty, encouraging economic growth or otherwise. It is not to be forgotten though that the original reason of why the targets came about was after the Kyoto Protocol which makes it a policy drawing example, and the extent of its contextualization needs to be better understood. Though the real answer can only be known in due time, the tight time frame of 4–6 years —when legislation was just announced in 2015 and 2016—leaves little expectation for the targets to be achieved.

8 Conclusion

So how far is the GoE from achieving the 2020 targets? Currently, it is exactly halfway and that takes into account the hydropower jump start. The next question becomes, is Egypt going to make it? Unfortunately, not. On the positive side, however, progress is being made, and the targets become a motivation to utilize an environmentally and economically sustainable energy resource. The question then becomes: What is the purpose of having RE targets for the Egyptian context? The RE targets within the Egyptian context can be considered a policy drawing antecedent of the international protocol on integrating renewables in power generation. The fact that renewables are primarily associated with electricity generation rather than the whole of the energy mix sets a limitation—in a way—to the perseverance of the renewables agenda on the national level. In addition, the non-binding legal nature of targets makes the government the official responsible entity for implementing the goals while de-emphasizing the roles of the private sector, non-governmental organizations, and end users to realize the goals.

That being said, the main theoretical contribution of this research is to fill a timely knowledge gap on the barriers faced by the current RE electricity sector institutions on their road to restructuralization and market liberation, giving an in-depth insight into challenges they face. Secondly, the study provides a practical application of the NGT and TM theories and suggests a merged analytical framework to explain the governance concepts of RE transition. Moreover, the research conceptualizes a reversed framework for studying institutional barriers—inspired by the institutional analysis and development framework (IADF)—proving the conceptual hypothesis between barriers and the RE targets achievability. Nonetheless, the shortcomings of this research are the lack of quantitative indicators to back up the in-depth insights into institutional barriers. The study points out for a much-needed revision of the selected indicators to evaluate the institutional barriers of political commitment, participation, and institutional culture. Firstly, the political commitment indicator should have a focus on the implementability of strategies and policies and not only a checkmark of their existence. The 2030 strategy provides an abstract vision for the electricity sector and an outline for the reforms that should happen—or is happening in case of the legislative reform and institutional restructuring. However, as pinpointed by the strategy itself is there is still a need to write-up a hands-on energy strategy which optimizes the energy mix—for both the electricity and petroleum sector. Secondly, the stakeholder participation indicator is need of contextual reframing as well, for lack of participation eliminates the policy-making exercise of both the private sector and end users totally. The participation indicator, especially in an early transition context, should exhibit an understanding of effective participation in RE generation which is an outcome of the governance structure, societal level of awareness, and market conditions. Moreover, the research had undercut the qualitative indicators chosen for transparency and experimentation and instead formulated the "degree of" measurability on qualitative argumentation. However, the qualitative arguments presented in this study provide a solid foundation to expand the research population and withdraw quantitative results.

In answering the main research question: *To what extent does the current institutional barriers of the RE electricity sector in Egypt influence the achievement of the proposed 2020 Targets? It is with the significant extent that the institutional barriers influence the achievement of the RE targets.* Validating the conceptual framework of this research, an evident interlinkage between the barriers does exist in which the institutional barriers do not have an equal impact on the targets achievement, as explained earlier. Summing up the findings, it was found out that the degree of political commitment was evaluated as poor after analyzing the strategic, fiscal, and institutional framework root causes. Moreover, the participation of the private sector and the end users in the policy-making is non-existent; however, the participation of the private sector in RE generation is on the rise. Moving on to the institutional culture, the degree of experimentation and transparency is on that of a low level. Reconceptualising the theoretical link between institutional barriers on the RE targets, it is evident that the degree of impact differs. In conclusion, the complexity of policy-making governing the RE transition in Egypt is due to the experimental nature of the whole energy generation process as well as the long learning curve for the policy cycle, and the existing interdependencies of governance networks. Questionable political commitment, semi-transparent administrative communication, non-flexible regulations to conduct business as usual, and non-participatory policy-making, all impact the competencies of current institutions to reach the 2020–2022 targets.

Acknowledgements Indebted to the supervision of Dr. Ogenis Brillante, Somesh Sharma, and all faculty at the Institute of Housing and Urban Development Studies, Erasmus University, Rotterdam, for supervising the outcome of the original version of this work, M.Sc. thesis entitled Governing the Transition of RE Electrification in Egypt, for the year 2016.

References

Bottoms, I. e. a. (2016). 80 gigawatts of change: Egypt's electricity pathways. North Africa - Tunis: Egyptian Center for Economic and Social Rights; Heinrich Boll Stiftung.

Davies, M., Elmatbouly, S., El-Mahgouzy, D., Schellekens, G., & Ahmed, S. (2015). Developing renewable energy projects: A guide to achieving success in the middle east. Egypt: EverSheds, Shahid Law Firm, Pwc.

Economic Ministerial Committee. (2015). Strategy: Egypt's five year macroeconomic framework and strategy FY 14/15 - FY 18/19. Egypt Economic Development Conference (A Joint Work of the Economic Misterial Committee), Sharm el-Sheikh.

Egyptian Electricity Holding Company. (2014). Egyptian electricity holding company: Annual report (2013/2014). Cairo.

El-Markabi, M. S. (2015). Addressing Egypt's electricity vision.

El-Refai, M. L. (2015). Evaluating egypt's renewable energy and energy subsidies policies. (Unpublished M.P.P). The American University in Cairo, Cairo.

Freris, L., & Infield, D. (2008). Renewable energy in power systems. UK: John Wiley & Sons.

Hassan, F., Penglis, E., Seferiadis, G. N., & Araya, M. K. (2015). MENA energy series: Transparency and social accountability in the egyptian power sector. (No. 93936-EG). Washington DC: International Bank for Reconstruction and Development/The World Bank.

Kieffer, G., & Couture, T. D. (2015). Renewable energy target setting. UAE: International Renewable Energy Agency (IREA).

Klijn, E., & Koppenjan, J. (2016). Governance networks in the public sector (1st ed.). New York: Routledge.

Koster, A. M., & Anderies, J. M. (2013). Institutional factors that determine energy transitions: A comparative case study approach. In E. Michalena, & J. M. Hills (Eds.), Renewable energy governance: Complexities and challenges (pp. 33–61). London: Springer.

Loorbach, D. (2007). Transition management: New mode of governance for sustainable development. Rotterdam: Dutch Research Institute for Transitions (DRIFT).

New and Renewable Energy Authority (NREA). (2016). Retrieved from http://www.nrea.gov.eg/english1.html.

Osman, S. H. (2015). Overview of the electricity sector in egypt. 2nd Capacity Building Reporting Methodologies: How to Collect Data and Monitor Regulated Entities - Presentation Notes, Milan.

Ostrom, E. (2010). Beyond markets and states: Polycentric governance of complex economic systems. Transnational Corporations Review, 2(2), 1–12.

Painuly, J. P. (2001). Barriers to renewable energy penetration: A framework for analysis. Renewable Energy, 24(1), 73–89.

RES4MED. (2015). Country profiles: Egypt. Italy: Renewable Energy Solutions for the Mediterranean (RES4MED).

Voß, J., Smith, A., & Grin, J. (2009). Designing long-term policy: Rethinking transition management. Policy Sciences, 42(4), 275–302.

Contemporary Urban Centres as a Mean Towards Sustainability

Sara Mamdouh Farouk Elrawy, Hany Louis Atalla, and Ebtesam M. Elgizawy

Abstract

Urban centres were affected by contemporary trends and stakes of the various forces that drive the city to change in sequence and lead to growth and development. Especially, after the emerge of many movements that were related to improve the built environment such as: The New Urbanism, Smart growth and the Compact city. Those movements involved the mixed usages as their primary principle in the construction, planning and real estate development sector. In return, the urban construction of Egypt inevitably was affected by the impact of new trends especially in new urban areas. Many of those movements and trends that advocate mixed-use, believed that such as this contemporary concept will lead to more sustainable lifestyles and they emphasize that urban centres will become more attractive, viable and safer places to live and work. Thus, it is important to look at the relationship between this contemporary trend and the concept of sustainability and its dimensions, to reach more awareness and understanding. The main objective of this paper is to provide an analytical narrative for the concept of mixed-use development, the movements and advocacies as well as the benefits associated with it and the discussion of the relationship between this contemporary trend and sustainability. This study is expected to contribute to the literature enrichment with respect to the lessons learned which assumes a strong relationship between the trend of contemporary urban centres and sustainability.

Keywords

Urban centre • Contemporary movements • Mixed-use development • Sustainability

S. M. F. Elrawy (✉) · H. L. Atalla · E. M. Elgizawy
Architecture Department, Faculty of Fine Arts, Helwan University, Cairo, Egypt
e-mail: smamdouh.eng@gmail.com

1 Introduction

As a consequence of the urban center evolution, especially after World War II, advocate to suburban and new cities and move towards the separation of land uses and the return to their integration (Grant 2002; Coupland 1997). And a focus for serving the needs of those who live and work, promoting self-containment, providing higher-density housing in projects with a suitable scale for walking, combining different activities and facilitating access to public transport are the primary justifications for many contemporary movements to achieve environmental, economic and social benefits, such as New Urbanism, compact cities, smart growth. The concept of mixed-use development has been established as a significant model in the worlds of planning and development, and has emerged as a major planning tool for compact city, smart growth and New Urbanism movements. At the end of the twentieth century, these three similar movements emerged, which are based on the belief that returning to traditional development and searching for new alternative cities are essential in restoring functional and sustainable communities. They have become very influential amongst the eyes of planning theorists and urban designers due to their focus on contemporary issues and adopted land use integration as a key principle (Gleeson 2006). In addition, it has been noted as the key to creating vibrancy in otherwise dull urban environments, whether the development of new urban areas, or the re-consolidation of existing urban spaces (Downs 2001; Kelbaugh 2002; Daniels 2001; Talen and Knaap 2003; Congress for the New Urbanism 2001).

New Urbanism is a standout amongst the most critical fields of urban design, which prospered in the United States in the past two decades; sometimes, the term (neo-traditional design) is used as a replacement of the new Urbanism movement (Talen 2005). Spatial ordering of uses was one of the most aspects that have been listed by the New Urbanism movement and the emphasis on its effects on our social, economic and civic life. New Urbanism is beginning to have

C. Alalouch et al. (eds.), *Advanced Studies in Energy Efficiency and Built Environment for Developing Countries*, Advances in Science, Technology & Innovation, https://doi.org/10.1007/978-3-030-10856-4_2

widespread impact, especially in town centre, which has a design methodology that favouring the pedestrian over the automobile and adopt some principles of mixed land use projects with the business office, retail building and light industrial located in urban centres (Rowley 1996a).

Smart growth is another movement and one of the latest important sustainable urban planning paradigms to inhibit the effects of distribution, and is based on the incorporation of land uses and buildings compressed, which leads to high density with less environmental impacts. It supports growth in the walkable urban centres. It also encourages compact, transit-oriented and mixed-use development (Porter 2002).

The idea of compact city is fundamentally the same as "smart growth"; it is an urban planning and design concept, which advances generally higher residential density with mixed land uses. In 1990, the European Commission urged the use of high-intensity mixed-use, and the transformation of former infrastructure sites rather than development in open green areas (CEC 1990). The idea emphasizes that urban activities should be nearer to each other to ensure better access to facilities and services, also to make them more efficient. Where the greater reliance on public transport with walking and biking, as well as the concentration of individuals and activities and reduced passenger traffic all contribute in reducing energy consumption; hence, it would also give a series of environmental, social and economic benefits (Jenks and Jones 2010). The implicit image of the compact city movement has proved to be a highly influential interpretation of what a sustainable community should be carried, through the importance of urban form with relatively mixed land uses (Jenks and Jones 2010; Jenks and Dempsey 2005). Those contemporary movements agreed to take a series of considerations: see Fig. 1.

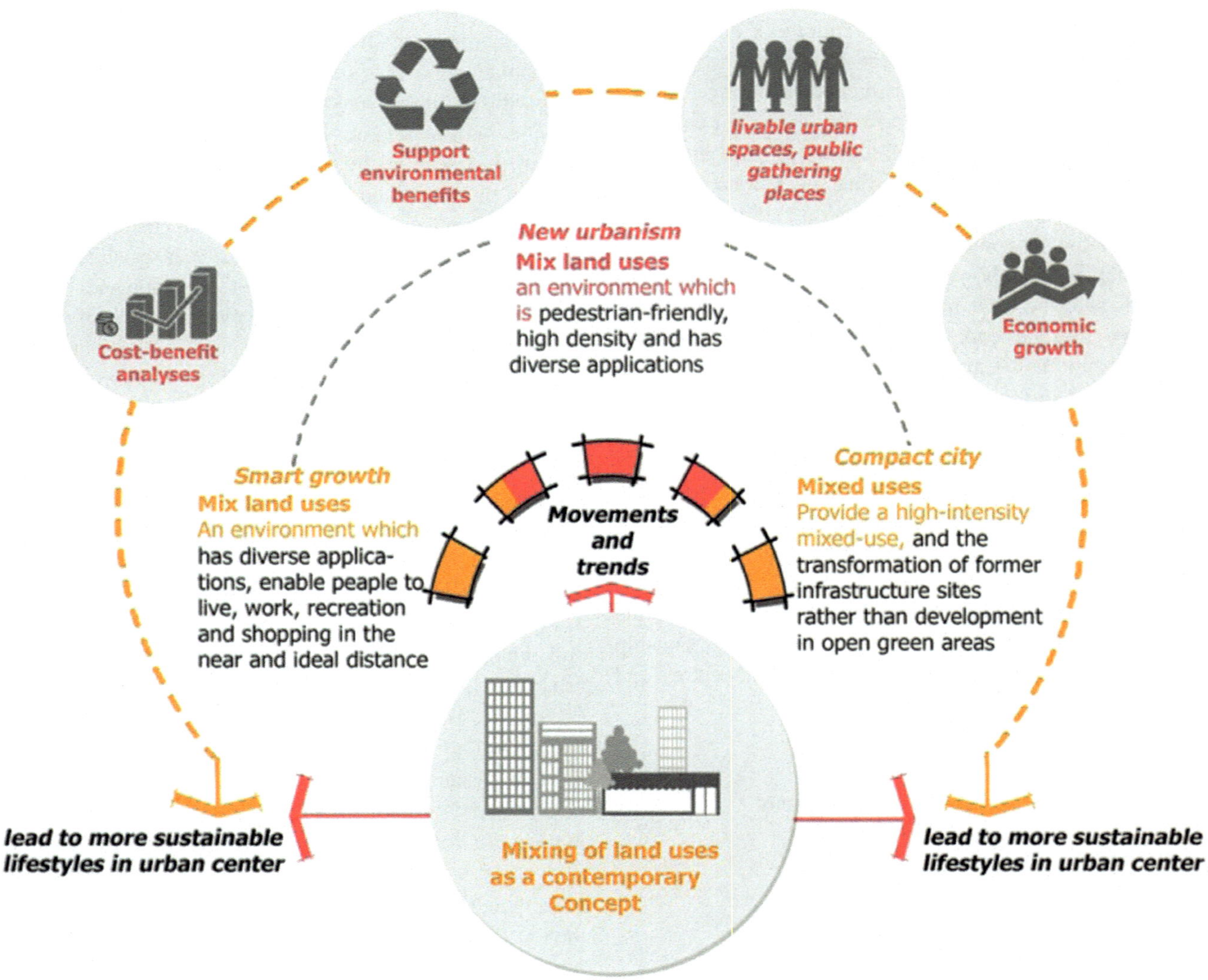

Fig. 1 Mixing of land uses as a contemporary concept, adapted by the author

- Mixing land uses to reach greater densities, responding to changing consumer demands, and taking advantage of synergies arising from the integration of complementary uses and internal travel on foot (Rabianski et al. 2009). Provide livable urban spaces with public gatherings and a variety of shops, restaurants and entertainment, as well as more vibrant commercial areas that provide retail and services to beneficiaries.
- The trend of economic growth and employment focuses on increasing the number of low- and medium-sized firms, focusing on knowledge-based industries and services, and other types of corporations that are aware of the importance of creating an environment that facilitate innovation through competition and knowledge sharing (Porter 1998; Hall 1998). This provides another reason for aggregating urban activity in mixed-use projects.
- Mixed-use refers to the use of different uses such as office, commercial and residential in a building, or integrated project with the adequate functional relationship and coordinated physical design have been combined (Gregory 2003).
- Cost benefits of transportation and infrastructure projects through time savings, especially in business travel, as well as taking into account that the collection of activities needs a less expensive infrastructure than distributed in different places.
- Bolster environmental advantages through dense urban structures which encourage advanced sewage systems, material reusing, waste collection and diminished individual interest for possessing land and endeavour to greatly reduce the energy and material utilization related to urban communities that negatively affect the environment (Grant 2002).
- In this light, the contemporary concept of mixed-use development has become a new trend that has received significant attention in pursuit of more sustainable communities.

1.1 Objectives and Methodology

The aim of this paper is to present an analytical narrative of the concept of mixed-use development in new urbanization. The research is based on a relative benefits index according to sustainability categories, in the analysis of the concept of contemporary mixed-use centre, to discuss the relationship between that contemporary trend and sustainability, and highlight the benefits associated with it.

The study below is following the strategy of the qualitative methods which in fact enables the researcher to show the link between the theoretical thought and the output of the induction, analysis and description through developing a broad base of data and information.

The research begins with a broad analytical narrative of the literature to determine the nature and scope of issues surrounding mixed-use urban centres and sustainability as well as to provide insights into the present condition of knowledge on the subject. Then the study comes to call for an investigation the relation between mixed-use development and sustainability by exploring the sustainability encounter and presents the perceived relationship between its basic dimensions and mixed-use development, then discusses the position of those dimensions with regard to mixed-use development. At long last, the paper concludes with a summary of the main findings.

2 Mixed-Use Development as a Contemporary Trend for Urban Centres

Demystifying the concept of mixed-use development and determining a common theoretical framework that forms the practiced requirements of it, need to understand both of its core components and the strategic goals of the theory behind the concept. Questions remain as to whether the concept is to be consistently understood or there are amount of underlying complexity; thus, there are many questions that should be answered to clear and full understanding of all aspects and meanings related to the mixed-use development concept (see Fig. 2).

The term mixed-use arose in urban planning circles over as a tool for urban revitalization, especially in large-scale projects (Miller and Miller 2003). But, it has some ambiguity. It has different connotations for different authors addressing the lack of clarity surrounding the topic. Coupland mentioned, "The term 'mixed-use' or 'mixed-use development' are widely used, but seldom defined" (Coupland 1997). Rowley view that, "Mixed-use development is an ambiguous, multi-faceted concept" (Rowley 1996b). Rabianski sight that "The definition of mixed-use is not as precise as for a single property type" (Rabianski et al. 2009). In addition, Grant's opinion is "Although the term appears frequently in the planning literature, it is rarely defined" (Grant 2002). Furthermore Hoppenbrouwer said, "The term mixed-use development has frequently appeared in the planning literature. However, this term is hardly defined" (Hoppenbrouwer and Louw 2005).

2.1 Defining the Mixed-Use

The ambiguity in defining mixed-use lies in the fact that there are several ways of mixing land uses and almost infinite

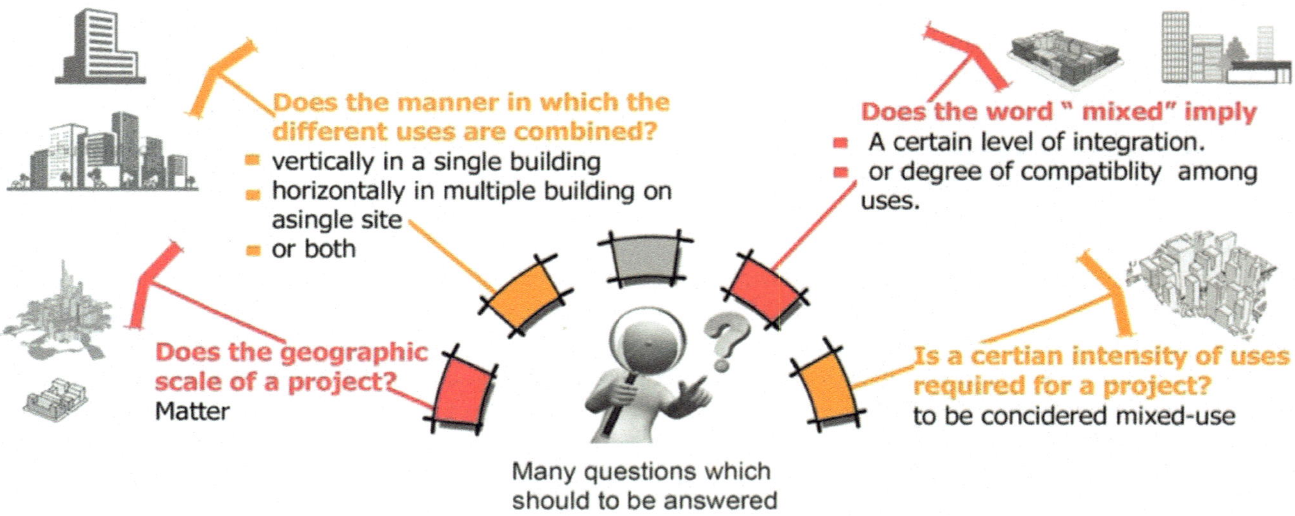

Fig. 2 Mixed-use development concept

forms of development, scale, density and the degree of mix the most famous ways (Marsh 2006). Thus, there are many institutions in different countries which have fashioned definitions of mixed-use in order to come a clear understanding of what it actually is. The Urban Land Institute[1] (ULI) in the United States has been developed the most widely definition of the mixed-use term. The original definition, developed in 1976 which published on the topic of their first book entitled Mixed-Use Developments is: "A mixed-use development is a real estate project with planned integration with some combination of retail, office, residential, hotel, recreation or other functions. It is pedestrian-oriented and contains elements of a live-work-play environment. It maximizes space usage, has amenities and architectural expression, and tends to mitigate traffic and sprawl" (Niemira 2007).

The ULI's definition characterized by three important factors consists of: First, three or more important uses supports each other and produce revenue (e.g., retail, entertainment, residential, offices, hotel, and/or can be divided into civil, cultural, recreation uses). Then, physical and functional integration of the project elements within the coherent and intensive use of the land and considerable pedestrian connectivity. Finally, compatible planning and development with a well-knit plan and taking into account the scale and type of uses, allowed densities, and related things (Schwanke et al. 2003).

[1]*Definition* is the product sponsored by four professional organizations in the real estate industry: the National Association of Industrial and Office Properties (NAIOP), the Building Owners and Managers Association International (BOMA), the International Council of Shopping Centers (ICSC), and the National Multi Housing Council (NMHC).

Additionally, ULI specifies that mixed-use must include at least three integrated uses and each of them must be substantive enough to draw a substantial market in their own right, which excludes the uses that serve as amenities for a fundamental use (Schwanke et al. 2003). Therefore, projects that contain several uses, but the lack pedestrian connectivity, such as master planned communities in the suburbs that require an automobile to traverse, do not match their demands for mixed-use.

Upon inspection, the ULI's definition of mixed-use was first conceived in 1976, a time when large-scale mixed-use developments were applied as an instrument for urban revitalization in the United States (Miller and Miller 2003). This clarifies that the development trends of the time play a substantial role in the how mixed-use developments are perceived.

The Industry survey definition, a survey done in 2006 by many national and international associations, The purpose of the survey was to identify the fundamental characteristics of mixed-use development and the results were synthesized to produce the following definition; It was defined as a real estate project with a combination of some function such as retail, office, residential, hotel and recreation in planning integration with some characteristics; It is pedestrian-oriented and contains components of a live-work-play environment. As well, it has amenities and architectural expression and tends to ease traffic and spread with maximized space utilization (Niemira 2007). The industry definition implies that, only two uses are required in the development for considering it as a mixed-use (Hoppenbrouwer and Louw 2005).

The Planning Portal is the UK Government's online planning that defines the mixed-use as a combination of complementary uses such as; residential, community and

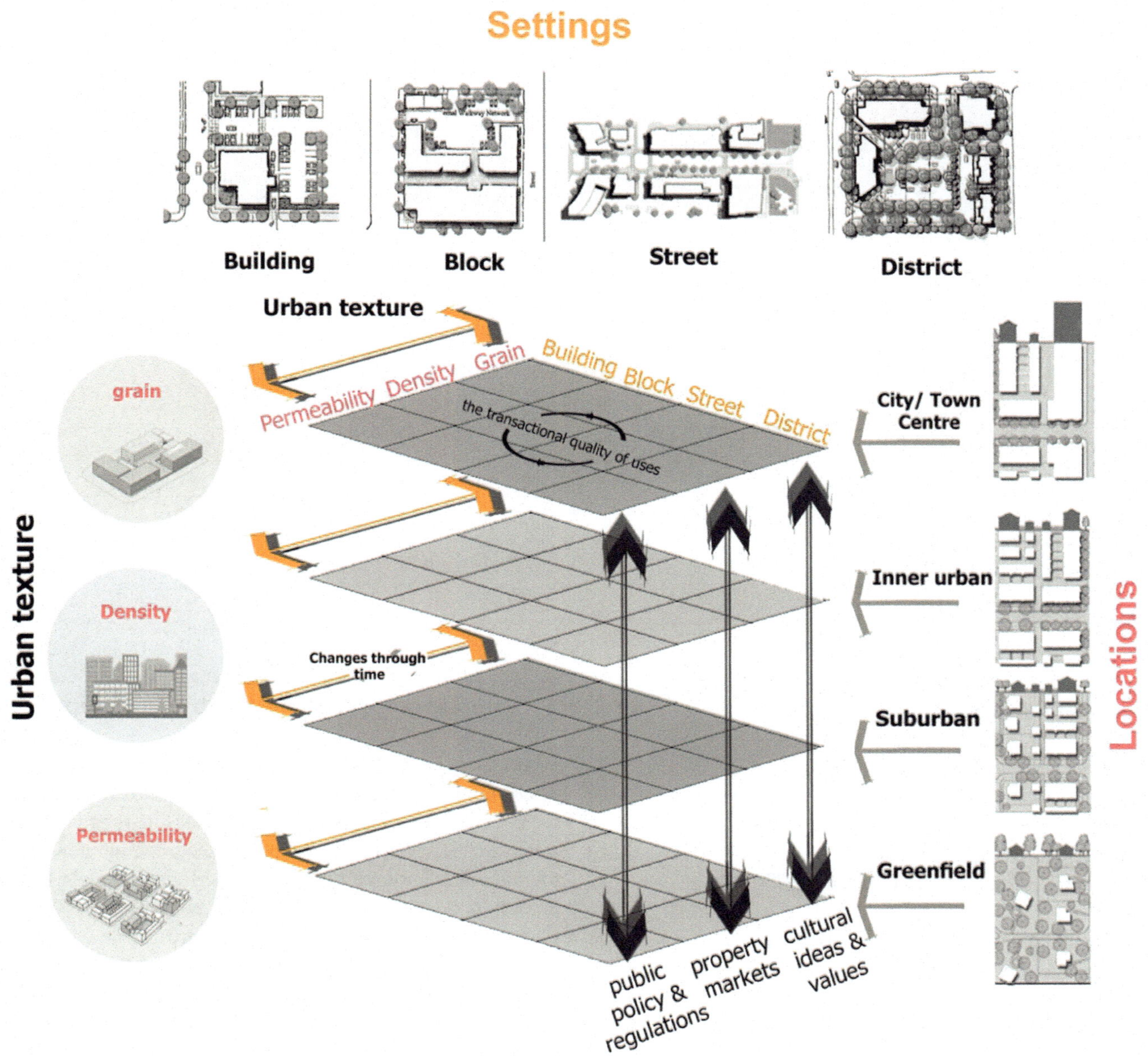

Fig. 3 Alan Rowley's model, adapted by author

recreational uses in a particular location or region (Planning Portal 2009).

In Australia, in particular, there is growing concern about the ability of the development sector to adequately respond to the challenges of urbanization, which has led to the strengthening of mixed-use projects by government and local authorities, and therefore many definitions have been developed for many government institutions. Thus, some local governments like Parramatta City Council: "Mixed-use development means a building or place comprising two or more different land uses" (PCC 2009). And The City of Sydney: "Mixed-use development means a building, or buildings, in which two or more uses are carried out" (COS 2005).

Both Parramatta City Council and City of Sydney Council take very wide definitions for mixed-use development. Compared to the ULI, which maintains that three uses is the minimum number to be considered as a mixed-use development, they agree that mixed-use development is any mix of two uses or more. In addition, both councils maintain that a mixture of uses within a single building is considered mixed-use.

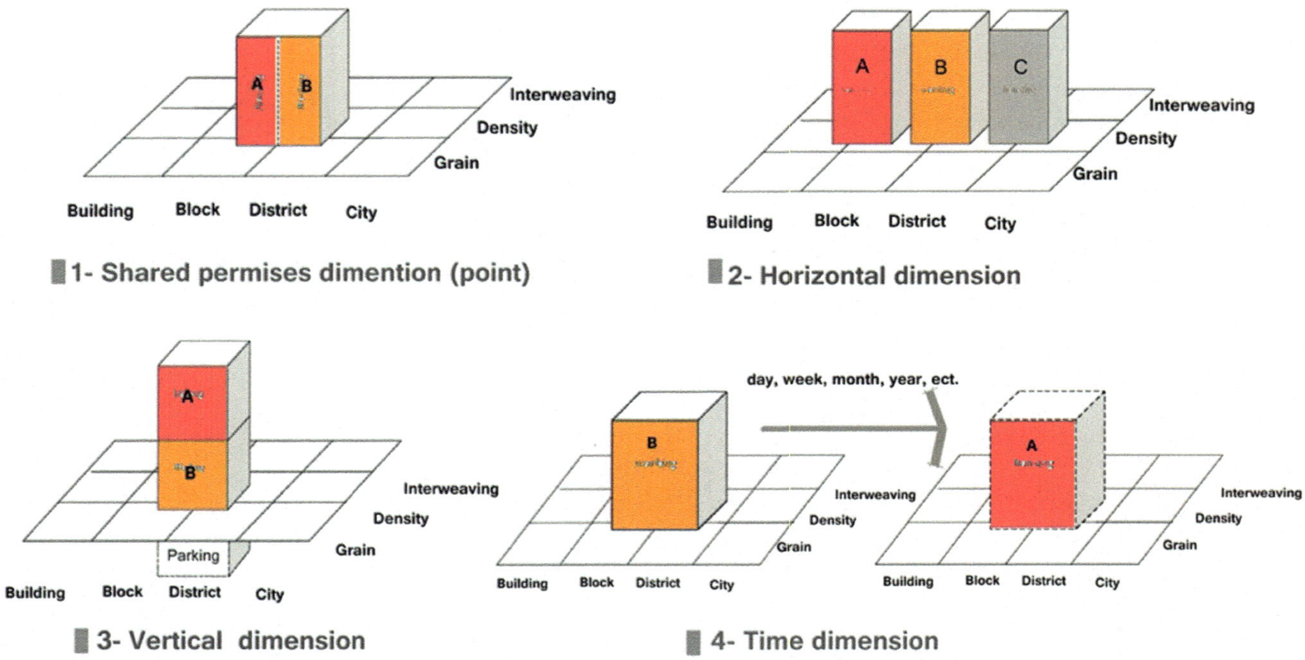

Fig. 4 Hoppenbrouwer and Louw's mixed-use model

2.2 Conceptual Analysis of Mixed-Use Developments

In order to better define mixed-use, it is important to understand the primary elements that influence the texture of an urban area. Furthermore, conceptualizing the diversity of ways that mixed-use projects can be exhibited in physical space besides a series of consideration:

The function of the individual land uses.
The manner and form in which multiple uses can be combined.
The scale at which a mixing of uses can occur.

Thus, there are two conceptual models which are consistently referenced in the literature. The first typology was developed by Rowley (1996b) (Table 1).

And the second was made by Hoppenbrouwer and Louw (2005) in an attempt to expand upon Rowley's ideas (Hoppenbrouwer and Louw 2005) (Table 2).

Consequently, there are assortments of factors that effect in various ways the function and form of urban Centre's land uses. The idea of mixed development can vary in use, as far as the nature and combination of uses, the scale in which the blend of uses happens, the dimension in which the uses are mixed, and the urban texture that is shaped inside the project as well as throughout the surrounding area. Thus, there is an almost infinite amount of possible mixed-use configurations and properties.

3 Contemporary Urban Centres in Egypt

Egypt has adapted many contemporary experiences by establishing new cities and communities in urban regions. In the Greater Cairo region, attention has recently begun to new trends in the real estate development sector within new cities to raise and improve community performance as well as attract individuals until these new communities. Consequently, the construction of urban centres in new communities also has inevitably been affected by the impact of new trends, mixed urban centres have been steadily increasing. To illustrate the features of the contemporary centre in the Greater Cairo area, some urban centres will be selected briefly, highlighting their features and description them according to the characteristics already mentioned (Fig. 5).

The Cairo Festival City is a mixed-use urban community strategically located in the new city of Cairo. The project includes a number of uses such as: restaurants, shopping and entertainment areas, residential areas of villas and luxury apartments, as well as main office space, along with hotels and educational institutions. In addition, many parks, paths and squares combine areas to promote and enhance gatherings for family and friends as well as various amenities, places and garages.

Arkan is located on the main commercial and retail centre, which houses several projects by the largest and most prominent developers in Egypt. The project comprises of

Table 1 The first conceptual model (Alan Rowley's typology)

Alan Rowley's typology	The internal texture of settlements is a primary aspect on the conceptual model of mixed-use development created by Rowley (1996a, b). This model concentrates on mixed-use in the horizontal dimensionor between contiguous buildings, and submit that the physical form of mixed-use development is a function of urban texture, setting and location. See Fig. 3
Urban texture	Rowley submits that the urban texture is the result of three things: grain, density and permeability – Grain: The grain of a project refers to the manner in which the various elements of a project are mixed together in space. Level of integration means the level intensity of the mix uses of the grain, a fine grain refers to a project where similar uses are widely dispersed amongst non-similar uses, While a coarse grain indicates project were extensive areas of one use are separated by extensive areas of another use. As well Level of transition: from one use to the next. Moreover, an abrupt transition from a cluster of similar uses till non-similar uses is referred to as a sharp mix and gradual transition is referred to as a blurred mix – Density: Level of compactness (horizontal to vertical). It is a measure of the intensity of land uses, and refers to the amount of space or count of units contained within a particular area – Permeability: Level of distinctions between buildings, blocks, streets and districts. Permeability refers to the layout of the roads and the number of possible ways that the pedestrians choose from as they move through a certain area. The corresponding size and shape of the blocks, and the placement and design of the buildings and public spaces within each block
Setting	Rowley also inserts setting or spatial scale into his model, making distinctions between buildings, blocks, streets and districts
Locations	Furthermore, he differentiates between four types of locations where mixed-use development: – City or town centres, including the commercial and civic core of towns and cities – Inner-city areas on sites including derelict, vacant or built-up land needing revival – Suburban or edge-of-town locations – Greenfield sites out past the urban fringe

Table 2 The second conceptual model (Hoppenbrouwer & Louw's typology)

Hoppenbrouwer and Louw's Typology:	Hoppenbrouwer and Louw's typology expands upon Rowley's and integrates the components in a more systematic way (Rabianski et al. 2009). They recognize the need to consider the vertical dimension, as well as a mix of uses within a single structure and different uses of a space over some period of time. They include other elements in his conceptual model and concentrates on the spatial perspective like function, dimension, scale and urban texture. See Fig. 4
Function	The function on mixed-use development indicates to the single land uses that are being mixed. However, their typology used housing and working for the purpose of simplicity, the model is flexible enough to be extended to any other combination of uses
Dimension	The dimension component consists of four elements: The shared premise dimension The horizontal dimension, The vertical dimension The time dimension
Scale	The method of subdivision is different in Hoppenbrouwer and Louw's model, they divided the mixed-use development to levels (the building, block, district and city)
Urban texture	Urban texture was taken from Rowley's model, but they used the interweaving of functions instead of the permeability, while retaining with grain and density

office space and in addition, retail space and eateries that are accommodated for the maximum variety of tenant options and an assortment of activities. The territory has just turned into a business centre point for the city and is expected to be one of the key business areas in the Middle East in the coming years.

Galleria 40 is a mixed-use commercial complex in Egypt, featuring a luxury shopping area and green office buildings. It has been designed as a forward-thinking, innovative development and ensuring lasting investment value. Its parts are within walking distance of all business facilities on site and are surrounded by pedestrian areas.

Katameya Downtown, located at the tip of the East 90 Road, the bloodline of New Cairo, sited in a prime location which is easily accessible via a core network of roads from the major Cairo districts. Katameya Downtown consists of retail shops, cafes and restaurants, office spaces with business visitors and shoppers daily.

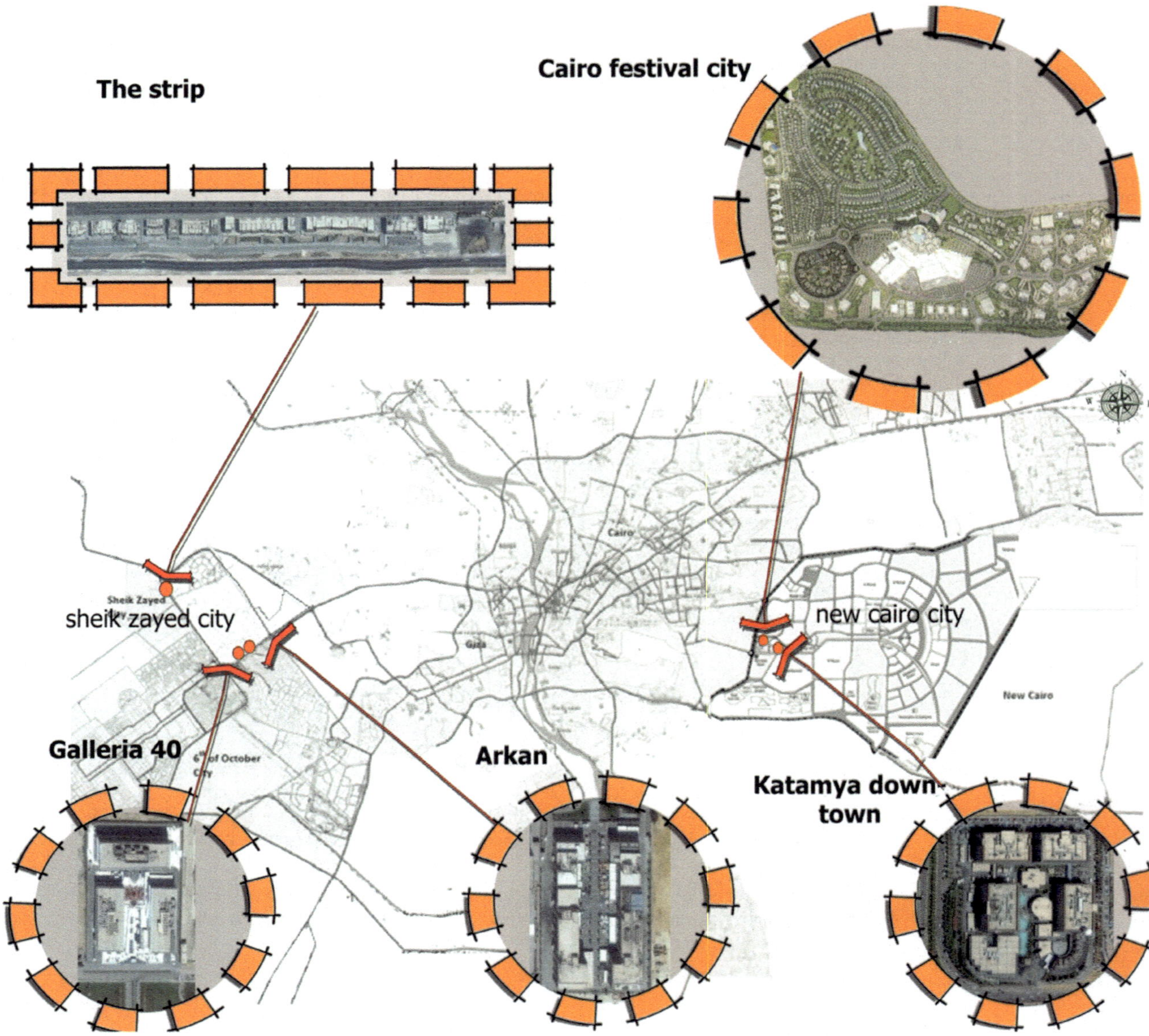

Fig. 5 Map for the Greater Cairo area shows the examples

The Strip is located directly on Dahshur road, linking the Cairo–Alexandria Desert Road in Juhayna Square. Strategically located in a growth zone with excellent transport connections, it is located at the intersection of two main streets which located in the heart of Cairo's west. This Strip provides high-end retail space, and shops for a full range of retail and services. Also, parking spaces for visitors (Table 3).

4 Goals Underlying Contemporary Urban Centres

Mixed-use development can contribute to a variety of goals around the desire to change current patterns of urban centres' growth and correcting adverse effects. Moreover,

Table 3 Description the contemporary urban centre in Egypt, according to the mixed-use topology

Contemporary urban centre examples	Level of integrations (Urban texture)			Scale and setting				Location				Mixed-use scenarios		
	Horizontal	Vertical	Both	Building	Block	Street	District	City/town centre	Inner urban	Suburban	Green field	2	3	More
The strip	√					√				√		√		
Arkan		√			√			√					√	
Galleria 40		√			√			√					√	
Cairo festival city			√				√		√					√
Katameya Downtown			√		√			√					√	

Table 4 Agreed benefits underlying contemporary centre, adapted by the author

The expected benefits	The contemporary movements, trends and advocates								Target
	1	2	3	4	5	6	7	8	
Introduce cost savings in infrastructure, land and energy		√	√		√		√		Economic
Allow shared parking and transport infrastructure to be used more efficiently	√		√			√			Economic
Decrease the economic costs of travel time		√	√	√	√	√	√		Economic
Lower household expenditure on transport and more options of transport	√		√	√	√				Economic
Helping in concentrate the knowledge and innovation activities in the urban centre		√	√						Economic/Social
Saving green spaces in conjunction with specific sorts of urban advancement	√	√	√			√			Environmental
Reduce emissions to the air and atmosphere (reducing the need to own a car)	√		√		√		√		Environmental
To help encourage greater physical activity, with consequent health benefits	√					√			Social
Promote vitality through activity and uses' diversity during the day and in the evening, which can reinforce each other	√	√	√	√	√	√	√	√	Social/Economic
Offer people convenience, choices and opportunity which lead to a sense of personal well-being	√		√	√	√	√	√	√	Social/Economic
Increase the viability and more vibrant commercial areas	√		√	√	√	√	√	√	Social/Economic
Enhance walkability and pedestrian connectivity with surrounding areas	√	√				√			Social/Economic/Environmental
More attractive and better-quality urban Centres	√		√	√		√	√	√	Social/Economic
Be associated with lower crime and more personal safety			√	√			√	√	Social/Economic
Component integration	√	√	√			√			Economic/Environmental

1 New Urbanism, *2* Smart growth, *3* Compact city, *4* Coupland (1997), *5* Hoppenbrouwer and Louw, *6* Alan Rowley's, *7* Grant Jill, *8* others: Jane Jacobs, Joseph Rabianski (Congress for the New Urbanism 2013; Daniels 2001; Hofstad 2012; Coupland 1997; Hoppenbrouwer and Louw 2005; Rowley's 1996a, b; Grant 2002; Jacobs 1961; Rabianski et al. 2009). √Means agreed

it "forms part of a strategy for sustainable development as well as a theory of good urban form, with the objectives of economic vitality, social equity, and environmental quality" (Grant 2002). The goals can then be divided into social, economic and environmental objectives. Information has been sourced from published journals and books, web pages and unpublished theses that illustrate the benefits of mixed-use integration (Table 4).

5 Sustainability

Today, the word "sustainability" has become very common and used in everyday life by most of us. We tend to talk about sustainable energy, sustainable ecosystems, sustainable development and so on to concern for the environment and the planet in general. Sustainability is a phenomenon of multi-disciplinary theories about sustainable development, which has been shaped by individuals and organizations for a long time and developed on the basis of different points of interest (Giddings et al. 2002). There has been growing literature on sustainability over the past three decades in urban studies. Many interpretations, definitions and practices have been presented in each of them along with a range of criticisms. Progress towards sustainability is now a path littered with many definitions, indicators and principles. We can see that, sustainability is a term that has evolved from the idea of "sustainable development", defined as realizing the development needs of all people without sacrificing the earth's ability to sustain life. Sustainability means balancing the human and absorptive effects of the natural world that can last indefinitely. Sustainability takes into account three interrelated components: the environment, the economy and the social order. The balance between these elements will require the adoption of a new ethic, a new way of life and new expectations to ensure our collective survival. With the advent of "sustainability" in developmental science, it has similarly prompted the use of advanced concepts of "sustainability" in contemporary debates on how to renew, revitalize and reform urban areas and urban communities by the planners. On the other hand, "sustainability" is seen as the correct means or the best possible end to urban improvement. A wide range of authors proposed different principles and methodologies for urban sustainability, after realizing the importance of urban sustainability in the overall practical improvement of sustainable development.

The sustainable city conference in Rio de Janeiro in 2000 expressed that the concept of sustainability, as applied in the city, is the ability of the urban area to continue to work at the levels of quality of life sought by community without limiting the options available to present and future generations, And causing adverse effects beyond urban boundaries (Brebbia and Florez-Escobar 2015). Sustainability is therefore the only way to sustain and achieve the end product required for a lifestyle. Sustainability in urban areas concentrates on the human community as a multidimensional space, shaping and affecting sustainable development at the local and global levels; a space that is manifested not only in the form of physical infrastructure, but also in terms of social, economic and environmental systems situated in communities (UN Habitat 2009).

Move towards sustainability is also requires a series of actions and a range of approaches which have been pursued to measure it, the most influential ones have been those dedicated to developing sets of sustainability indicators. They have prompted much debate in the literature as a result to the tensions between expert-led and citizen-led models in their development. However, the indicators were the most effective instrument of measurement at all. Sustainability is divided into a set of dimensions, each of which has a combination of indicators to measure as will be shown in the next section.

5.1 Dimensions of Sustainability

The most common division of sustainability includes three dimensions: the social, the economic and the environmental (Crilly 2000).

The Economic Dimension:
"Economic sustainability" means a system of production that satisfies present consumption levels without compromising future needs. Sustainability sought by economic sustainability is the sustainability of the economic system itself. Holmberg (1992) states that an economically sustainable system must be able to consistently produce better services and avoid harm to products. Additionally, Jacobs and Šlaus (2011) call attention to that an economically sustainable system is to enhance human economic well-being in disposable personal income, employment, equality in income distribution, energy efficiency education and net household savings.
Economic sustainability is an important for two reasons (Jonathan et al. 2001): First, the most businesses do not seek to sustainability unless they view the financial benefits back from it. Second, financial wealth is a significant criterion affecting the quality of life. On this light, the economic sustainability should yield many solutions with long-term financial viability and the goal of financial profit should not be inconsistent with social and environmental targets.
In sum, the basic concept of economic sustainability is how to increase the income and the revenue while maintaining or reducing asset stocks, as well as the possibility of determining economic sustainability in terms of maximizing welfare over time (Munasinghe 2007).
The Environmental Dimension:
Indeed, this dimension stems from concern about the environment and the degradation of natural resources. Similar to the meaning of sustainability, it is characterized as meeting the services and resource needs of present and future generations without bargaining the strength of the ecosystems that give them (WCED 1987). Most importantly, environmental sustainability could be described as a state of balance, strength and cohesion that allows the human community to meet its needs while not exceeding the

boundaries of its supporting ecosystems to maintain the regenerating of important services to address those needs and not through our actions that reduce biodiversity and have consequences pose a direct threat to society and the economy. These results incorporate an unnatural climate change, asset exhaustion, waste disposal, air contamination, land degradation, ozone consumption, diminished biodiversity and water contamination (Jonathan et al. 2001). The environmental dimension pays attention to regulating services, the benefits that are obtained from the management of ecosystem processes, the conservation of natural resources and maintain the environmental conditions for people (eg. Clean water and air, a suitable climate).

The Social Dimension:

Social sustainability is the key element of sustainability (Jonathan et al. 2001). Environmental and economic factors are imperative; however, they are mandate not only to themselves but also in achieving social sustainability. It is indeed working towards social sustainability. Social sustainability is concerned with ensuring the sustainability of the diverse social relationships found in healthy communities. In the context of the built environment, this involves the creation of physical, cultural and social spaces that promote the prosperity of individuals and promote the sense of community.

Urban regions have empowered social change and integration through some remarkable sustainability criteria that address social difficulties, for example, poverty, misuse, weakness (mental and physical), an absence of essential services, common distress, wrongdoing and depression (Roseland 2000; Colantonio 2007; Ndeke 2011; Jacobs and Šlaus 2011). Jenks and Jones (2010) argue that the socially sustainable system must include certain characteristics such as: protection of mental well-being, physical health, promotion of society, fair treatment of all individuals and provision of basic services to all individuals. As well as complementary needs (business, entertainment, infrastructure to bolster social and cultural life, social comforts, frameworks for national engagement and space for individuals and places to develop) without affecting future generations. This is consistent with the goals of sustainable development.

There is an increasing concern with establishing a means to assess progress towards sustainability as well the advocacy has increased to measure it. Therefore, numerous analysts in colleges, environmental associations, think tanks, national governments and worldwide organizations have upheld the development of indicators for sustainability measurement as the dominant and the best method (Ling 2005; Munier 2005). Accordingly, the indicators to be addressed in this paper are particularly appropriate and adapted to urban areas from a very wide range of sustainability indicators (Table 5).

6 Discussion

This research was intended to provide a comprehensive analysis of the relation between mixed-use contemporary centre and achieving sustainability across three categories: economy, society and environment. Thus, the research tries to delve into how to reverse this relationship through scores to facilitate the investigation process.

Given the subsequent tables that illustrate the relationship between indicators of sustainability dimensions and the expected benefits of contemporary urban centres. And considering these benefits as elements or criteria, the impact of each element on sustainability indicators was measured. An effective relationship between the element and sustainability indicators is fixed by a (√) sign, then each sign was calculated by a value equal to (1). The goals of urban centres have been divided into environmental, social and economic benefits compared to the dimensions of sustainability. The relation value in this study is the relative benefit index which reflects the correlation level of each benefit with categories of sustainability (Table 6).

It is found that the relative environmental benefits index equal to 72% which means there is strong relation between the contemporary centre and environmental category of sustainability (Table 7).

Furthermore, it is found that the relative economic benefits index equal to 63% which means there is also med-high relation between the contemporary centre and economic category of sustainability (Table 8).

In addition, the relative Social benefits index equal to 74% which means the highest relation is between the contemporary centre and social category of sustainability (Chart 1).

Finally, the relative results show that there is a strong relationship between the contemporary centre for mixed-use and sustainability. First of all, contemporary mixed-use centre received the highest score in achieving social sustainability, followed by environmental and economic sustainability.

Table 5 Sustainability dimensions

Dimensions	Indicators	Descriptions, Purpose
Environment	Air quality	Increase air quality while avoiding air pollution
	Water quality	Increase water conservation, including water-sensitive urban design
	Green/open space	Provide sufficient green/open space with good accessibility, helping mitigate the effects of climate change and urbanization
	Waste	Reduce solid waste with disposing it in an environment-friendly way, Increased use of recycled or sustainable sourced materials
	Transportation	Create pedestrian/bicycle-friendly community and less car dependency
Economy	Business and economic diversification	Supporting ventures and private businesses to get better service and a wider choice of goods as well to create competitive advantages
	Employment	Provide of adequate neighbourhood employments and relating data for nearby inhabitants
	Infrastructure/energy	Provision of sufficient infrastructure (water, sewage, electricity and telecommunication). Introduce a system for conserving and reusing waste water, improve energy-efficient building design and layout Enhance facilities that utilize sustainable power sources, or decrease the utilization of non-renewable energy source for their operations and transport needs
	Public transportation	Transport systems make a positive contribution to the economic sustainability. It offers choices of transport mode, affordable, operates efficiently and supports a vibrant economy, decrease the cost value of losing time in traffic congestion.
	Income and well-being	Improve household income. The benefits reach well beyond the traditional and limited measurements of economic performance and improves human well-being and overall welfare well, services and goods.
Social	Social equity	People of all ages, races, cultures, sexes and abilities are given access to services (ensure affordability of services and amenities and adequate communication and fair access to social services and goods and access to property ownership and security of tenure and improve community transport).
	Social capital	Increase opportunities for social interaction and provide a focus for the community. Social capital refers to the institution, relationships and norms that shape the quality and quantity to community's social interactions such as (working, family, friend and neighbour's relationship)
	Community stability	The longer the population wants to survive in the long run and invest in the community, the more socially sustainable the community becomes, and the more it will be attractive
	Sense of place	Enhance a sense of place and create unique places that feel authentic and possess a variety of distinctive characteristics that draw pedestrians back again and again
	Safety and security	Increase public safety (neighbourhood, pedestrian and bicycle) with less crime/accident, security of property

Table 6 Relation between sustainable environmental dimension and contemporary mixed-use centre environmental goals

Sustainability dimensions	Indicators	Contemporary mixed-use Centre environmental benefits (CMUCENB)				
		A	B	C	D (Shared)	E (Shared)
Environment	Air quality	√	√	√	√	√
	Water quality	√			√	√
	Green/open space	√	√	√	√	√
	Waste				√	√
	Transportation		√	√	√	
Total score	5	3	3	3	5	4
Percentage		0.6	0.6	0.6	1	0.8

A Saving green spaces in conjunction with specific sorts of urban advancement and conservation of natural ecosystems within urban areas and protect the integrity of the natural environment, *B* reduce emissions to the air and atmosphere, *C* enhance walkability and pedestrian connectivity with surrounding areas, *D* component integration, *E* introduce savings in infrastructure, land and energy

Table 7 Relation between sustainable economic dimension and contemporary mixed-use centre economic goals

Sustainability dimensions	Indicators	Contemporary mixed-use Centre economic benefits (CMUCECB)						
		D (Shared)	E (Shared)	F	G	H	I (Shared)	K (Shared)
Economic	Business/economic diversification	√	√	√	√	√	√	√
	Employment						√	√
	Infrastructure/energy	√	√	√		√	√	
	Public transportation		√		√	√		
	Income and well-being		√		√	√	√	√
Total score	5	2	4	2	3	4	4	3
Percentage		0.4	0.8	0.4	0.6	0.8	0.8	0.6

D Component integration, *E* introduce cost savings in infrastructure, land and energy, *F* allow shared parking and transport infrastructure to be used more efficiently, *G* decrease the economic costs of travel time, *H* lower household expenditure on transport and more options of transport, *I* helping in concentrate the knowledge and innovation activities in the urban centre, *K* promote vitality through activity and uses' diversity during the day and in the evening, which can reinforce each other

Table 8 Relation between sustainable social dimension and contemporary mixed-use centre social goals

Sustainability dimensions	Indicators	Contemporary mixed-use Centre Social Benefits (CMUCSOB)						
		I (Shared)	J	K (Shared)	L	M	N	O
Social	Social Equity	√	√	√	√	√		
	Social Capital	√		√	√	√	√	√
	Community stability	√	√		√	√	√	√
	Sense of place	√			√	√	√	√
	Public safety/crime		√	√			√	√
Total score	5	4	3	3	4	4	4	4
Percentage		0.8	0.6	0.6	0.8	0.8	0.8	0.8

I Helping in concentrate the knowledge and innovation activities in the urban centre, *J* to help encourage greater physical activity, with consequent health benefits, *K* promote vitality through activity and uses' diversity during the day and in the evening, which can reinforce each other, *L* offer people convenience, choices and opportunity, *M* increase the viability and more vibrant commercial areas, *N* more attractive and better-quality urban centres and *O* be associated with lower crime and more personal safety

Chart 1 Contemporary mixed-use centre benefits

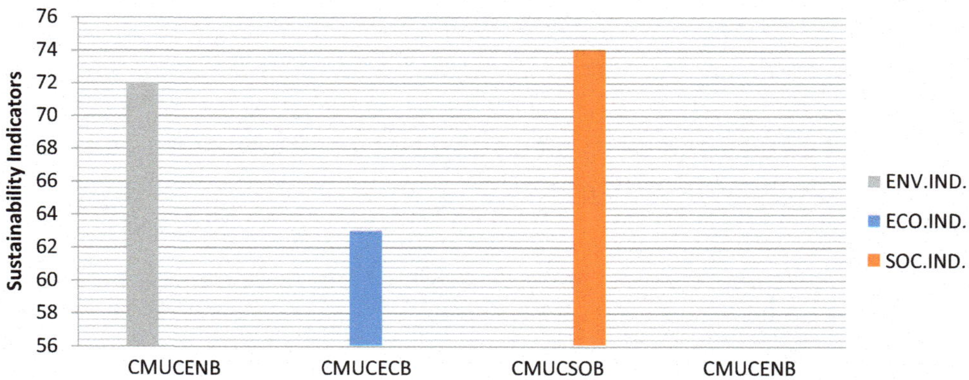

7 Conclusions

Evolution of the Urban Centres and being influenced by contemporary trends and movements, such as New Urbanism, smart growth and compact city, the tendency to adopt the concept of mixed-use as a common principle in recent times. As these movements gained great momentum, influential writers inspired planners around the world to advocate mixed-use developments and incorporate many of its benefits and the persistence of this concept as facilitators of the social, economic and environmental sustainability objectives of the urban form. Egypt has been affected by many experiments, especially in the new urban areas of the Greater Cairo region, characterized by a legacy of mixed-use of the contemporary urban centres.

It was the purpose of this study to provide a narrative analysis of the benefits associated with mixed-use as a main towards achieving sustainability. The study began by looking at literature and understanding the concept of mixed-use and its vocabulary and characteristics, and to extract a range of benefits and classify them into social, economic and environmental benefits, while shedding light on some examples in Egypt. In addition, the study was built on the understanding that currently there is a debate about the importance of sustainability and identifying its concept, its dimensions, and its indicators that were categorized and placed within groups. Through discussing the relationship between a trend of the contemporary urban centre and the sustainability, there are advantages to bringing together different types of activities within a relatively mixed-use and diversified urban areas.

In the end, the results of this research yielded a link between the trend of contemporary urban centre as a mixed-use and sustainability, and the results indicate a strong relationship. Consequently, the results can be used to develop sustainability policies and action plans for urban development, and to develop and take into account guidelines that will help create successful urban environments.

References

Brebbia C, Florez-Escobar W. *The Sustainable City X*. UK: WIT Press; 2015.

Colantonio A. Social Sustainability: An Exploratory Analysis of its Definition, Assessment Methods, Metrics and Tools, *OISD (EIB) WP* 2007; **1**.

Commission of the European Communities (CEC). *Green Paper on the Urban Environment.* Brussels: Commission of the European Communities; 1990.

Congress for the New Urbanism. *Charter Of The New Urbanism.* McGraw-Hill Education; 2013.

Congress for the New Urbanism. *New Urbanism: Comprehensive report & best practices guide*. NY: New Urban Publications; 2001.

COS. *Sydney Local Environmental Plan 2005 Appendices*. City of Sydney Council, Sydney; 2005. http://www.cityofsydney.nsw.gov.au.

Coupland A. *Reclaiming the City: Mixed Use Development*. London: E & FN Spon; 1997.

Crilly M. Dimensions of Sustainable Urbanism. *PhD International Ecotechnology Research*. School of Industrial and Manufacturing Science, Innovation and Technology Assessment Unit, International Ecotechnology Research Centre. Cranfield University; 2000.

Daniels T. Smart growth: A new American approach to regional planning. *Planning Practice and Research* 2001; **16** (3/4): 271–279.

Downs A. What does smart growth really mean. *Planning* 2001; **67** (4): 20–25.

Giddings B, Hopwood B, O'Brien G. Environment, economy and society: Fitting them together into sustainable development. *Sustainable Development* 2002; **10** (4):187–196.

Gleeson B. Towards a new Australian suburbanism. *Australian Planner* 2006; **43** (1): 10–13.

Grant J. Mixed Use in Theory and Practice: Canadian Experience with Implementing a Planning Principle. *Journal of the American Planning Association* 2002; **68** (1): 71 – 84.

Gregory C Dale FAICP. Current Planning Issues: Smart Growth. *Planning Commissioners Journal* 2003; **50**. Accessed June 2017, available at: http://plannersweb.com/wp-content/uploads/2012/08/182.pdf.

Hall P. *Cities in Civilization*. New York: Pantheon; 1998.

Holmberg J. *Making Development Sustainable: Redefining Institutions, Policy, and Economics*. Washington DC: Island Press; 1992.

Hoppenbrouwer E, Louw E. Mixed-use development: Theory and practice in Amsterdam's Eastern Docklands. *European Planning Studies* 2005; **13** (7): 967–983.

Jacobs G, Šlaus I. Human Capital and Sustainability. *Sustainability 2011*; **3**: 97–154.

Jacobs J. *The death and life of great American cities*. London: Pimlico; 1961.

Jenks M, Dempsey N. *Future Forms and Design for Sustainable Cities*. Oxford: Architectural Press; 2005. P. 287.

Jenks M, Jones C. *Dimensions of the Sustainable City*. Springer Science+Business Media B.V.; 2010.

Jonathan M, Timothy M, Kevin P, Neva R. *A Survey of Sustainable Development: Social and Economic Dimensions*. Global Development and Environment Institute, Tufts University. Washington D C: Island Press; 2001.

Kelbaugh D. The New Urbanism. In S Fainstein and S Campbell (Eds.), *Readings in urban Theory*. Oxford and Malden: Blackwell; 2002. p. 354–361.

Ling G. *Sustainability and Cities: Concept and Assessment*. Singapore: Institute of Policy Studies; 2005.

Marsh A. Mixed Management: Mixed-Use Property Managers Juggle Varying Tenant Needs. *Commercial Property News* 2006; **20** (14): 23.

Miller N, Miller J. Defining Mixed Use Development. *Design Center for American Urban Landscape*, Minneapolis; 2003.

Munasinghe M. *Making Development More Sustainable: Sustainomics Framework and practical applications*. 1st edition. MIND Press; 2007.

Munier N. *Introduction to Sustainability, Road to a Better Future*. Netherlands: Springer; 2005.

Ndeke E. A critical review of the development of sustainability indicators for the City of Cape Town: A focus on environmental and socioeconomic sustainability; Master Thesis, *Stellenbosch University* 2011. http://hdl.handle.net/10019.1/18078.

Niemira M P. The Concept and Drivers of Mixed-Use Development: Insights from a Cross-Organizational Membership Survey," *Research Review* 2007; **4** (1): P. 54.

PCC (2009) Draft Local Environmental Plan 2009. Parramatta Council, Sydney. https://www.cityofparramatta.nsw.gov.au.

Planning Portal. *Planning Portal*. United Kingdom; 2009. http://www.planningportal.gov.uk/england/genpub/en/.

Porter M. Clusters and the New Economics of Competition. *Harvard Business Review* 1998; **76** (6): 77–90.

Porter, D. *Making smart growth work*. Washington D C: Urban Land Institute; 2002.

Rabianski J, Gibler K, et al. Mixed-Use Development: A Call for Research. *Journal of Real Estate Literature* 2009; **17** (2): 205–230.

Roseland M. Sustainable community development: Integrating environmental, economic, and social objectives. *Progress in Planning* 2000; **54** (2): 73–132.

Rowley A. Mixed-use Development: ambiguous concept, simplistic analysis and wishful thinking. *Planning Practice & Research;* 1996b. **11**(1): 85–98.

Rowley, A. *Mixed-use Development: Concept and Realities*. London: Royal Institution of Chartered Surveyors; 1996a.

Schwanke D, Phillips P, et al. *Mixed-Use Development Handbook*, 2nd ed. Washington D C: ULI - The Urban Land Institute; 2003.

Talen E, Knaap G. Legalizing smart growth: An empirical study of land use regulation in Illinois. *Journal of Planning Education and Research* 2003; **22**: 345–359.

Talen E. *New Urbanism and American Planning: The Conflict of Cultures*. New York: Routledge; 2005.

UN-Habitat. *Planning Sustainable Cities*. United Nations Human Settlements Programme (UN-Habitat); 2009.

World Commission on Environment and Development (WCED). *Our Common Future*. Oxford: Oxford University Press; 1987.

Hofstad H. Compact city development: High ideals and emerging practices. *European Journal of Spatial Development* 2012; **49**.

Geographic Information System as a Support Tool for Planners to Choose the Best Suitable Location for the Development of New Economic Cities

Marwa Adel Elsayed and Samah M. Elkhateeb

Abstract

In Egypt, the challenge posed by current urban situation accompanied by the continuous increase in population is currently of crucial importance. The total land area of Egypt is about one million square kilometres, where only 7.6% of its total area is cultivated and inhabited. Egypt 2030 urban development plan aims to spread development over 40% of Egypt's area to exploit available natural resources and provide around 20 million job opportunities. Sinai Peninsula is one of the most promising destinations for development. However, development in such a virgin area and unique environment as Sinai should be cautious. Hence, the research paper aims to adopt a pilot method to support planners in choosing new cities locations using geographic information system as a scientific tool. The results have been shown through implementing a pilot analytical methodology to finally reach suitable city locations through combining the planning matrix with GIS technology to query the overlay of the target area of interest. Four different sets of criteria have been proposed accordingly to plan for economic city base founded on natural and physical resources.

Keywords

GIS • New cities • Sinai • Sustainable development • Sustainable location • Geographic information system

1 Introduction

Choosing new cities location are considered one of the major planning issues that should be based upon the particular database, for the planner to be able to choose the best sustainable site. Egyptian Government has adopted the policy of establishing three generations for the new cities (Hafez 2015). The Egyptian new cities as El-Shrouk, El-Obour, and 10th of Ramadan have been established according to the different aims (economic, social and environmental situations) (Hafez 2015). Most of these cities in the planning stage have been established upon GIS database taking the potential and constraints of their location into consideration. Hafez (2015) has reviewed the development bases of the new cities in some Arab countries and the influence of the real estate investment on developing the new cities and finally recommended the sustainable resilience alternative for the new cities in the presence of current potentials and constraints.

Building sustainable cities using different planning strategies are analysed by Martos et al. (2016), as they adopted ways for sustainable development not only for decision-makers and designers but also for the city users' and citizens' (Ding et al. 2015). Martos et al. (2016) designed a model "Trinity of Sustainable Cities" from spatial, logical and time dimensions, and principles for sustainable development assessment of cities in developing countries. Garcia-Ayllon and Miralles (2015) presented a GIS model with 50 indicators implemented in the territorial analysis, they developed a tool called a GIS retrospective analysis, and this tool helps the planner in decision-making and to make future advance diagnostics in territorial management. Limited researchers have addressed the use of GIS in selecting the best locations for development of new cities. This study aims to choose best city sustainable locations in Sinai using pilot meta-analysis methodology that could be applied along in Egypt to finally ensure choosing best

M. A. Elsayed
Department of Architecture, The British University in Egypt, Cairo, Egypt

S. M. Elkhateeb (✉)
Faculty of Engineering, Ain Shams University, Cairo, Egypt
e-mail: samah.elkhateeb@eng.asu.edu.eg

S. M. Elkhateeb
College of Architecture & Design, Effat University, Jeddah, Saudi Arabia

C. Alalouch et al. (eds.), *Advanced Studies in Energy Efficiency and Built Environment for Developing Countries*, Advances in Science, Technology & Innovation, https://doi.org/10.1007/978-3-030-10856-4_3

locations with economic bases grounded on the area resources.

1.1 Development Plan of Egypt 2030

Egypt's population is about 100 million inhabitants which is the 16th largest population among world population. Cairo is the largest city in Egypt with a population over 9 million. The next largest city, Alexandria, has about half of that number, with 4 million inhabitants (Hafez 2015). Egypt put its vision of the development of 2030 to achieve the sustainable development of the country that serves the current and future population of the country. In Egypt Vision 2030, there are 12 pillars in Egypt sustainable development strategy (SDS); one of them is the urban development pillar. This pillar is dealing with urban development issues, regarding current and new urban development plans. The new urban development plan aims to double the urban space, geographically redistributing development and citizens to capitalise on the use of available resources; as well as replacing and developing slum areas, and improving the quality of life. Double the urban space means planning more cities and redistributes development and citizen among the whole area of the country.

1.2 Sustainable Development Definition

Sustainable development is the overlay between, economic, environment and social sustainability. Sustainable development is the future of the planet, and it has been a keyword for most of the future development plans to cities and urban areas (Packalén 2010; Nurse 2006; Duxbury 2001).

Obviously, sustainable development is not the end goal only, but it is the process as well. The final goal is an equitable society that moves in manners and does not exhaust the planet's capability to maintain its resources. Achieving sustainability goal is a multi-stakeholder process that includes actors at different levels of government and throughout disciplines. Sustainable development in city planning and new cities should start from an early step, from selecting the location of the city and the economic base as well. Choosing the economic base of the community will act a significant role on the future sustainability of the community and the city.

1.3 Land Suitability Analysis

The land is an area of the earth, and its natural resources are valuable to all human. Land has an economic value varied based on the use and purpose, land for housing different than education, public services, transport, mining or recreation use. This means that land is a valuable resource in the urban area because more than half of world population lives in the urban area, this also in developing the area, half of its population lives in urban areas. This continuing urbanisation estimated to add 2.5 billion people to the urban population by 2050, with almost 90% increase concentrated in Africa and Asia (Madurika and Hemakumara 2017).

The importance of the land raised the importance of thinking of suitability techniques for land to inform decision-maker about suitable opportunities. Many valuable decisions can be taken when land suitability techniques implemented (Steiner et al. 2000).

Land use suitability analysis is identifying the most suitable location for future land uses based on specific requirements, predictors of some activities or preferences. The geographical information system (GIS) is one of the tools that can be used for land use suitability mapping and analysis for urban, mining, agriculture and land use projects (Collins et al. 2001). GIS has been used as a tool for land use suitability, especially for ecological use for plant species and animal habitat (Store and Kangas 2001). GIS also can be used in landscape evaluation, private and public property planning and analysing geographical favourability (Cambell et al. 1992; Eastman 1997; Church 2002). Additionally, GIS used in regional planning by Shin HuoyTerhKaiCao in 2018, as GIS is an essential tool in many planning activities.

Land use suitability definition is different based on the intended purpose of the land future use. For example, the agriculture use, it means the suitability of the land for cultivation of various crops, for urban planning it can be the suitability of the land for building various uses of the city. There is a difference between site search problem and site selection problem. Site search problem prearranged site for suitability analysis. On the other hand, site selection is the identifying of a specific location for a suitable activity based on the potentials of the site like size, location and other attributes. In the site search and suitability analysis, there should be a given site and this site should be identified through set of observations and data. Land suitability analysis is not only a GIS-based procedure; it can be used to allocate the most appropriate location for a significant use or future project. This process can be very important for urban planner and land developer of master plans and land use plans (Joerin et al. 2001).

2 Methodology

The research adopted a pilot analytical methodology, where it synthesise the results from overlaying multiple studies to determine the best city location across the surrounding potential and constrain, using GIS as a tool to support in decision-making. The researchers collected qualitative data concerning all the possibilities and constrained of the target site, and set three phases for the research developed after Robson 2011, which are data filtering phase, data integration phase and data deduction phase, as shown in Fig. 1.

a. Data filtering phase: this phase considered the research concrete stone, as the researchers made much effort in collecting all the maps of the target site in different formats and converting these maps to an optimum format to be imported into GIS database.
b. Data integration phase: after having a complete database, all the feature classes are now ready to union them in one feature class holding the attribute data for all the features.
c. Data deduction phase: in this phase, the researchers have designed queries to be able to refer to the database attributes, to select the best city location, according to the site potential and constraints.

2.1 Study Area

Sinai lies in the north-east of the Egyptian border; it is a triangular peninsula about 60,000 km^2. Sinai surrounded on the north by the Mediterranean Sea, and at the south, it lies between the Suez Gulf and the El Akaba Gulf. Approximately, 66,500 people live in South Sina, and 314,000 live in North Sina. The population of Port Said is 500,000 people

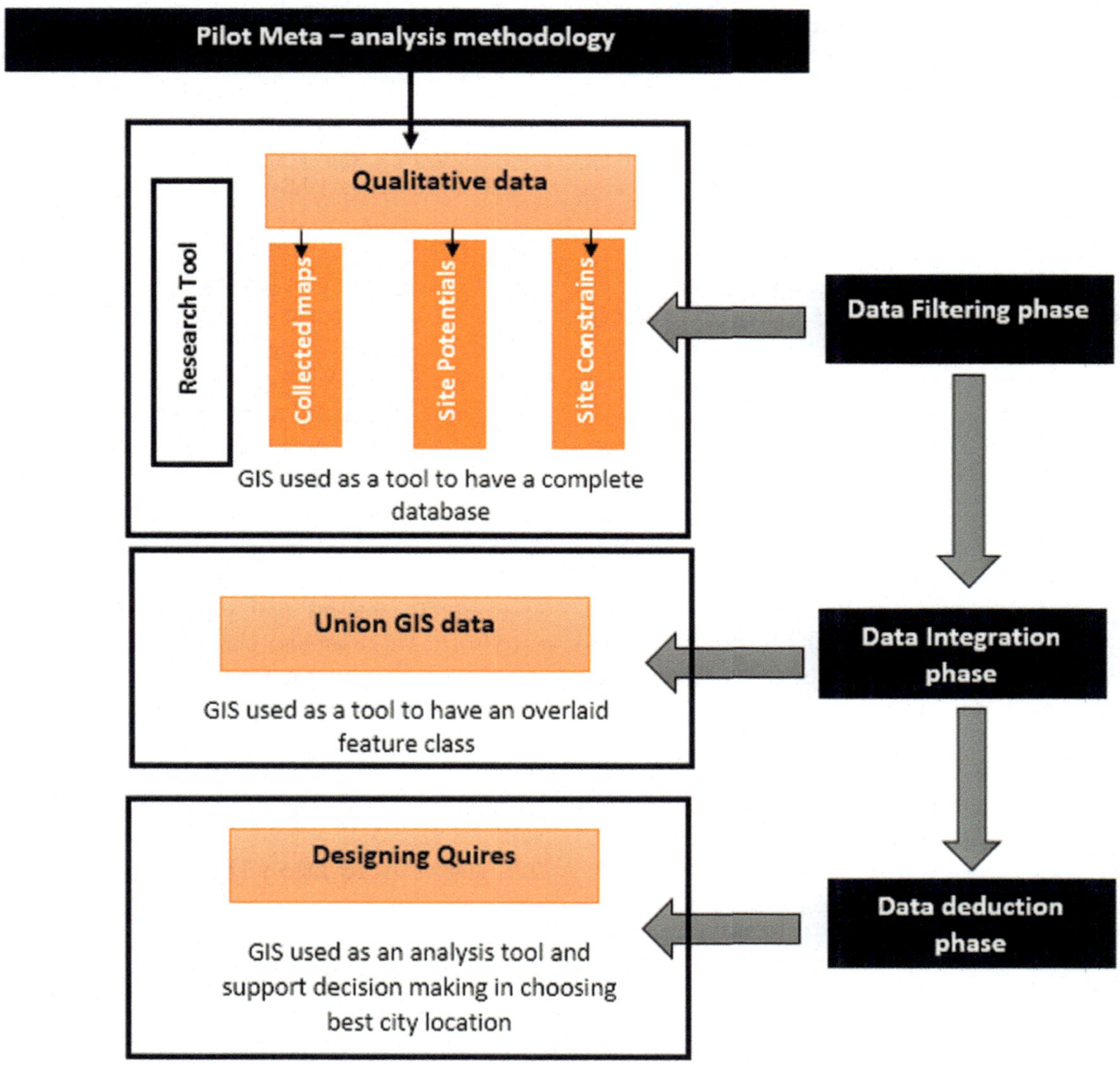

Fig. 1 Suggested methodology (Researchers 2017)

(Central Agency For Public Mobilization & Statistics 2016). Portions of the populations of El-Suez and Al-Ismailia live in Sinai, while the rest live on the western side of the Suez Canal. The total population of El-Suez and Al-Ismailia governorates is about 1.3 million, in fact, part of this population live in the Sinai, but the majority lives in the western part of Suez Canal.

The Camp David accord was signed 17 September 1978, by Egyptian President Anwar El Sadat and Israeli Prime Minister; Menachem Begin after thirteen days of secret negotiations at Camp David. The consequences of this accord are dividing Sinai into three sectors (A, B and C); each of these sectors has its conditions concerning the urban development (Camp David Accord 1978).

The first sector of Sinai Peninsula according to Camp David agreement starting from Port Said to Sharm El Sheikh. The area is chosen as a one unit part of Camp David agreement. This area has been chosen as a pilot project that can be repeated on the national level ("Egypt's Vision 2030" 2015). This area has been chosen for different reasons:

a. Spread of urban development that appears within various activities (Agriculture—Industry—Trade—Tourism—Residential) without the existence of a planned strategy for this area that organises this growth.
b. The existence of significant investment in the area that needs a kind of organisation and planning to reach the optimum methodology to manage the area sources.
c. This area is considered the most reliable central location among the other developed areas in the world, that is because of its unique location along the Suez Canal (new and old one), and in between the Red and the Mediterranean Sea.
d. This area passes beside the highest density region and highest population and resources as well, and it is adjacent to the eastern delta, Ismailia, and El Suez which is highly populated with people and labour which can be used in the development process in the chosen area (Fig. 2).

2.2 Data Filtering Phase

The area of interest is 5,374,684.454 acre. It lies in four governorates (Ismailia, Suez, Port Said, North of Sinai and South of Sinai). After stabilising at the study area, the researchers have gathered maps and information to set the potential and the constraints. These maps (survey map, climate maps, building material map, metallic and nonmetallic resources map, lithology map, land use map, conservation map for flora and fauna, petroleum wells map) were available in a raster format, and the following steps were applied to change its format to vector format;

e. Determining the coordinate system WGS 1984 UTM Zone 36N was set to project all the maps using this system.
f. Allocating area of interest (AOI) coordinates on a projected UTM map with the presence of Egypt map layers in the background.
g. Adding the raster maps to the GIS, and then using the Georeferenced tool, each map is adjusted.
h. The raster maps are digitised concerning the area of interest related to its target feature class.

The database is structured of many data sets; each contains GIS feature classes that are created or imported by ArcGIS and organised systematically and logically into the data sets so that can be found when needed. The pilot test case consists of 1 geodatabase that contains ten feature data sets which illustrated as follows climate data set (max and min temperature, max and min humidity, max and min wind speed in January and July), conservation data set (flora and fauna), geological data set (lithology, metallic and non-metallic areas, building materials areas, and geology of the soil), hydrology data set (underground water level and type in the soil), infra-structure data set (roads, railroads and utilities), land use dataset (vacant and current urban), regional data set (all Egypt regional data from roads, railroads, government and regional boundaries), sun duration data set (duration of sun in each month and its intensity), and topographic data set (contour lines and railways).

After importing the required feature classes within the previously mentioned datasets, the next step is to prepare all the feature classes to be ready to overlay in one union feature class. Buffer analysis tool creates buffer polygons to a specified distance around the input features (lines-points). An optional dissolve can be achieved to remove overlapping buffers, and the distance is identified in linear units like; (metres, feet, etc.) geodesic buffers will be generated (Fig. 3).

2.3 Data Integration Phase

There are two methods for performing overlay analysis feature overlay (overlaying points, lines or polygons) and raster overlay. Some types of overlay analysis provide themselves to one or the other of these methods. Overlay analysis to find locations meeting specific criteria is often best done using raster overlay. Of course, this also depends

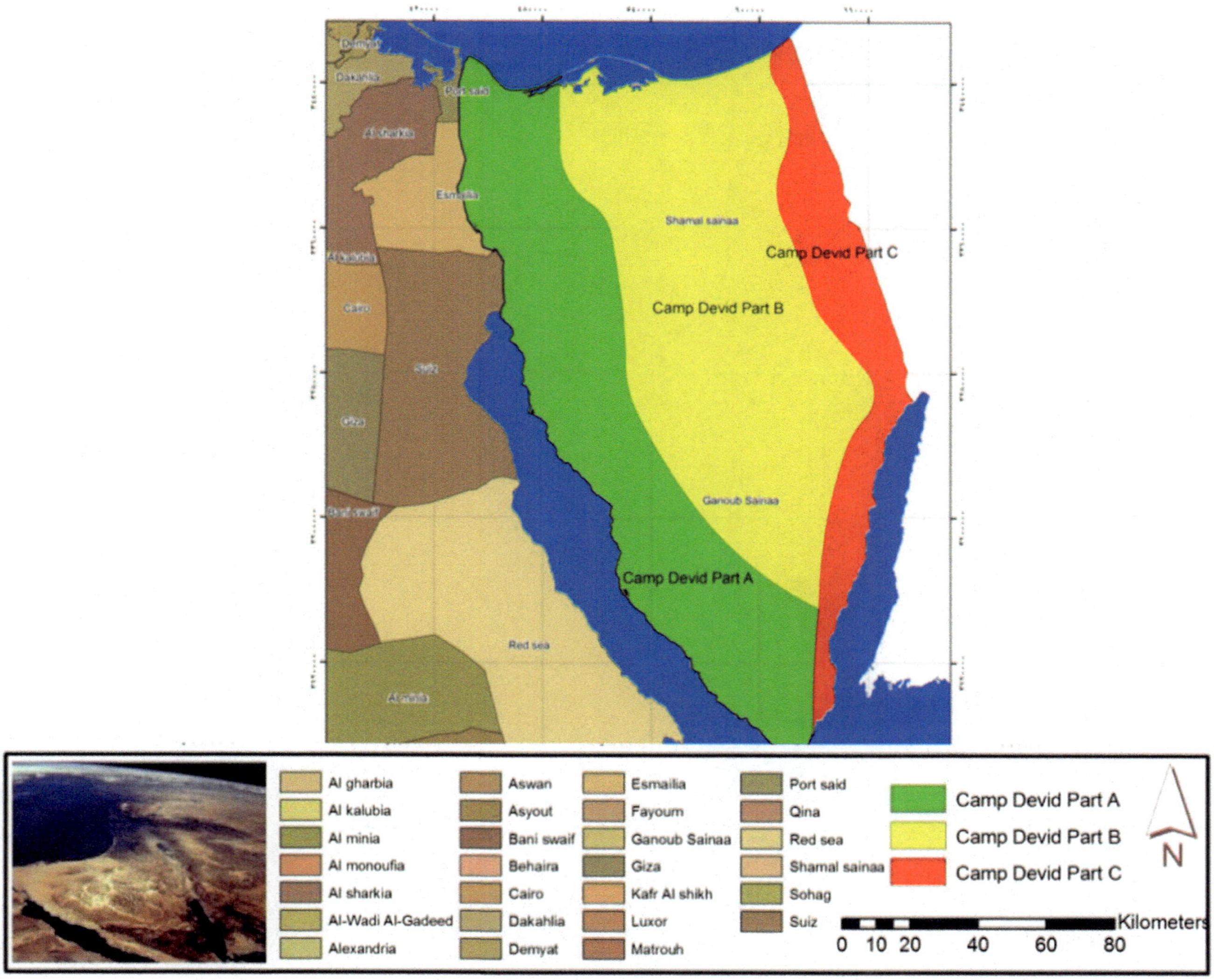

Fig. 2 Camp David accords (Researchers 2017). Adapted from Camp David Accord (1978)

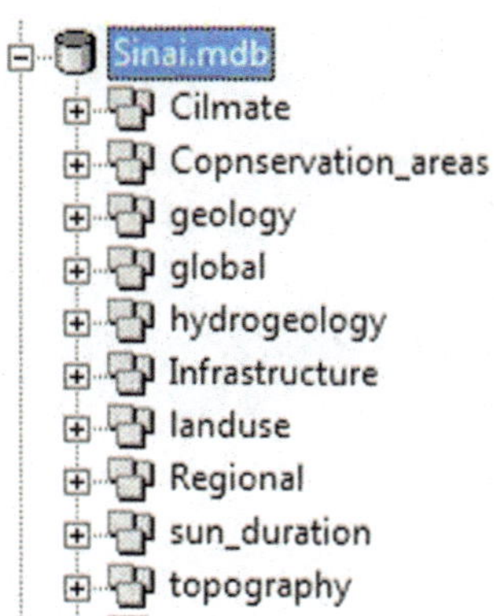

Fig. 3 GIS database (Researchers 2017)

on whether the data is already stored as features or raster. It may be valuable to convert the data from one format to the other to perform the analysis, as the data available for the case study of this article is a vector, so it has been converted to raster to reach the best and accurate results and this will be shown in details in the result and discussion section.

An overlay operation not only a simple merging of polygons; in facts, all attributes of the features that take part in the overlay are carried through, polygons are overlayed (using the Union tool) to create a new polygon layer. The parcels are split where the polygon boundary crosses them, and new polygons created.

Exact locations or areas that have a specific set of attribute values can be found easily according to the criteria that are specified and the queries that are designed with the help of the correlation matrix. This approach is used to find locations that are suitable for a particular use or are susceptible to some risk. After reaching the complete overlay feature class and be sure that all the silver gaps (the errors that resulted from the overlay) have been removed, now the database is ready for the next step (Fig. 4).

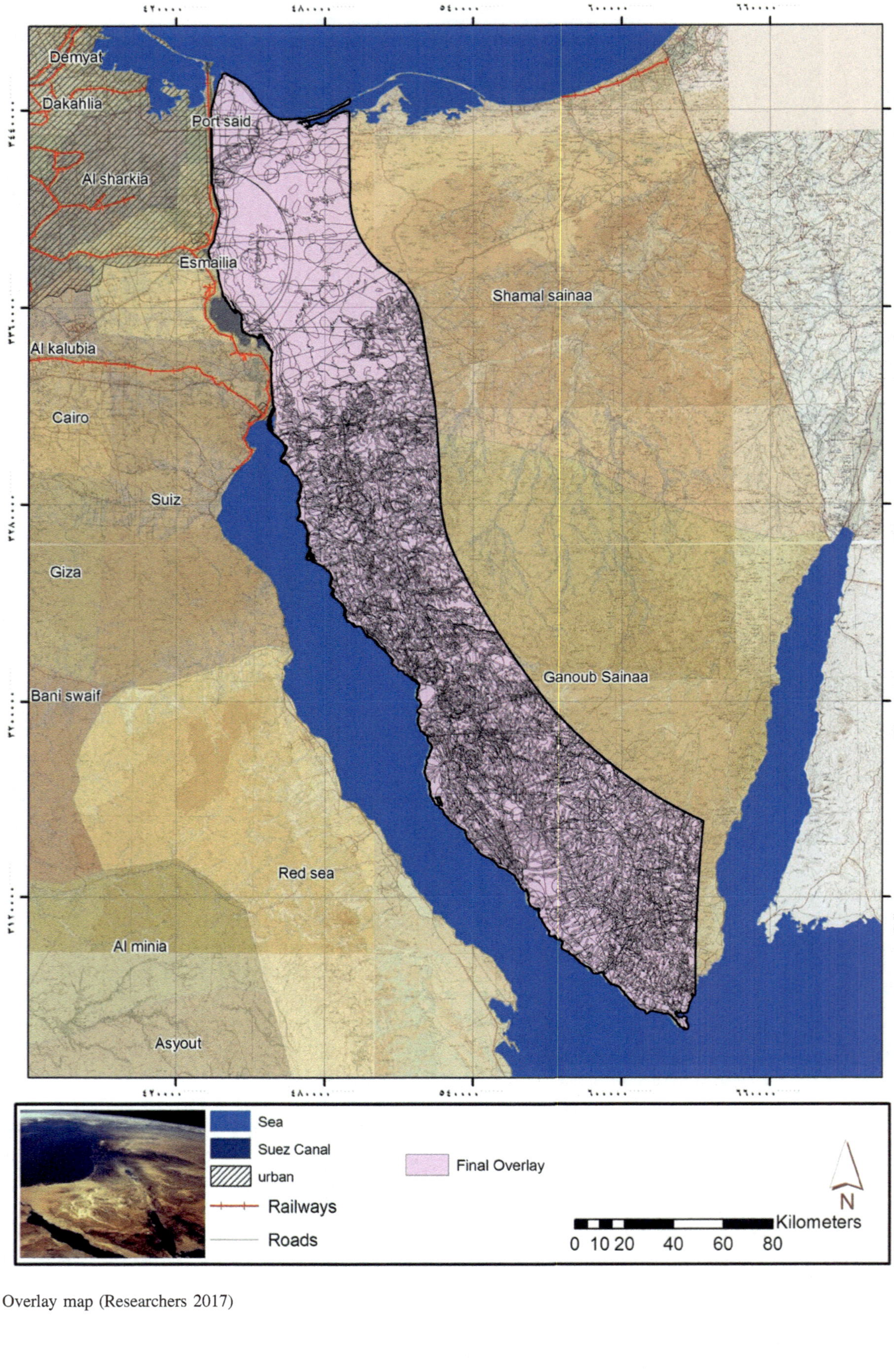

Fig. 4 Overlay map (Researchers 2017)

2.4 Data Deduction Phase

The SQL query tool was applied on the developed geodatabase and the results are represented in layer related to a table view. After reaching the overlay and determining the alternatives strategy, now it is time to query the database to reach the most suitable area to develop the new cities. A query is generated, according to the potentials and constraints of the area of study.

New cities with agriculture or trade economic base are applicable;

- The intergranular productivity ranges from moderate to low productivity.
- Contour ranges from 100 to 400 m above sea level.
- Soil type is dissected limestone plateau with Lithosols, Gravelly and Gravelly Sand Beaches, Loamy sand to sandy Loam Soils with Lithosols, Rugged Rock Land Mainly of the Basement Complex, Sandy Soils and Gravelly Lithosols of the Desert Plains with Rocky Hills.
- Temperature between 8 and 20° in winter and between 25 and 37 in summer, humidity not more than 60%, global sun duration between 3 and 7.5 kWh/m^2/day, solar Radiation from 7 to 13 h daily.
- Served by a road buffer can be located in an area of the electric field, areas served by canals or the River Nile, and finally can be located in significant fault lines buffer and rain ways buffer where both can be treated and do not affect city safety.

New cities with agriculture or trade economic base are not applicable where

- Contour level more than 400 m above sea level.
- Not applicable in areas of metal extraction buffers, areas of building materials buffer, conservation areas, major fault line buffer, oil gas pipeline buffer, soil type is sabkha or dunes, significant rain way buffer that forms main lakes and valleys, and
- Finally, areas that can be exposed to flood risks.

New cities location with industrial base is applicable where

- Areas of metal extraction buffer.
- Areas of building materials buffer, the intergranular productivity ranges from moderate to low productivity.
- Contour ranges from 100 to 400 m above sea level.
- Soil type is Dissected Limestone Plateau with Lithosols, Gravelly and Gravelly Sand Beaches, Loamy sand to sandy Loam Soils with Lithosols, Rugged Rock Land Mainly of the Basement Complex, Sandy Soils and Gravelly Lithosols of the Desert Plains with Rocky Hills.
- Temperature between 8 and 20° in winter and between 25 and 37 in summer, humidity not more than 60%.
- Global sun duration between 3 and 7.5 kW, solar from 7 to 12 h daily.
- Served by a road buffer can be located in an area of the electric field, areas served by canals or the river Nile, and finally can be located in significant fault lines buffer and rain ways buffer where both can be treated and do not affect city safety.

3 Results and Discussions

After determining the areas of new cities location as shown in the following figure, the researcher determined different strategies to allocate the new cities economic bases according to its location potential and constraint (Table 1). Four main approaches have been settled:

a. Labour force coming from delta as the primary constraint to choose city location was the easy access from delta to a new city, so El Salam bridge, El Fardan bridge, and Ahmed Hamdy tunnel had been chosen to be the easy access to the new city location.
b. El Salam Lake as the primary source of water and electricity, so the primary constraint to choose city location was the city distance from El Salam Lake.
c. The Red sea shore as an international trade path between Asia and Africa, so the primary constraint to choose city location was the city distance from the seashore.
d. The proposed bridge between Saudi Arabia and Egypt, this bridge is proposed along the Red sea as the main gate for international trade, and also this alternative also depends on labour force coming from the Red Sea, El Minia and Beni Sweif governments.

Table 1 Table shows different city locations according to the previously mentioned strategies (Researchers 2017)

First alternative for city location 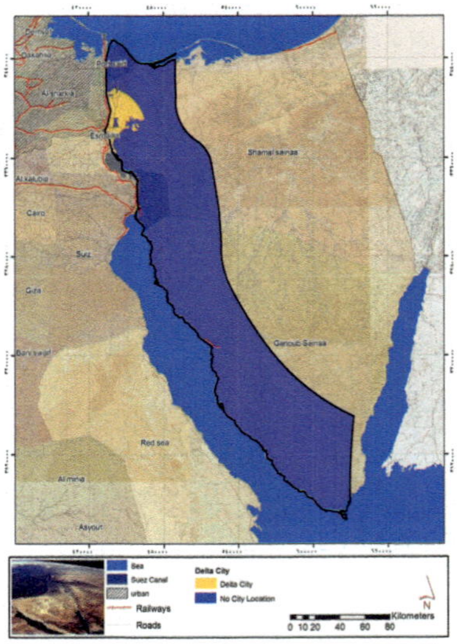 Regarding the potential and constraints, it has been found that this location is the best for agriculture, livestock grazing activities due applicability of the soil, availability of underground water, suitability of temperature and educational, handcrafts activities due presence of technology valley that can be developed	Second alternative for city location 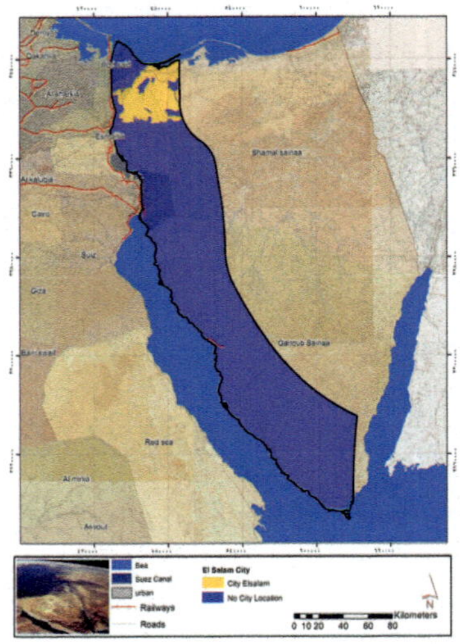 After choosing the second city location and studying its potential and constraints, it has been found that this location is the best for agriculture, the livestock grazing activities due applicability of the soil, availability of water from the canal and suitability of temperature
Third alternative for city location 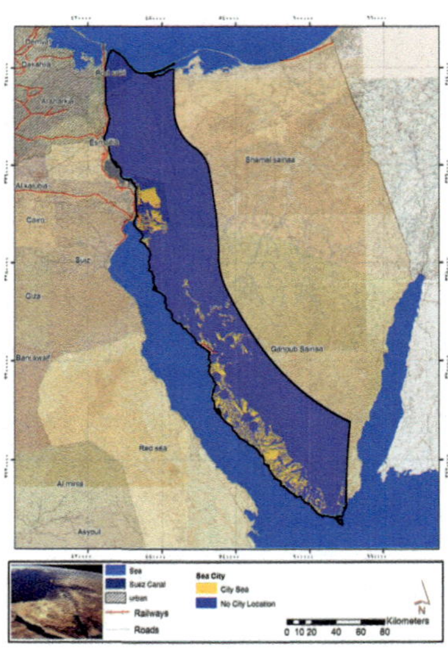 After choosing the city location and studying its potential and constraints, it has been found that this location is the best for logistic international trade activity and constructing international port	Fourth alternative for city location  After choosing the city location and studying its potential and constraints, it has been found that this location is the best for international trade activity, constructing land port and industrial cities

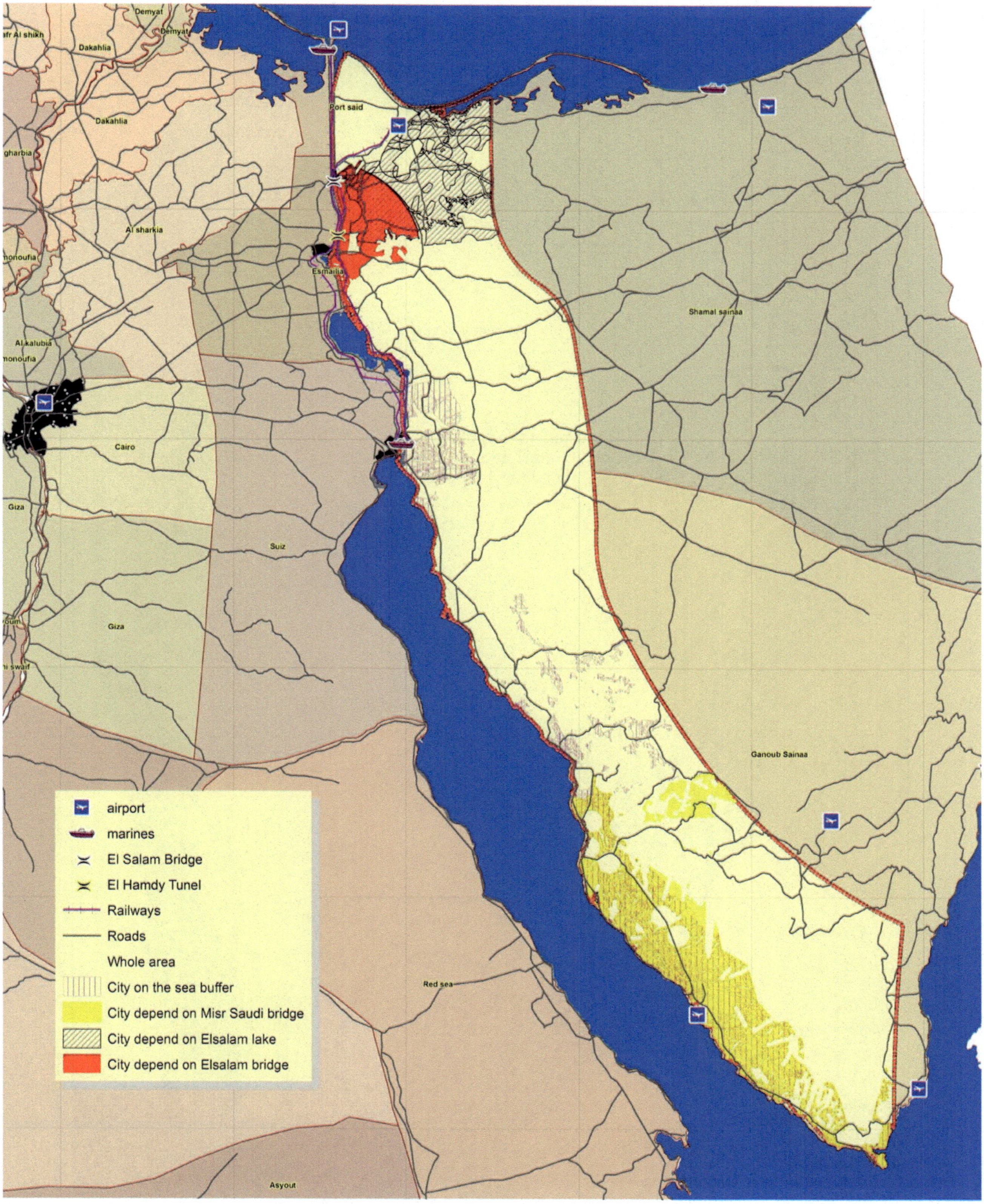

Fig. 5 Overlaid city chosen location (Researchers 2017)

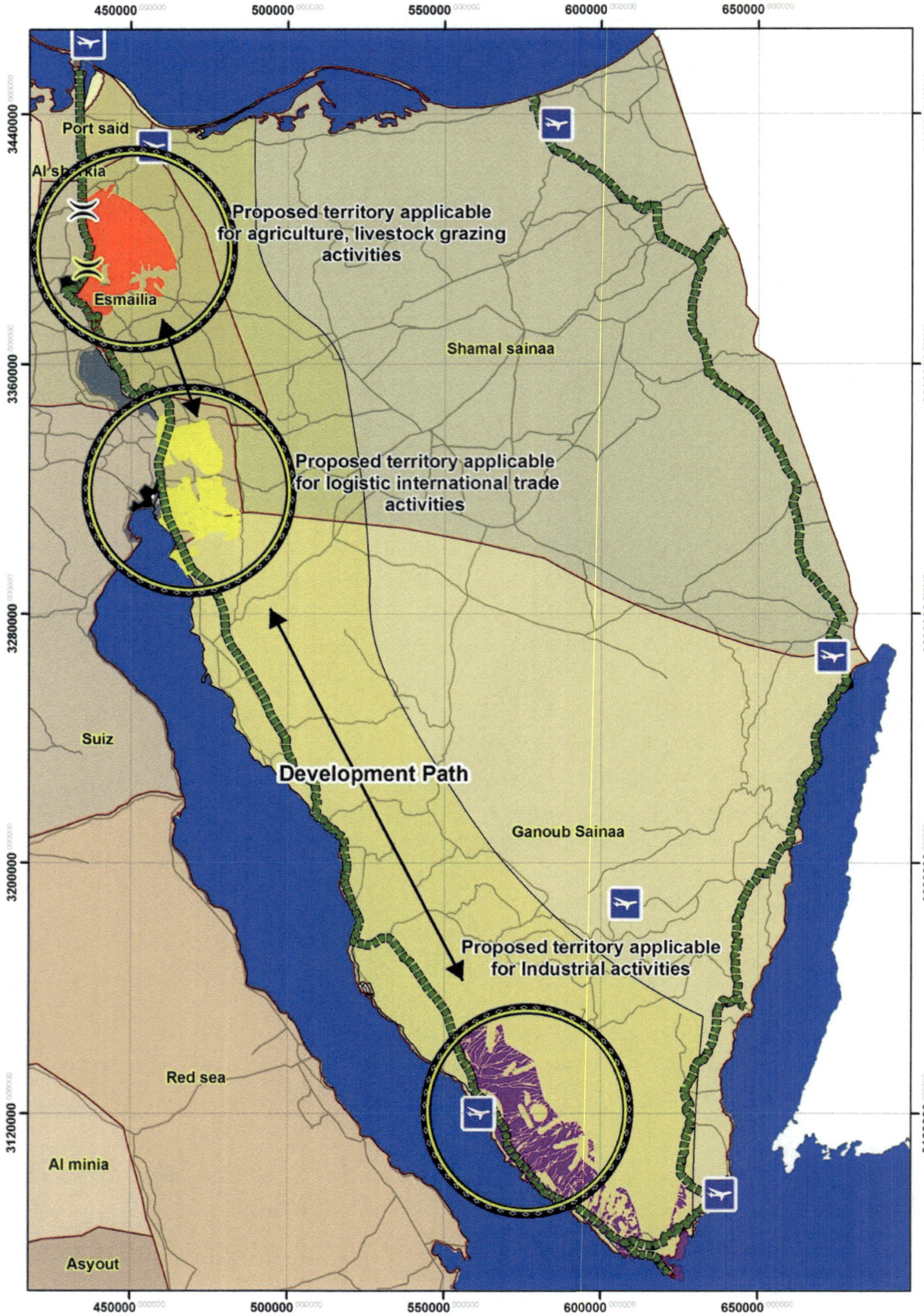

Fig. 6 Suggested locations for new city development (Researchers 2017)

4 Conclusion

Applying a pilot analytical methodology to allocate the optimum location for the new city can be a supporting tool for decision-makers. This tool is depending on using GIS that builds up a site analysis system through taking into account the potential, constraints and the consideration of the spatial resources as well.

Regarding the previous table and four main strategies that have been settled by the researchers, Fig. 5 shows the overlay of the four chosen locations.

According to Egypt 2030 development strategy and the results that have been concluded from the database, the researchers suggested three locations for new city development (Fig. 6).

a. A location of a central city has been chosen east of Ismailia with economic base depend mainly on agriculture, livestock, grazing, and scientific research as the presence of Technology Wadi in this area will encourage a lot developing researchers in renewable energies, agriculture, mining and industry.
b. Location of small cities has been chosen west of El Suez city with economic base depends mainly on international logistic trade and this location matches the government 2020 strategy for Sinai.
c. A central city south of Elkea valley location has been chosen mainly due to the availability of raw materials in this area, and it is close to the suggested bridge between Egypt and Saudi Arabia. Accordingly, its economic bases will be mainly industry and trade.

References

Cambell, J. C., J. Radke, J. T Gless and R. M. Whirtshafter 1992. An application of linear programming and geographic information systems: cropland allocation in antique. Environment and Planning A, Issue 24.

Camp David Accord, (1978). Camp David accords: the framework for peace in the Middle East.

Central Agency For Public Mobilization & Statistics, (2016). No Title. Population census. Available at: http://www.capmas.gov.eg/.

Church, R.L., 2002. Geographical information systems and location science. Computers and Operations Research Volume 29, issue 6.

Collins, M. Steiner F and Rushman M., (2001). Land-use suitability analysis in the United States: historical development and promising technological achievements. Environmental Management Volume 28.

Ding, X. et al., (2015). An inclusive model for assessing the sustainability of cities in developing countries – Trinity of Cities' Sustainability from Spatial, Logical and Time Dimensions (TCS-SLTD). Journal of Cleaner Production, Volume 109, Available at: http://www.sciencedirect.com/science/article/pii/S0959652615008732 [Accessed August 11, 2016].

Duxbury, N. (2001) "Exploring the role of arts and culture in sustainable urban development", Table d'Hôte on Building Sustainable Communities: Culture and Social Cohesion, Available at: http://www.cultureandcommunities.ca/downloads/Duxbury-urban-sustainability-2001.pdf [Accessed September 16, 2017].

Eastman, J., 1997. Idrisi for Windows, Version 2.0: Tutorial Exercises, Graduate School of Geography—Clark University, Worcester, MA.

Egypt's Vision 2030, (2015), Report, the official publication of the country of Egypt, Sustainable development strategy, Egypt's Vision 2030, Egypt economic development conference.

Garcia-Ayllon, S. & Miralles, J.L., (2015). New Strategies to Improve Governance in Territorial Management: Evolving from "Smart Cities" to "Smart Territories." Procedia Engineering, issue 118. Available at: http://www.sciencedirect.com/science/article/pii/S1877705815020512 [Accessed February 4, 2016].

Hafez, R.M., (2015). New cities between sustainability and real estate investment: A case study of New Cairo city. HBRC Journal. Available at: http://www.sciencedirect.com/science/article/pii/S1687404815000188 [Accessed February 16, 2017].

Joerin, F., Thériault, M., and Musy A., (2001). Using GIS and outranking multicriteria analysis for land-use suitability assessment. Geographical Information Science Volume 15.

Madurika, H., Hemakumara, G., (2017) GIS-Based Analysis For Suitability Location Finding In The Residential Development Areas Of Greater Matara Region, an international journal of scientific & technology research, volume 6, issue 02. Available at; http://www.ijstr.org/final-print/feb2017/Gis-Based-Analysis-For-Suitability-Location-Finding-In-The-Residential-Development-Areas-Of-Greater-Matara-Region.pdf.

Martos, A. et al., (2016). Towards successful environmental performance of sustainable cities: Intervening sectors. A review. Renewable and Sustainable Energy Reviews, Issue 57. Available at: http://www.sciencedirect.com/science/article/pii/S1364032115014781 [Accessed January 10, 2016].

Nurse, K. (2006) "Culture as the Fourth Pillar of Sustainable Development", Commonwealth Secretariat Marlborough House Pall Mall London, Available at: http://www.fao.org/sard/common/ecg/2785/en/Cultureas4thPillarSD.pdf [Accessed September 16, 2017].

Packalén, S. (2010) "Culture and sustainability. Corporate Social Responsibility and Environmental Management Corp. Soc. Respons-sib. Environ. Mgmt", Vol. 17.

Shin HuoyTerhKaiCao, (2018), GIS-MCDA based cycling paths planning: a case study in Singapore. (2018). Applied Geography, 94, 107–118. https://doi.org/10.1016/J.APGEOG.2018.03.007.

Steiner, F., L. McSherry, and J. Cohen. (2000). Land suitability analysis for the upper Gila River watershed. Landscape and Urban Planning Volume 50.

Store, R. and Kangas J., (2001). Integrating spatial multi-criteria evaluation and expert knowledge for GIS-based habitat suitability modelling. Landscape and Urban Planning Volume 55(2).

The Energy Poverty Situation: A Review of Developing Countries

Ifeoluwa Garba, Kamila Nieradzinska, and Richard Bellingham

Abstract

This paper reviews the inadequacies in energy systems in the three most deprived regions across developing countries: South Asia, Sub-Sahara Africa and East Asia. It highlights the weaknesses in energy systems contributing to energy poverty at the national levels as well as discusses overlooked issues which translate to energy poverty at household levels. Based on observations, as a starting point, we suggest a new definition for energy poverty in these regions. Finally, we recommend some policies to address highlighted issues. Adequate energy services are a crucial element for sustainable development and are vital for ensuring a good quality of life for people: a fact which has led to increased interest and attention in the area of energy sectors. However, previous strategies and approaches to eradicate energy poverty have failed to accomplish set goals. One of the arguments for this has been the lack of understanding of the problem and inefficient policies. To combat this problem, which the International Energy Agency (IEA) predicts will increase by 2030, if not effectively addressed, a better understanding of the energy systems in the most deprived regions is necessary. In addition to this, more effective energy policies which encourage adequate, reliable and affordable energy services need to be designed.

Keywords

Energy poverty • Developing countries • Sustainable development

I. Garba (✉) · K. Nieradzinska · R. Bellingham
University of Strathclyde, Glasgow, UK
e-mail: ife.garba@strath.ac.uk

1 Introduction

Energy is critical to addressing many global development challenges, including but not limited to education, poverty, inequality and health.

The relationship between energy and economic development has been extensively researched in past years (Kebede et al. 2010; Karekezi et al. 2012; Omri 2013) while the role of energy in attaining the Sustainable Development Goals (SDG) is becoming more realised. Yet, current approaches to eradicate energy poverty are failing in terms of progress and scale (Energy Sector Management Assistance Program (ESMAP); Global Alliance for Clean Cookstoves (Alliance) 2015). Indeed, if current trends continue, more people will be living in energy poverty by 2030 than currently (International Energy Agency 2006).

To address this prediction, target goals and benchmarks are not enough, there is a crucial need for full engagement from global political, academic and industrial players. And although the international community has set a goal for universal energy access, the lack of understanding of the situation will seriously impede the realisation of such goals. A review of the situation of energy systems in the most deprived countries is thus beneficial. By reviewing current inadequacies in energy systems, this paper seeks to unearth the underlying reasons for lack of reliable access to energy in the South Asia, East Asia and Sub-Sahara Africa regions; providing an insight into deficiency areas that can be addressed through optimised policies and investments. It is worth highlighting that although access to clean cooking fuels for cooking and heating is vital, and the current trend is presented for completion purpose, it is beyond the scope of this paper. As such, focus is placed on access to electricity.

The remainder of the paper is organised as follows: Sect. 2 analyses the current situation and trend of energy poverty in the world then emphasises on the most deprived regions. Section 3 examines the relationship between energy and sustainable development while Sect. 4 examines the

C. Alalouch et al. (eds.), *Advanced Studies in Energy Efficiency and Built Environment for Developing Countries*,
Advances in Science, Technology & Innovation, https://doi.org/10.1007/978-3-030-10856-4_4

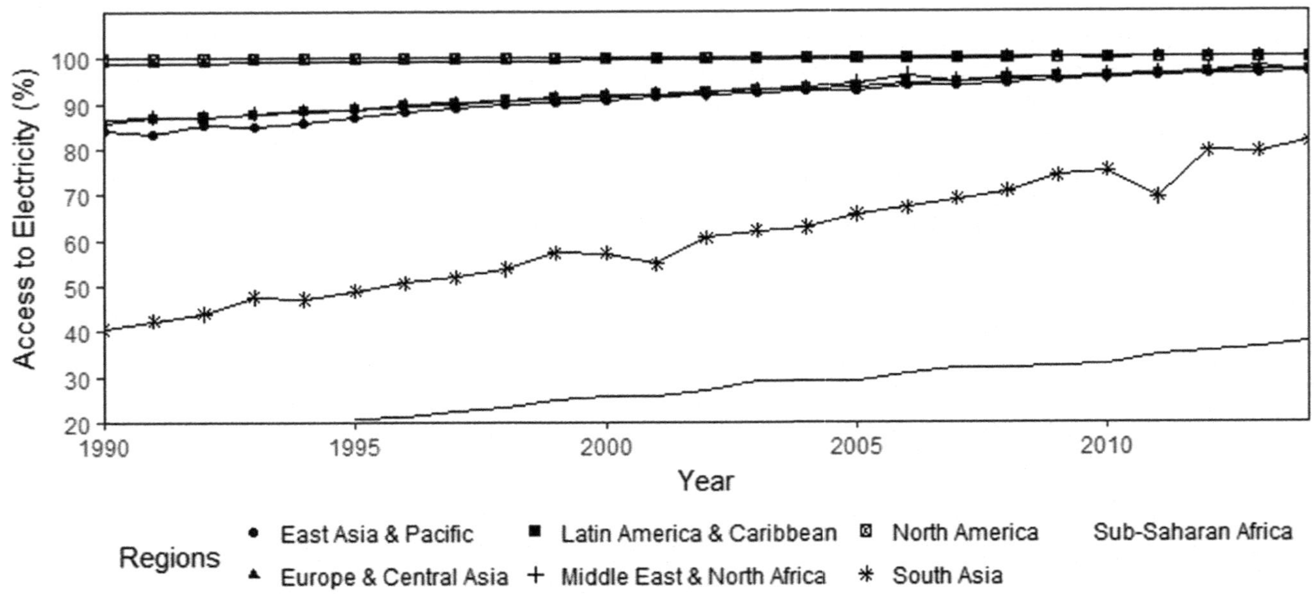

Fig. 1 Trend of access (%) to electricity across the world. *Source* World Bank (World Bank Group 2016)

energy systems in the most deprived regions in the world. Section 6 provides some policy recommendations based on observations while Sect. 7 discusses and concludes.

2 Current Situation and Trends

Over the past few decades, access to 'modern' energy systems[1] has been rising. Yet, although significant progress has been made, about a third of the world's population still lack access to electricity, while just above half the population have access to clean fuels and technologies to satisfy basic needs like cooking and heating (World Bank Group 2016).

Existing literatures have provided excellent insights into the progress towards universal access on a global scale. However, most fail to illustrate the actual distribution and/or severity of energy poverty issue across the various regions in the world.

To address this, we demonstrate the progress trend across the various regions across the world.[2] A trend analysis for the different regions is performed and the results presented in Figs. 1 and 2. Moving averages of access across the regions were also compiled and are presented in Tables 1 and 2. By calculating the moving averages, we are able to further understand and demonstrate the overall progress trend across the regions of interest.

Figure 1 shows access to electricity by region from 1990 to 2014. The trend illustrates an improvement in access to electricity across all regions, with the exception of 'Europe and Central Asia' and 'North America' which already had universal access in 1990. By 2014, the 'Latin America and Caribbean' and 'Middle East and North Africa' regions had progressed towards universal access, with less than 10% of the populations in these regions lacking access to electricity. The biggest improvement is seen in 'South Asia', where over two decades, even with the dips in trend, a growth rate of over 30% is observed. The worst growth rate is observed in the 'Sub-Saharan Africa' region where in 2014, just over 30% of the population have access to electricity.

To further demonstrate the growth rates in these regions, 3, 5 and 10 years, moving averages (MA) of access to electricity over two decades across all regions were computed and the results presented in Table 1.

Figure 2 shows the trend of access to clean cooking fuels by regions. However, unlike access to electricity, the progress towards access to clean cooking fuels has been slower; with only one region having achieved universal access. Across 'East Asia and Pacific', 'South Asia', and 'Sub-Saharan Africa', less than 50% of the population in these regions have access to clean cooking fuels. Even in South Asia, where tremendous progress was made in relation to electricity access, access to clean cooking fuels across the region is seriously lacking.

Considering the poor performance across all regions, this suggests that perhaps across all regions, less attention is given to access to clean cooking fuels or perhaps it is due to an assumption that with access to electricity, these needs can be met without the need for clean cooking fuels.

Nonetheless, what is more striking is the growth rate in 'Sub-Saharan Africa', where over the period of a decade, there has been a growth rate of less than 5%. As of 2014, less than 20% of the population in 'Sub-Sahara Africa' had access to clean fuel for cooking and/or heating. In addition,

[1]Modern energy systems here refer to electricity and clean cooking fuels.

[2]For this purpose, we have used the World Bank region classification method.

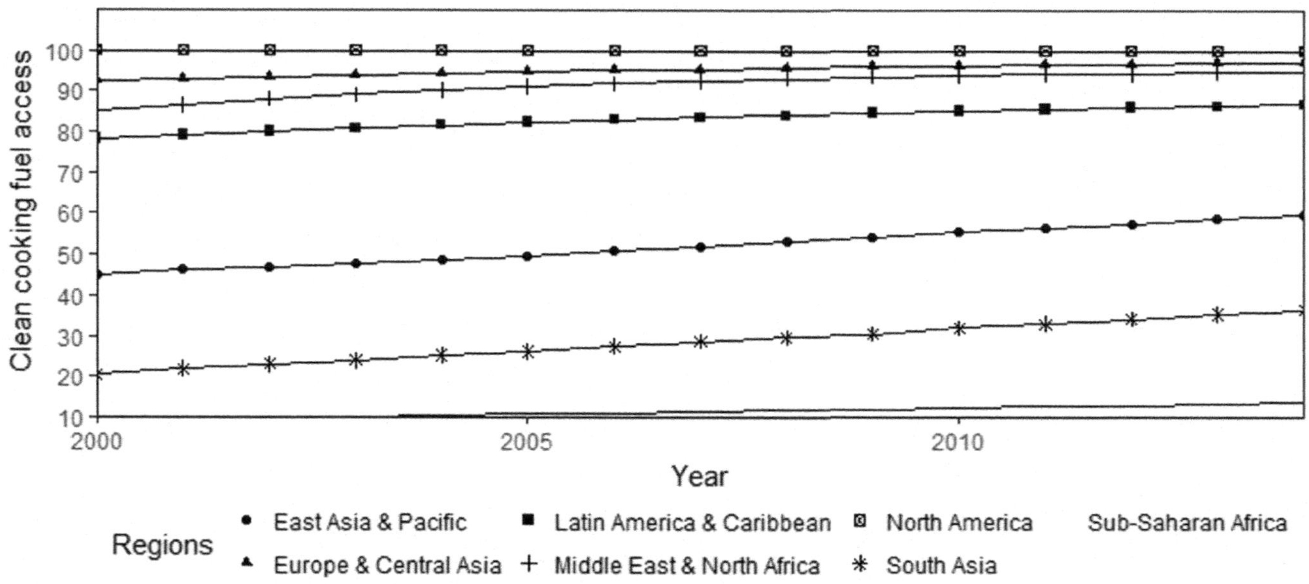

Fig. 2 Trend of access (%) to clean cooking fuels across the world. *Source* World Bank (World Bank Group 2016)

Table 1 Access to electricity—moving averages

Region	Year	3 years MA	5 years MA	10 years MA
East Asia and Pacific	1993	82.58	82.89	
Europe and Central Asia		98.50	98.50	
Latin America and Caribbean		87.01	86.98	
Middle East and North Africa		86.72	86.71	
North America		100	100	
South Asia		45.64	45.67	
Sub-Saharan Africa		20.63	20.90	
East Asia and Pacific	2003	90.99	91.00	91.22
Europe and Central Asia		99.43	99.43	99.45
Latin America and Caribbean		92.55	92.52	92.74
Middle East and North Africa		91.86	91.94	92.23
North America		100	100	100
South Asia		58.90	59.33	60.88
Sub-Saharan Africa		28.86	28.03	28.26
East Asia and Pacific	2013	96.11	96.08	
Europe and Central Asia		99.99	99.99	
Latin America and Caribbean		96.67	96.62	
Middle East and North Africa		96.59	96.50	
North America		100	100	
South Asia		75.70	76.49	
Sub-Saharan Africa		35.97	34.93	

across the 3, 5 and 10 years moving averages, a trend of less than 2% is observed, confirming there has been little to no progress in the region with regards to access to clean cooking fuels.

The poor performances across the regions are further emphasised in the moving averages results, presented in Table 2. From the results obtained in Tables 1 and 2, as well as the trend behaviours shown in Figs. 1 and 2, it becomes

Table 2 Access to clean cooking fuels—moving averages

Region	Year	3 years MA	5 years MA	10 years MA
East Asia and Pacific	2000	45.62	46.60	49.19
Europe and Central Asia		93.75	94.08	94.83
Latin America and Caribbean		78.79	79.45	80.98
Middle East and North Africa		89.54	90.37	92.29
North America		100	100	100
South Asia		23.50	24.25	26.01
Sub-Saharan Africa		11.35	11.47	11.74
East Asia and Pacific	2005	50.71	51.77	54.42
Europe and Central Asia		95.31	95.59	96.20
Latin America and Caribbean		81.93	82.51	83.93
Middle East and North Africa		93.53	94.21	95.44
North America		100	100	100
South Asia		27.09	27.76	29.45
Sub-Saharan Africa		11.91	12.01	12.31
East Asia and Pacific	2010	56.02	57.08	
Europe and Central Asia		96.63	96.82	
Latin America and Caribbean		84.79	85.34	
Middle East and North Africa		96.42	96.68	
North America		100	100	
South Asia		30.47	31.14	
Sub-Saharan Africa		12.46	12.61	

evident that although global progress towards universal energy access is improving, there exists an uneven distribution in progress being made across regions. What's more, in the case of access to electricity, across all regions, with the exception of 'Europe and Central Asia' and 'North America', although the trend results show an improvement in access to electricity across all regions, the progress in the regions are discontinuous (as demonstrated in the moving average results in Table 1).

It is important to be cognisant of the fact that due to the aggregate nature of these analyses, these figures do not illustrate the differences in countries within the regions. They do not reflect the differences in provinces within the countries and more importantly, do not highlight the significant differences in modern energy access levels between urban and rural areas. However, evidences show that about 70% of the population without access to modern energy services resides in rural areas (World Bank Group 2016; Foell et al. 2011).

3 Energy and Development

Several literatures have investigated the linkage between energy consumption and economic development (Nussbaumer et al. 2012; Omri 2013); primarily focusing on the relationship between energy consumption and GDP. Yet, the impact of lack of energy in developing countries goes well beyond just economic development and growth. Social and human developments are vital areas that need accurate representations. Hence, there exist momentous gaps regarding the impact of energy on sustainable socio-human development (see Fig. 3); particularly in developing countries (Culver 2017).

The relationship between energy and sustainable development can be complicated in positive and negative ways. On one hand, it is becoming widely accepted that services enabled through the access of energy such as lighting, better cooking possibilities, income generation process, contribute to economic and environmental improvements which are crucial to sustainable development in developing countries. Yet, the conventional production and distribution of energy do contribute to the pollution of the environment and in some cases, disruption of ecosystems (Dincer 2000). To assess sustainability, two major conceptual frameworks have been widely utilised in literatures: substantive and procedural sustainability (Jain and Jain 2013; Komiyama and Takeuchi 2006). To demonstrate how energy impacts on development and more importantly, sustainable development, we adapt the first of the two frameworks.

Our decision is based on the fact that a substantive framework can be used to evaluate the improvements in

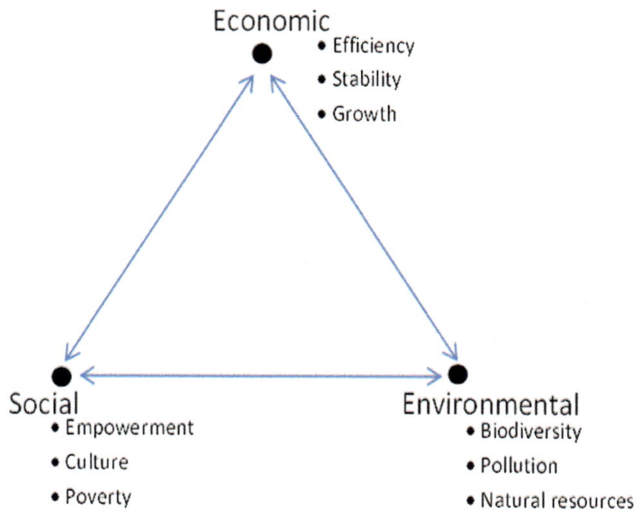

Fig. 3 Impacts of energy on sustainable development. *Source* Munasinghe and Shearer (1995)

economic, social and environmental conditions of a population as a result of a project or action while procedural frameworks generally aim to evaluate the acceptance or feasibility of a project or action.

Combining two of the three major approaches used in substantive sustainability literature; triangular and constant-capital approaches, we illustrate the relationship of energy and development in Fig. 3. Table 3 further demonstrates the performances of the regions across some development indicators.

In terms of economic impacts, access to energy can be associated with several development opportunities facilitated through longer working hours, use of electrical appliances, amongst others. To name a few, it supports the possibility for higher productivity and even possible employment opportunities. In itself, the process of creating or expanding accessibility of energy services could contribute to provision of employment for the unemployed by means of creating jobs such as sales, installation and maintenance jobs, within the energy sector. It could also serve as a supplementary source of employment for the already employed, transfer workers from agriculture and farming, to the energy sector and other sectors, thus creating diversification of industries as well as income generation.

Likewise, the shift from poor fuel sources such as firewood and charcoal to modern energy services created by access to modern energy helps abate deforestation, forest degradation and biodiversity loss (United Nations Environment Programme 2012). Thereupon, increased water availability, economic income as well as soil conservation, through active forestry activities can contribute to the reduction of poverty, hunger and environmental degradation. Finally, the incomplete combustion of solid fuels in traditional and/or open stoves is attributable to emissions causing greenhouse effects, thus contributing to global warming effects (Panwar et al. 2011). Hence, access to modern energy services would contribute to the mitigation of greenhouse gas effects (Mensah and Adu 2015; Smith et al. 1993).

The provision of access to modern energy services can also contribute to the empowerment of women and girls in developing countries (Habtezion 2013). By freeing up time otherwise spent on gathering and preparing fuels for tasks, such as cooking and/or heating, available time can be spent on pursuing education or employment, thus contributing to equality.

4 Overview of Energy Systems in Deprived Regions

In the past couple of decades, international organisations such as the World Bank (WB) and United Nations (UN) have realised that to promote economic growth as well as achieve the sustainable development goals, emphasis have to be placed on expanding and improving the energy sectors in developing regions. This section discusses the weaknesses in the energy sectors of the three most deprived regions: South Asia, East Asia and Sub-Sahara Africa, with the aim of unveiling some of the reasons for inadequate energy supply in these regions.

South Asia

In South Asia, although the region's energy demand has expanded two-and-a-half times in the past two decades (International Energy Agency (IEA) 2013), the energy situation has significantly improved.

Nonetheless, the region still has issues needing addressed. One of the biggest challenges in the South Asian region is the unreliability of energy services, due to its limitation in generation capacity. As a result of high energy demands and limited generation capacity, the region struggles with providing reliable energy services to the population. In addition to energy unreliability, available energy access is not uniformly distributed (especially in India and Bangladesh). Energy access is mostly concentrated in urban areas, leaving the population in rural areas without access (World Bank Group 2016). More to this point, more than 40% of the rural population in India and Bangladesh lack access to electricity (World Bank Group 2016) and considering that most of the population in these countries reside in rural areas: 67 and 66% in India and Bangladesh, respectively, although the

Table 3 Energy and development indicators—2014

Region	Electricity consumption (kWh)	HDI (per capita)	Life expectancy (years)	CO_2 per capita (t)	GDP per capita (US $ 000's)	GDP annual growth (%)	Poverty head-count ratio $ 1.90 (%)	Youth literacy rate (%)
East Asia and Pacific	3679.76	0.70	75.10	6.30	9.66	3.41	3.54	99
Europe and Central Asia	5372.44	0.76	77.21	6.92	26.14	1.28	2.15	99.71
Latin America and Caribbean	2128.82	0.75	75.06	3.06	10.05	0.21	5.40	97.92
Middle East and North Africa	2875.32	0.69	73.12	6.16	8.44	1.19	2.10	92.70
North America	13,243.18	0.73	79.06	16.35	54.19	1.60		99
South Asia	707.50	0.59	68.21	1.46	1.50	5.65	15.09	83.48
Sub-Saharan Africa	480.57	0.50	59.40	0.84	1.81	1.81	40.99	71.41

Source World Bank (World Bank Group 2016)

overall reported figures across the region might appear tolerable, in reality, about one in four people in India and Nepal live without electricity while about one in three people in Pakistan and Bangladesh live without electricity (Saracini et al. 2014). As a result of these high-level problems within the region, inaccessibility and unreliability of energy services at household levels are a common issue amongst the population.

5 East Asia and Pacific Islands

East Asia

Across East Asia, there has been a tremendous progress towards universal access, with only four countries having less than 90% electrification rates: the poorest rates notably being in Myanmar and Cambodia. Due to the population distribution in these parts of the region, about 80% of the population live in rural areas (International Energy Agency (IEA) 2013); urbanisation driven economic growth and notably, dedicated government interventions, have been essential elements in the increased access to modern energy services. Nonetheless, these countries do have their challenges, some of which are: heavy reliance on fossil fuel such as coal for energy generation, heavy reliance on energy importation especially in Thailand, Cambodia and Vietnam (Asian Development Bank 2016): leading to high energy cost for consumers as well as lack of efficiency at both generating levels and consumption levels. As a result, at the high level, energy security, efficiency, sustainability and demand management inadequacy are issues faced by these countries, all of which can be argued to translate into energy poverty in terms of non-sustainability and non-affordability for consumers as well as other issues such as local air pollution.

Pacific Islands

With the exception of Australia and New Zealand, lack of access to electricity is deemed common in the Pacific islands. One of the main factors to the situation is the topography of the islands. For example, in the Solomon Islands, the population spans over more than 300 populated islands while in Kiribati, less than 80,000 people live on 33 widely scattered islands which span over 80 km^2 of land. As a result of this spread in habitations, the utilisation of a national grid is rendered impossible. Yet, the delivery of fossil fuels are expensive and often unreliable for the majority of the population that live in the outer islands or in the more remote areas of the larger islands (Urmee et al. 2009). However, the problem of lack of access to electricity is not limited to only the small islands, on 'large' islands, access to some villages is often a problem due to the absence of roads as a result of the dense jungles and rugged terrain (Flint 2008). Thus, as a result of lack of infrastructure, at household levels, inaccessibility to modern energy services is a common issue in this region.

Africa

The African energy sector is termed the most impoverished part of the global energy system due to the under-developed nature of the system and studies have long established that the development of the region's energy system is a crucial element to the region's overall development (International Energy Agency (IEA) 2014). However, although the continent as a whole crucially needs energy system and socio-economic development, the severity of the energy situation drastically varies across the region.

However, the problem of energy poverty resulting from absence or inadequate electricity supply in the region is multifaceted: to state the least, it varies from lack of generating capacity, decrepit existing stock, limited transmission and distribution infrastructures, as well as the lacking expertise crucially needed in maintaining and expanding the system, to the corruption issue(s) in most of the countries. What's more, some of the countries in the region have the highest electricity prices by world standard (International Energy Agency (IEA) 2014). The African Development Bank (AfDB) found that consumers pay as much as 20–50 cents per kilowatt-hour against a global average of approximately 10 cents (African Development Bank 2013). These coupled with the high poverty levels in the region (see Table 3) poses a serious case of energy poverty.

Thus, these high-level problems need to be addressed to facilitate the reduction of energy poverty at household levels. At household levels, non-affordability, inaccessibility, unreliability, inefficiency, pollution and non-sustainability of energy services are common major aspects of energy issues in this region.

6 Policy Recommendations

Energy poverty at the household level goes beyond the accessibility factors; as revealed in these regions, factors such as efficiency, reliability, affordability, security, health and safety are all aspects of energy poverty that need better recognition by stakeholders in the energy sector. As a starting point, developing countries need to establish a better and more comprehensive definition to the problem. The governments in these countries also need to officially address the phenomenon in government frameworks; to date, only a few developing countries address energy poverty in their national frameworks (African Development Bank 2013).

Across all the regions discussed in Sect. 4, improvement of the energy sectors needs to be prioritised. Consequently, investment in generating plants, transmission lines, distribution systems as well as maintenance of existing infrastructure needs to be crucially emphasised in national and regional polices. However, in the case of Africa (especially Sub-Sahara Africa) and the Pacific islands, where finance for expansion and maintenance of infrastructures is a major challenge, the governments within these regions need to look into designing and implementing policies which aid international instruments and treaties. Incentives which could entice the public and private sectors' into participating in developing, operating and maintaining their energy sectors, should also be considered.

In regions where generation capacity is the challenge, potential renewable energy resources can be explored in order to improve the issue of energy security and energy affordability at both the higher and lower levels. In addition, the use of renewable resources, where available, could allow for distributed generation, which could provide multifaceted benefits. Firstly, this could serve as a solution to the issue of accessibility of remote, rural communities (as is the case in parts of the Pacific islands and South Asia). Another benefit of this approach could be its ability to help avoid further energy system complications which could result from the expansion of already unreliable and in some cases, perilous grids. In addition, this could also allow for the promotion of energy distribution systems at the community level which in some cases, could save on capital costs, would promote community involvement, could promote co-generation and/or even support micro-funding and/or joint funding programmes.

Furthermore, governments in developing countries can implement consumer demand management control measures. Through improved efficiency measures, generating plants can be used to serve more consumers. One of the ways to accomplish this could be through the inclusion of awareness programmes on efficiency information and best practices, in government policies and plans. Finally, research and development measures need to be included in policies. Although currently severely lacking in developing countries, by investing in human resources, development of personnel who can promote the development and maintenance of infrastructures can be aided. On a final note, developing countries require a comprehensive strategy which can be used to effectively monitor and evaluate the progress of implemented energy schemes, to ensure the desired goals are being achieved.

Nonetheless, it is important to remember that policies are not judicially enforceable and have no binding power hence if progress is to be truly enforced, there might be needs for legislations which reflect these goals.

7 Discussion and Conclusion

Poverty is a reality for millions of people across the world and energy poverty is both a cause and result of it. Billions of people across the world do not have access to energy infrastructures and some who do, cannot afford to pay for it.

Energy poverty has crucial implications, ranging from the inability to use energy services to satisfy basic needs such as cooking and heating to the inability to obtain appropriate medical services, education, reduced productivity, amongst others. As such, the eradication or at the least, the reduction of energy poverty is vital to support sustainable development in developing countries.

To eradicate energy poverty, a better understanding of the problem and the causes is paramount. However, there are gaps in the understanding of the reasons behind energy poverty in developing countries. In this paper, we have tried to address this by reviewing the current stage of energy poverty in developing countries, as well as the intensity and distribution of the phenomenon across the most deprived regions. Following this, we have reviewed the current state of energy systems in these regions, with an aim of understanding what aspects of the inadequacies in energy systems in the regions are translating to energy poverty at household and societal levels. In reviewing and understanding the problem, we argue that accessibility to energy services is not the only contributing factor to energy poverty and as such, a more comprehensive definition of the problem is essential in tackling this issue.

As a starting point, we propose that in place of current definitions which focus on accessibility only, energy poverty in these regions should be defined as 'The inability of an individual or group of individuals to safely acquire sustainable, clean, efficient and reliable energy services, which adequately satisfies their basic needs such as lighting, cooking and heating/cooling,[3] at affordable costs.[4]'

By adequately representing the problem, stakeholders are better provided with relevant information to help effective decision-making. More importantly, investments can be directed towards addressing the problem root causes, resulting in more effective investment strategies and solutions.

One of the main conclusions of this review is that a better understanding of the root causes through thorough evaluation of the energy systems in these regions is crucial in the process of eradicating energy poverty. Otherwise, there are risks of investments being utilised in inefficient way. In addition, specific policies and programmes need to be implemented in these regions if the goal of eradicating energy poverty is to be realised. However, in order to critically address energy poverty, policies alone will not be enough. The design of effective legislations might be needed to tackle this momentous problem.

[3]With adequate heating and cooling meeting the WHO standardised level of acceptable temperatures in a household.

[4]In terms of monetary or labour.

References

African Development Bank (2013). Energy sector policy of the AfDB group. Technical report, African Development Bank Group.

Asian Development Bank (2016). *Achieving Universal Electricity Access in Indonesia.*

Culver, L. C. (2017). The Stanford Natural Gas Initiative. Energy Poverty: What You Measure Matters.

Dincer, I. (2000). Renewable energy and sustainable development: a crucial review. *Renewable and Sustainable Energy Reviews*, 4 (2):157–175.

Energy Sector Management Assistance Program (ESMAP); Global Alliance for Clean Cookstoves (Alliance) (2015). The state of the global clean and improved cooking sector. Technical report, Energy Sector Management Assistance Program (ESMAP); Global Alliance for Clean Cookstoves (Alliance), Washington DC.

Flint, A. (2008). *Trade, poverty and the environment: the EU, Cotonou and the African-Caribbean-Pacific bloc*. Springer.

Foell, W., Pachauri, S., Spreng, D., and Zerriffi, H. (2011). Household cooking fuels and technologies in developing economies. *Energy Policy*, 39(12):7487–7496.

Habtezion, S. (2013). Gender and energy. *United Nations development programme*, (4):1–6.

International Energy Agency (2006). Energy for Cooking in Developing Countries. Technical report, International Energy Agency.

International Energy Agency (IEA) (2013). Southeast Asia Energy Outlook. Technical report, International Energy Agency.

International Energy Agency (IEA) (2014). Africa Energy Outlook: A Focus on Prospects in Sub-Saharan Africa. Technical report, International Energy Agency (IEA), Paris.

Jain, P. and Jain, P. (2013). Sustainability assessment index: a strong sustainability approach to measure sustainable human development. *International Journal of Sustainable Development & World Ecology*, 20(2):116–122.

Karekezi, S., McDade, S., Boardman, B., and Kimani, J. (2012). Energy, Poverty, and Development. Technical report.

Kebede, E., Kagochi, J., and Jolly, C. M. (2010). Energy consumption and economic development in Sub-Sahara Africa. *Energy Economics*, 32(3):532–537.

Komiyama, H. and Takeuchi, K. (2006). Sustainability science: building a new discipline. *Sustainability Science*, 1(1):1–6.

Mensah, J. T. and Adu, G. (2015). An empirical analysis of household energy choice in Ghana. *Renewable and Sustainable Energy Reviews*, 51:1402–1411.

Munasinghe, M. and Shearer, W. (1995). *Defining and measuring sustainability.*

Nussbaumer, P., Bazilian, M., and Modi, V. (2012). Measuring energy poverty: Focusing on what matters. *Renewable and Sustainable Energy Reviews*, 16(1):231–243.

Omri, A. (2013). CO_2 emissions, energy consumption and economic growth nexus in MENA countries: Evidence from simultaneous equations models. *Energy Economics*, 40:657–664.

Panwar, N., Kaushik, S., and Kothari, S. (2011). Role of renewable energy sources in environmental protection: A review. *Renewable and Sustainable Energy Reviews*, 15(3):1513–1524.

Saracini, N., Krishnaswamy, S., and Dubey, S. (2014). Energy for Development in South Asia: addressing energy inequality sustainably. Technical Report 1, Christian Aid; Climate Action Network South Asia CANSA.

Smith, K. R., Khalil, M. A., Rasmussen, R. A., Thorneloe, S. A., Manegdeg, F., and Apte, M. (1993). Greenhouse gases from biomass and fossil fuel stoves in developing countries: A Manila pilot study. *Chemosphere*, 26(1–4):479–505.

United Nations Environment Programme (2012). Cost of deforestation in Kenya far exceeds gains from forestry and logging, UN joint study finds.

Urmee, T., Harries, D., and Schlapfer, A. (2009). Issues related to rural electrification using renewable energy in developing countries of Asia and Pacific. *Renewable Energy*, 34(2):354–357. Renewable Energy for Sustainable Development in the Asia Pacific Region.

World Bank Group (2016). World Development Indicators. Technical report, World Bank Group (WB), Washington DC.

The Recent Legislation for Eco-efficient Design in Quito, Ecuador: A Review and Case Study

M. J. M. Davis, David Jacome Polit, and J. L. Barros

Abstract

This paper sets out to review the recent legislation that was put into place for eco-efficient design in Quito, the capital city of Ecuador. It is argued that the resolution for Quito is better suited to ecological building design parameters than using international certifications imported from abroad, such as LEED or BREEAM. The background of the parameters used for the resolution is explored in depth in this paper. The results of an international review workshop are then given, followed by a summary of the resolution being put into practice through a brief case study of the water systems proposed for a building designed under the legislation. Quito´s urban sprawl poses fundamental challenges from an ecological and economic perspective, but reverting this trend also provides a unique opportunity to plan, develop, build and manage a city that is simultaneously more ecologically and economically sustainable and more resilient. The present paper analyses how the New Resolution in Eco-efficient Design for Quito helps out in this task and becomes a versatile instrument, part of a broader tool, in contributing to the goals of eco-efficient design, providing incentives to the construction industry and reducing the number of private vehicles coming into the city. The paper concludes that it is possible to construct an effective eco-efficiency code aimed at reducing the environmental footprint of buildings whilst improving people's quality of life. Furthermore, as the tool can be modified in future updates, research and better building practices are required to improve results and to adjust it to the ever-changing socio-economic contexts of the city. However, in order to be effective as the backbone of urban planning in Quito, the tool will need to be upgraded to the status of a municipal ordinance in future.

Keywords

Urban design • Legislation • Sustainable certification • Transport • Resilience

1 Introduction

Overall, work by Todd et al. (2013) shows how ecological rating systems for buildings allow for environmental assessments to be given and promote market transformations. Additionally, it is noted that tensions can be created in the application of the rating systems and subsequent building design requirements. The first metro line of Quito, amongst other solutions such as the Quito Cables or the extension of BRT coverage, "has been one of the most important projects to solve the problems of mobility in the city" (PMDOT 2015–2025 (iv), MDMQ). However, sustainable development as a concept with a broader scope requires looking at mobility, urban planning, land use and environmental issues together to propose a holistic tool. In 2016, the municipality of Quito introduced the Resolution No. 13-2016, which in Section VI laid out the parameters for real estate developers to be able to increase the height of their buildings if it complied with (a) being close to a mass public transport hub and (b) being in accordance with parameters laid out for ecological design. The resolution was then updated in April 2017. This legislation in eco-efficient design for Quito aims to be an instrument that can help pave the road to a better contribution to sustainable development. This paper sets out to cover three areas in order to review the resolution. First,

M. J. M.Davis
Evolution Engineering, Design and Energy Systems Ltd., 5 Silver Terrace, Exeter, EX4 4JE, UK

M. J. M.Davis
Department of Arquitecture, Pontificia Universidad Católica de Ecuador (PUCE), Av. 12 de Oct 1076 y Roca, Quito, Ecuador

D. Jacome Polit (✉)
Instituto Metropolitano de Planificación Urbana (IMPU)—MDMQ, 100 Resilient Cities, Quito, Ecuador
e-mail: djacomepolit@gmail.com

J. L. Barros
Private Researcher, Quito, Ecuador

C. Alalouch et al. (eds.), *Advanced Studies in Energy Efficiency and Built Environment for Developing Countries*, Advances in Science, Technology & Innovation, https://doi.org/10.1007/978-3-030-10856-4_5

the background to the legislation is laid out in the introduction, where it is compared to LEED and BREEAM certifications. The main section of the paper then begins by looking in further depth into the background for the weightings for each parameter. Second, the results of an international workshop on the resolution are presented, where feedback was given from international experts in the fields of resilience and sustainable urban design. Third, a case study presented a high-rise building that was designed in accordance with the legislation, showing the relation between theory and reality. The paper then finishes by bringing this all together in a review of the legislation in the conclusions section.

A background to the Resolution and a comparison to LEED and BREEAM.

Ecuador is located in Latin America and is characterized by having a mild climate with little seasonal variation, though the climatic conditions vary according to height more than the time of year. In the Andean Mountains, the temperatures tend to be between 12 and 25 °C, whereas in the Amazon and the coastal regions, the temperature often approaches 30° (Inamhi 2015). There is a negligible energy demand related to heating in the building sector when considering the context of the weather. Additionally, it means that international certifications such as LEED and BREEAM are not necessarily suitable to be used in the Ecuadorian context, given their focus on reducing the energy demand of the buildings to be certified (Fig. 1).

In contrast, national certifications have come about in Ecuador from public bodies and the private sector. Regarding the private sector, the environmental evaluation system (SEA) was introduced in 2013 by the Mutualista Pichincha. In the public realm, the Ministry of the Environment (MAE) brought the Punto Verde certification for ecological buildings into development in 2014. The parameters for both are shown in Fig. 2, where it can be seen how there is a shift from energy demands towards water resources, materials and residues.

The Resolution No. 13-2016 of the Secretary of Territory, Habitat and Housing of the Municipality of Quito (and its subsequent update in April 2017) offers a different perspective on eco-efficient design from LEED and BREEAM, where it

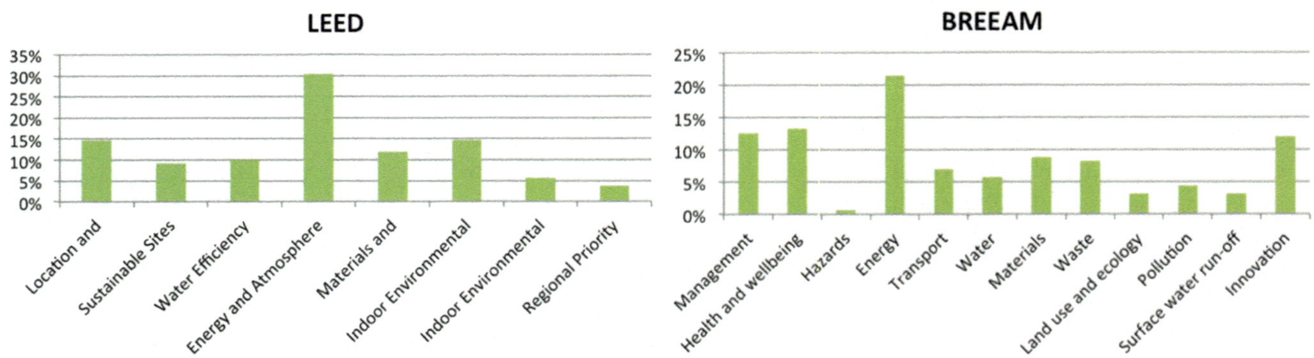

Fig. 1 Parameters for LEED and BREEAM certifications. Adapted from LEED (2016) and BREEAM (2016)

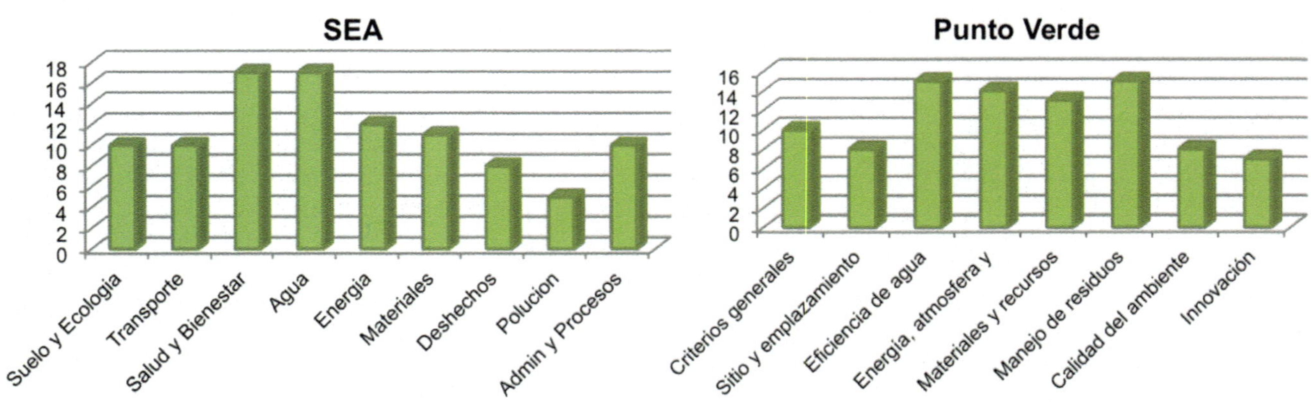

Fig. 2 Parameters for SEA and Punto Verde certifications. Adapted from SEA (n.d.) and MAE

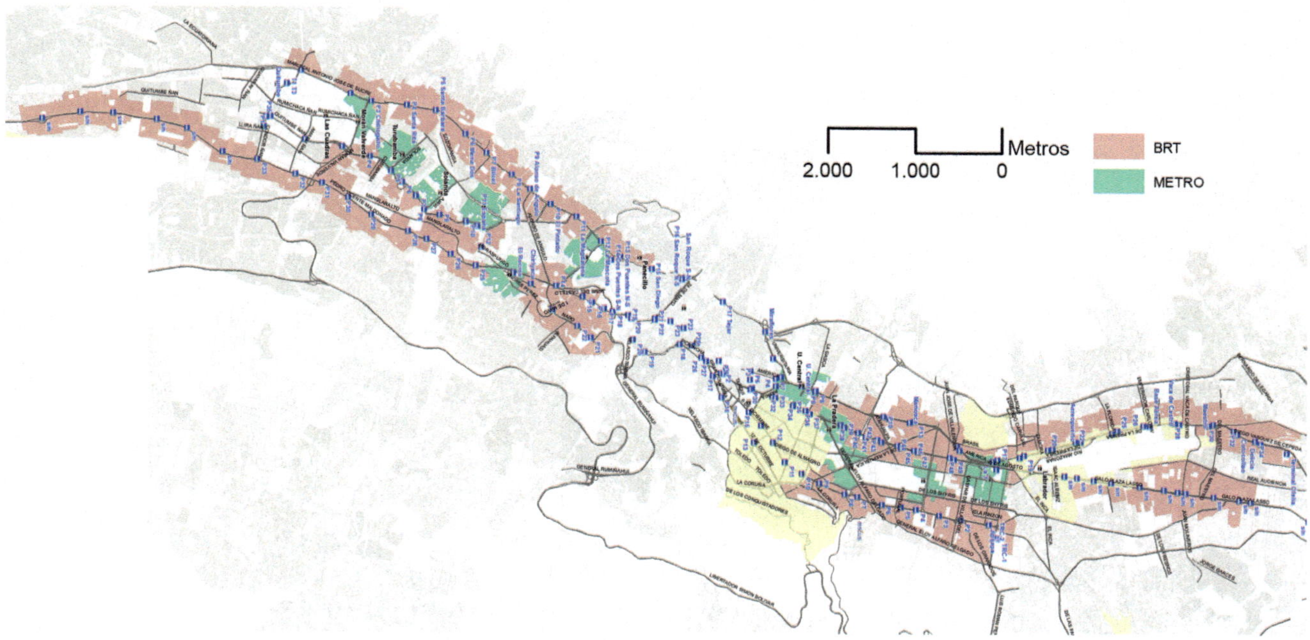

Fig. 3 Urban polygons showing 400 m within Quito's rapid bus transit system or planned metro stations

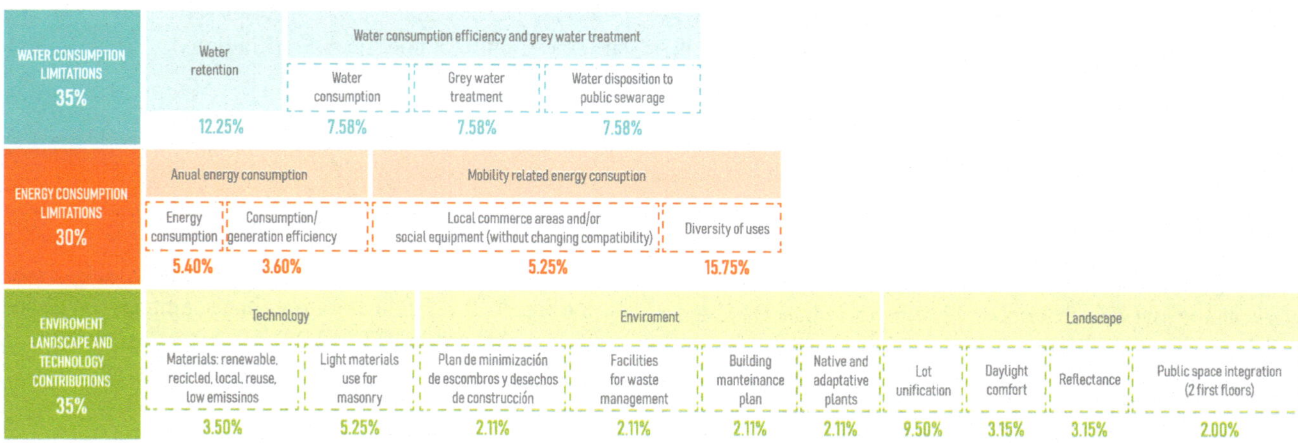

Fig. 4 Parameters for the Resolution 13-2016

builds on the parameters offered by SEA and Punto Verde. The aim was to encourage designers and the building industry to greatly reduce the environmental footprint of new buildings, in conjunction with moving away from private vehicles to mass public transport usage. In this context, the certification is aimed at the promotion of a compact city, with high-rise buildings located within 400 m to the mass public transport hubs (Fig. 3). In the instance of being within 400 m of a Rapid Bus Transit Station, it is possible to increase the building height by up to 50%. If the construction is within 400 m of a metro station, the building height can be increased by 100%. This offers an economic incentive for the building developers to comply with eco-efficient design, and to date, a significant interest has been shown already in the building sector.

The resolution dedicates 35% of the total percentage available to water (consumption, wastewater management and storm water control), compared to 10% and 6% for LEED and BREEAM, respectively (LEED 2016; BREEAM 2016). Additionally, of the 30% dedicated to energy, 21% is related to the promotion of diversification of uses to mitigate the need to travel large distances by car to meet shopping, work and leisure needs (Fig. 2). This is in relation to 42% of Ecuador's national energy demand being from the transport sector (MCSE 2015). Figure 4 shows the parameters for the

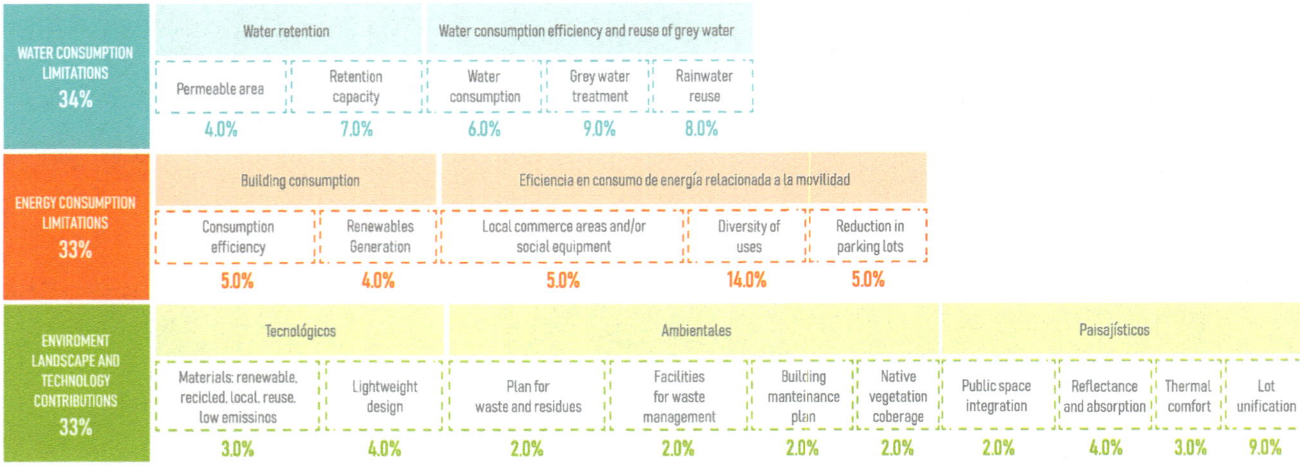

Fig. 5 Parameters for the Resolution 13-2016/07-April-2017

original Resolution No. 13-2016, and Fig. 5 for the updated version (Resolution No. 13-2016/07-April-2017).

2 Methodology

As mentioned in the introduction, this paper sets out to cover three areas in order to review the resolution. First, in the introduction a background to the legislation was given, where the differences to LEED and BREEAM certifications were laid out. The main section of the paper begins by looking in further depth into the background for the weightings for each parameter. The resolution was put to the test in an international workshop with experts in urban planning and resilience. The workshop was held on 5 May 2017 at the ICLEI 8th Global Forum on Urban Resilience and Adaptation in Bonn, Germany. Twenty-three experts attended with a ranging background of urbanists, resilient practitioners and technicians and decision-makers from different parts of the world. In the workshop, two key questions regarding the Resolution 13-2016 (prior to its update in April 2017) were posed: How can this tool be modified to contribute better to the proposed aim? What experiences from other parts of the world could be useful? The results of the workshop are summarized in this paper, and their relation to the updated version of the Resolution (2016/07-April-2017) is presented. Following this, the next section presents a case study of a high-rise building that was designed in accordance with the legislation and how the application of theory in reality is related to the need to have the resolution to be updated in 2017. The results are then summarized in the conclusion.

3 The Background to the Parameters Used in the Resolution

This section sets out to give a background to the weightings for each parameter used in the resolution, covering Agua (water), Energía (energy) and Aportes APT (environmental, landscape and technological contributions).

3.1 Agua (Water)

Quito is under an incredible strain regarding its wastewater treatment facilities, where currently wastewater from the city is discharged mostly untreated into surrounding rivers. According to the Public Metropolitan Company of Potable Water and Sanitation (EPMAPS) of Quito (2011), the sewerage discharge rates to the main rivers of the city are:

- Manchángara River = 3.48 m^3/s.
- Monjas River = 0.64 m^3/s.
- San Pedro River = 0.69 m^3/s.

Rainwater run-off also plays an important role in determining the capacity of the city's sewerage system. As the urban area of Quito expands, the amount of permeable ground of that previously unoccupied area is replaced by hard surfaces that lead to an increased surface run-off (Berndtsson 2009). This causes a sharp increase in the flow rate that is needed to be absorbed by the city's sewerage system. Estimates by Davis and Tapia (2016) for the increased loads in times of peak rainfall showed a 12-fold increase from 3.48 m^3/s into the Manchángara River to 55.32 m^3/s average. This puts an incredible strain on the sewerage system and diminishes the chances at all future attempts of pretreating the wastewater before discharge into local waterways.

In this context, the resolution sets out to:

1. Ensure the retention of rainwater run-off from new constructions and that this is not discharged directly to the sewerage system.
2. Promote a drastic reduction in the discharge of residual waters that would be produced by the new buildings.

Regarding point 1, the first section of the water parameters of the resolution covers the capability of the building retained rainwater. This is outlined in parameters that assess the percentage of permeable surface in the design and the ability to temporarily store any storm water run-off in peak precipitations. The updated resolution goes a step further in introducing a further parameter, where the storm water run-off then has to be reused in the building. Regarding point 2, the second section in the water parameters of the resolution focuses on efficiency in water consumption and the reuse of grey water. This is assessed in terms of the reduction in the use of potable water of the building before and after the eco-efficient design and the amount of grey water that is reused within the building. Consideration is also given to the pretreatment of blackwater before its discharge into the sewerage system. The idea is that blackwater can be pretreated for its reuse, say irrigation of green spaces. Otherwise, even if the water is discharged directly to the city's sewerage, the city should move toward a decentralized wastewater treatment system, where the requirement for large-scale, expensive infrastructure works is reduced.

3.2 Energía (Energy)

Buildings have connections and interactions with their surroundings: they have a place within the urban context through their users, they relate to mobility needs, and they serve as a channel to access, integrate and participate in the community on an urban scale (Anderson et al. 2015). Within this context, promoting sustainable urban transport enables a connection between individual buildings and the city where they are placed within, enhancing sustainability and helping to achieve reductions in energy consumption and environmental impacts (Larson et al. 2012). Furthermore, the promotion of renewable energies along with energy efficiency is considered the most widely spread solutions to control GHG (Özbuğday and Erbas 2015). Installation of renewable energies in buildings can also be linked to a decentralization of the energy system, enhancing independence, reducing energy consumption and increasing the resilience of the city in order to enhance urban sustainability (Kammen and Sunter 2016). As mentioned in Sadorsky (2014), renewable energy technologies address the need to meet the demands from growths in urban population and wealth, whilst also limiting GHG emissions and moving towards a more sustainable city. In the case of Ecuador and as previously mentioned in the introduction, the energy consumption for the building sector in Ecuador is far less than the international average, given that there is a negligible heating or cooling demand. According to the National Institute of Renewable Energies and Energy Efficiency of Ecuador INER (2016), Ecuador has six different climate zones. Quito is placed inside the climate zone three (3), which follows a climate criterion of CDD 10 °C for places ≤ 2500 masl. and HDD 18 °C for places ≤ 2000 masl. According to these characteristics, the need for cooling or heating in a building is not as important as targeting reductions in electricity demands, the promotion of sustainable mobility and water heating using renewable energies. The largest factor in energy demands is transport, which according to the Ministry of the Coordination of Strategic Sectors makes up 42% of total national energy demands (MCSE 2015). In Ecuador, residential buildings mainly need gas and electricity for cooking, water heating and electrical appliances, respectively. In terms of electrical energy, 42% was met by hydroelectricity in 2014 (MCSE 2015) and this is expected to increase dramatically with the Coca Codo Sinclair plant coming online (MEER, n.d.). As such, it is argued here that the electricity supply is to become renewable in future, and the aim is to ensure that the demands remain within the production capacity limits of Ecuador's hydropower plants. In relation to gas, this is mainly used for cooking and water heating. There has been a governmental push to replace three million gas cookers with electric induction technology (MEER 2015), which would mitigate the gas demand to electricity supplied from hydroelectric plants. In terms of water heating, however, it is not efficient to produce hot water with electricity (MacKay 2008), and as such solar water heating becomes of interest.

In this context, the resolution sets out to:

1. Limit the increase in energy demands, despite the increase in building heights permitted.
2. Produce hot water and/or electricity through renewable energies located on site.
3. Ensure commercial areas are included in the ground floor of the building.
4. Promote mixed-use buildings.

Regarding point 1, first section of the energy parameters examines the energy efficiency of the building. A base case for the building in a standard design is compared to the construction with energy efficiency measures put in place. In relation to point 2, the same first section goes on to stipulate that renewable energy needs to be produced on site, with a focus on solar water heating. The second section of the energy parameters looks at promoting mixed-use buildings and a reduction in transportation with private vehicles. Weighting is first given to having commercial areas on the ground floor of the building, in order to offer shopping, retail and food services to the building residents and public in general. A rather large weighting is then given if the building has a diversity of uses. The aim here is to greatly encourage buildings that are made up of residential dwellings, offices and shops. This in turn promotes the idea of compact cities, where one lives within walking (or even elevator) distance from the place of work and local amenities. A notable change between the original and the updated versions of the resolution is the inclusion of a substantial weighting for complying with minimal car parking requirements. The aim here is once again to displace the private car from the city centre.

3.3 Aportes APT (Environmental, Landscape and Technological Contributions)

Both LEED and BREEAM recognize the importance of environmental, landscape and technological contributions. LEED dedicates a total of 36% to parameters that are said to be related to this area (15% location and transportation, 9% sustainable sites and 12% materials and resources) (LEED 2016). BREEAM has a somewhat higher value of 37% dedicated (1% hazards, 9% materials, 8% waste, 3% land use and ecology, 4% pollution, 12% innovation) (BREEAM 2016). The resolution reflects the logic of LEED and BREEAM, with a weighting of 35% dedicated to this area. The parameters to be measured were inspired by both LEED and BREEAM, with adjustments made for them to be appropriate for the local context of Quito, Ecuador.

In this context, the resolution sets out to:

1. Encourage the use of local, renewable, lightweight materials for building constructions.
2. Transform the construction process and final building in use from being a polluting activity to one that transforms waste into nutrition and enhances local biodiversity.
3. Make the new constructions to have a positive contribution to local public space.

To this extent, the first section of parameters of the eco-efficiency matrix is aimed towards point 1 (materials that are local, recycled, renewable or with low VOC emission, plus lightweight materials in masonry and structures). Regarding point 2, the section is dedicated to closing cycles and promoting local biodiversity (plans to minimize residues from construction, plans to manage waste from the building in use, maintenance plans for the building installations and the use of native plants). Finally, in relation to point 3, the third section of parameters is related to improving public space (unifying allocations of land for construction), the shading cast by the new building (in the original Resolution 2016), reflectivity and heat absorption, thermal comfort (in the case of the updated resolution) and the integration of the building front (with public space). Overall, by taking inspiration from LEED and BREEAM and then adjusting the parameters to local necessities, the section of the certification aimed towards environmental, landscape and technological contributions should have a more positive impact than would be the case if the LEED and/or BREEAM parameters were applied directly.

4 Putting the Resolution to the Test: An International Workshop to Review the Legislation

4.1 The Workshop

As mentioned in the previous section titled "Methodology", the resolution was put to the test in an international workshop with experts in urban planning and resilience. The workshop was held on 5 May 2017 at the ICLEI 8th Global Forum on Urban Resilience and Adaptation in Bonn, Germany. In the workshop, two key questions regarding the Resolution 13-2016 (prior to its update in April 2017) were posed: How can this tool be modified to contribute better to the proposed aim? What experiences from other parts of the world could be useful? The results of the workshop are summarized in this paper, and their relation to the updated version of the Resolution (2016/07-April-2017) is presented.

4.2 The Main Observations

First, there was a general consensus between the members that there was a shortcoming in the resolution in addressing parking spaces allocated to each building. This was in relation to an international trend to reduce insofar as possible the parking spaces offered, in order to promote alternative forms of transport to private vehicles. Second, it was considered important that the legislation offered an incentive to the building industry to construct sustainable designs, as opposed to opting for a fine- or penalty-driven process. Third, it was observed that in reality the resolution was most likely aimed at an upper–middle class market sector, wherein the future improvements would need to be made regarding low-cost and informal housing. Fourth, it was deemed necessary to distinguish between the design parameters to be applied to housing versus office space versus commercial centres, as it is not necessarily the same sustainability drivers that would need to be applied. Fifth, in relation to this the idea of zoning the city was put forward. It was suggested, for example, that the poorer and less developed areas of the city should be given greater flexibility than the richer sectors, due to the amount of upfront finance initially required for ecological design. Finally, it was argued that a method for reviewing and carrying out sanctions was necessary if, for example, at the time of construction the building did not comply in reality to the coefficient design presented for the increase in permitted floors.

4.3 The Workshop Results in Relation to the Updated Resolution

The updated resolution came into force in April 2017. One of the main differences between the original and the updated resolutions was the inclusion of a limited number of parking spaces. This was reflected in the comments from the workshop, where there was a general consensus in the need to reduce urban parking spaces in Quito. The other observations are currently work in progress, where they will be assessed for the upcoming updates to the resolution.

5 Theory Versus Reality. A Case Study of a Building Designed Under the Resolution

5.1 The Building

The project consisted of a building called Tesla that sought to increase its height from 10 to 15 floors through being an ecological design. The company responsible for delivering the ecological design and obtaining the permission to increase the building height was Evolution Engineering and Design and Energy Systems Ltd., working for the client Tesla Buildings Inc. This section focuses on the water systems implemented in the building, as this was proven to be a reality check in terms of the costs incurred.

The building was designed under the original resolution, before it was updated in April 2017. The full percentage for water was reached, which in this case meant incorporating:

- Permeable surfaces and a storage tank to temporarily retain the rainwater run-off from the buildings.
- A grey water recycling system, which pretreated grey water from the showers and hand basins to be pumped and reused for toilet flushing, clothes washing and irrigation. The strategy is shown in Fig. 6. It can be seen how the shower and hand basin wastewaters are treated in a centralized plant, to be reused mainly in the toilets and washing machines. The wastewater from these is then treated in a biodigestor plant, to be reused in watering green areas.

5.2 Theory Versus Reality

The design complied with the requirements of the resolution and as such managed to obtain the special permission to increase the building height. When in the process of construction, however, it became clear that the infrastructure required to (a) pretreat the grey water destined for reuse, and (b) construct of over 500 m^2 of vertical garden needed for watering in accordance with the Resolution, was unfeasible in terms of a cost-benefit analysis of the project (the wastewater treatment plant would have taken up half a basement floor alone), and the whole system looked to need some 300,000 USD extra investment. For this reason, it was chosen to comply without the inclusion of the biodigestor or external vertical gardens (see Fig. 7).

5.3 The Reality in Relation to the New Resolution

It can be seen in the updated version of the matrix (Fig. 5) that these adjustments have been made to reflect this: the emphasis on not sending water to the sewerage system was replaced with a parameter that focuses on using rainwater as a water source. As mentioned previously, one of the main problems regarding Quito's sewerage system capacity is the inclusion of storm water run-off in times of peak rainfall. To this extent, the new resolution aims towards the management of this storm water run-off, through promoting the maximum usage of collected rainwater from the run-off from the building.

6 Discussion of Results and Conclusions

The recently introduced resolution for ecological design was the first step in sustainable certification in Quito, Ecuador. Like any other certification (such as LEED or BREEAM), the resolution needs to be updated to adjust requirements to a changing socio-economic context and to meet the market technical capabilities. To this extent, the resolution was updated in April 2017, where amongst other aspects greater emphasis was placed on storm water management and

Fig. 6 Original water management system proposed for the Tesla Building

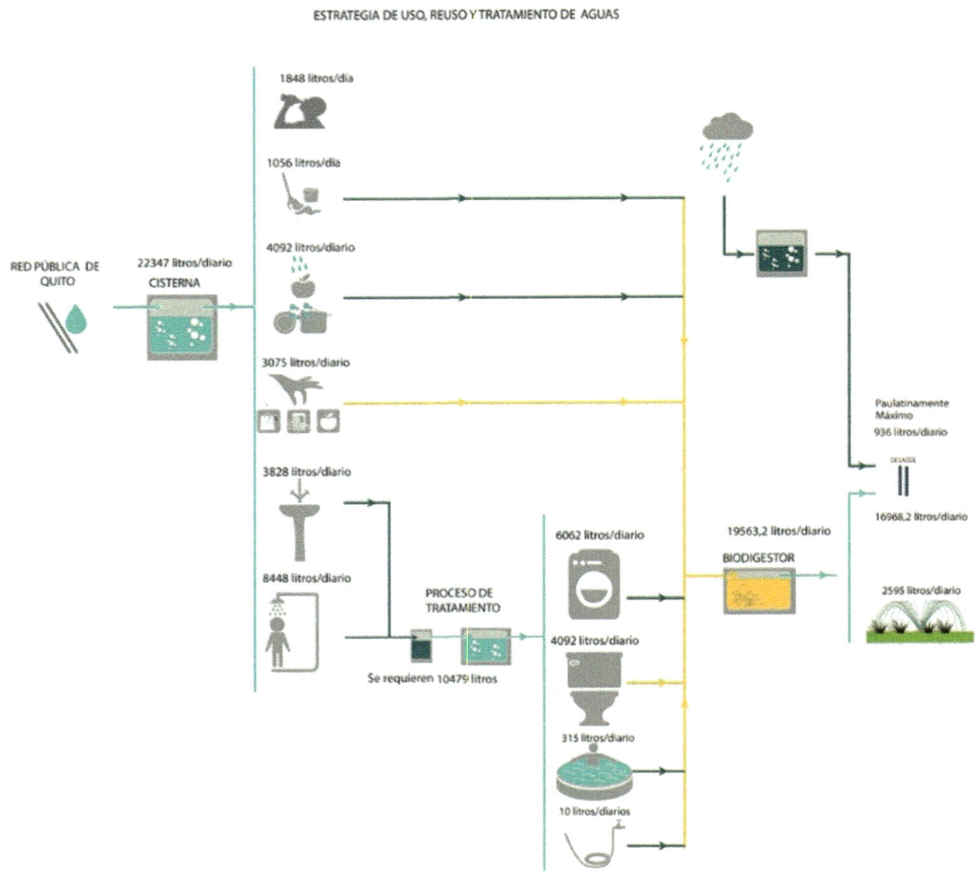

Fig. 7 Final proposal

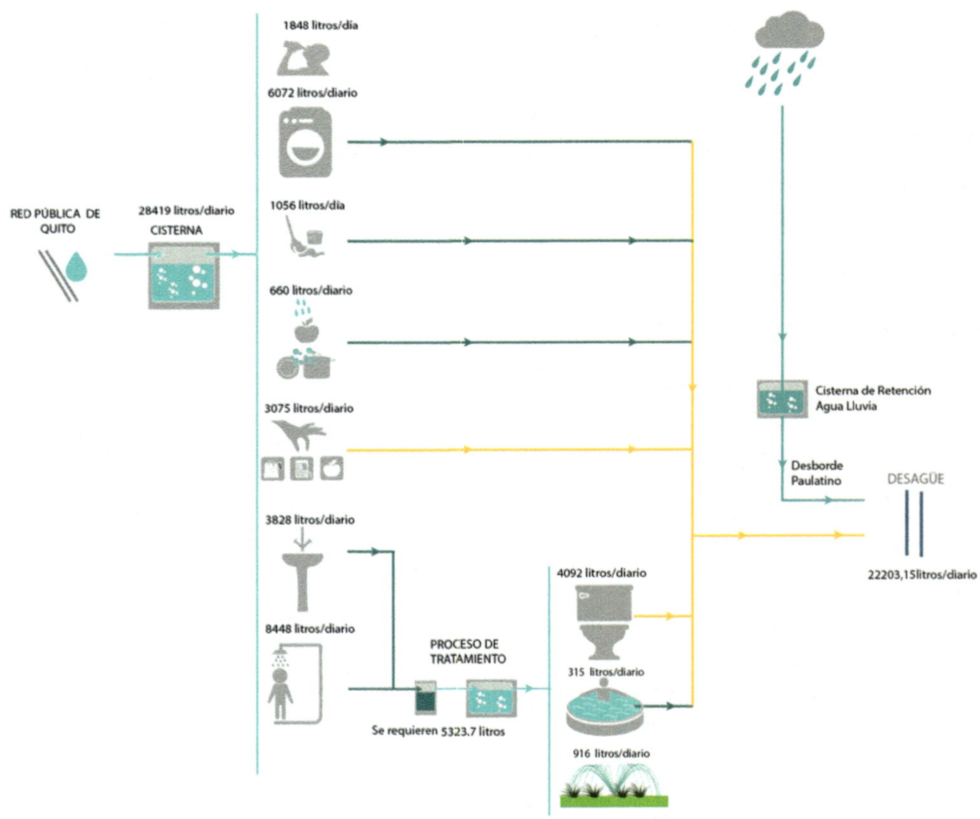

reduced the number of urban parking spaces. In future, it is hoped that the resolution can be developed further and climb up the legislation hierarchy to become an ordinance for the city. This would make it more resistant to the whims of political changes in future. The resolution was the first step with real impact on the real estate market of the city. In this sense, it has fulfilled its aim: reactivating the building sector. In future and with the introduction of an underground metro system, Quito stands to grow as a city. In this context, it is of paramount importance to ensure the building sector in the private sphere responds to the growth whilst not having a negative impact on the natural environment.

Overall, this paper highlighted the differences between the resolution designed for Quito and the international certifications such as LEED and BREEAM. It is argued in this study that the resolution is much more appropriate for local contexts and the climate of Ecuador. The resolution has less emphasis given to energy demands, with a greater focus on demand of resources in general and reducing the number of private vehicles on the road. In terms of international scrutiny, several valuable observations were made at an international workshop to discuss the resolution. One of these in particular was taken on board for the updated resolution: the need to reduce parking spaces in the city centre. Updates to the resolution also took into account the realities faced by the building sector, such as the space and cost requirements for on-site wastewater recycling.

Acknowledgements Special thanks need to be extended to Mauricio Masache, whose contributions to the energy section of this paper were invaluable.

References

Anderson, J. E., Wulfhorst, G., & Lang, W. (2015). Energy analysis of the built environment—Areview and outlook. Renewable and Sustainable Energy Reviews, 44, 149–158.

Berndtsson, J. C. (2009). Green roof performance towards management of runoff water quantity and quality: A review. ELSEVIER (Ed.), 3–4.

BREEAM. (2016). BREEAM International New Construction, Technical Manual SD233 1.0. Watford, UK.

Davis, M. J. M., and Tapia, A. (2016). The potential for Green roofs in Sustainable Urban Drainage Systems (SUDS) in Quito, Ecuador. Interaction between Theory and Practice in Civil Engineering and Construction; Komurlu, Ruveyda; Gurgun, Asli P.; Singh, Amarjit; and Yazdani, Siamak (eds.), 655 pp., ISEC Press - ND, USA, 2016, ISBN-10: 978-0-9960437-2-4. Available from: https://www.isec-society.org/ISEC_PRESS/EURO_MED_SEC_01/html/AW-7.xml.

Empresa Pública Metropolitana de Agua Potable y Saneamiento (EPMAPS). (2011). Programa de Saneamiento Ambiental para el Distrito Metropolitano de Quito. Retrieved from http://www.aguaquito.gob.ec/sites/default/files/documentos/plan_maestro_agua_potable.pdf.

Instituto Nacional de Meteorología e Hidrología (INAMHI). (2015). Anuario Meteorológico No. 52-2012. Quito, Ecuador. Available from: http://www.serviciometeorologico.gob.ec/wp-content/uploads/anuarios/meteorologicos/Am%202012.pdf.

Kammen, D. M., & Sunter, D. A. (2016). City-integrated renewable energy for urban sustainability. Science, 352(6288), 922–928.

Larson, W., Liu, F., & Yezer, A. (2012). Energy footprint of the city: Effects of urban land use and transportation policies. Journal of Urban Economics, 72(2), 147–159.

LEED. (2016). LEED v4 for Building Design and Construction – current version. USGBC. Washington, USA.

MacKay, D. (2008). Sustainable Energy-without the hot air. UIT Cambridge.

Ministerio Coordinador de Sectores Estratégicos (MCSE). (2015). Balance Energético Nacional 2015. Quito, Ecuador. Available from: http://www.sectoresestrategicos.gob.ec/wp-content/uploads/downloads/2016/01/Resumen-Balance-Energético-20151.pdf.

Ministerio De Electricidad Y Energía Renovable (MEER). (2015). Programa de eficiencia energética para cocción por inducción y calentamiento de agua con electricidad en sustitución del gas licuado de petróleo (GLP) en el sector residencial. Available from: http://www.energia.gob.ec/programa-de-eficiencia-energetica-para-coccion-por-induccion-y-calentamiento-de-agua-con-electricidad-en-sustitucion-del-gas-licuado-de-petroleo-glp-en-el-sector-reside-2/.

Ministerio de Electricidad y Energía Renovable (MEER). (n.d.). Coca Codo Sinclair. Avilable from: http://www.energia.gob.ec/coca-codo-sinclair/.

Mutualista Pichincha, (n.d.). Sistema de Evaluación Ambiental (SEA). Available from: http://www.sistemadeevaluacionambiental.com.

Özbuğday, F. C., & Erbas, B. C. (2015). How effective are energy efficiency and renewable energy in curbing CO_2 emissions in the long run? A heterogeneous panel data analysis. Energy, 82, 734–745.

Sadorsky, P. (2014). The effect of urbanization on CO2 emissions in emerging economies. Energy Economics, 41, 147–153.

Todd, J. A., Pyke, C., & Tufts, R. (2013). Implications of trends in LEED usage: rating system design and market transformation. Building Research & Information, 41(4), 384–400.

R.I.S.Q: Risk Assessment Tool for Quito

David Jacome Polit, P. Cubillo, D. Paredes, and P. Ruiz Villalba

Abstract

The Sustainable Development Goals and the New Urban Agenda require building resilience and effective risk management in order to achieve a better possible future. It is also true that no urban sustainability is possible without taking into account the risk dimension. In this sense, indices are increasingly important for emergency and sustainable development planning at different levels, particularly with respect to identifying vulnerable settlements and mapping disaster potential. This paper provides both a critical literature review and an empirical case study that highlight the importance of different types of decisions in the construction of risk index. However, during the exercise, it was clear that a risk index to be constructed with contextual sensibility can be challenging, especially around what indicators should be included, or how should they be aggregated, and if they should have differentiated weights. The purpose of this paper is to demonstrate a method of analysis, first based on existing information, second by proposing an aggregation method, and third proposing an overall risk index that can be graphically displayed. The case study focuses on the main natural hazards, exposure, and the social vulnerability of Quito, Ecuador. Overall results are calculated by applying Pareto ranking. The paper aims at providing an effective index to inform decision making and for effective urban planning in multirisks cities, with the final goal to reduce physical impacts and propose effective, sustainable, and resilient urban strategies.

Keywords

Risk index • Pareto • Hazards • Exposure • Vulnerability

D. Jacome Polit (✉)
Instituto Metropolitano de Planificación Urbana (IMPU)—MDMQ, Quito, Ecuador
e-mail: djacomepolit@gmail.com

P. Cubillo · D. Paredes · P. Ruiz Villalba
Private Researcher, Quito, Ecuador

1 Introduction

The UN's Sustainable Development Goal 11 establishes that cities and human settlements should be inclusive, safe, resilient, and sustainable. In this goal, the 11.b target states that by 2020, "substantially increase the number of cities and human settlements adopting and implementing integrated policies and plans toward inclusion, resource efficiency, mitigation and adaptation to climate change, resilience to disasters, and develop and implement … holistic disaster risk management at all levels." At the same time, the New Urban Agenda "Quito Declaration on Sustainable Cities and Human Settlements for All," in the Our principles and commitments section, literal c, establishes that to achieve our vision "building urban resilience, by reducing disaster risks and by mitigating and adapting to climate change," among other actions, are required to guide the proposed principles. However, in order to start working toward these goals and targets, informed decisions need to be taken, to plan, and to act.

In the recent past, vulnerability as part of risk analysis has been used widely based on interdisciplinary research with the aim to explain the degree to which economic, physical, and environmental systems suffer from different hazards. However, this practice has been based solely upon scientific simulation of economic loss on bio-physical conditions of the environment, depending on the physical preparedness of the communities and organizations affected by an adverse change (Ludena and Yoon 2015). When including other factors in the analysis, like socioeconomic, for instance, vulnerability depends on the sensitivity of an urban system to different hazards and how effectively affected population can act to reduce the detrimental effect of such impacts.

In a highly complex world, vulnerability is often an intrinsic characteristic of a system or element (UN/ISDR 2004; Cardona 2004, p. 37; Wisner 2002, pp. 12/7; Thywissen 2006 in Birkmann 2007), where most analysts acknowledge that vulnerability is a condition related only to

C. Alalouch et al. (eds.), *Advanced Studies in Energy Efficiency and Built Environment for Developing Countries*, Advances in Science, Technology & Innovation, https://doi.org/10.1007/978-3-030-10856-4_6

hazard exposure, e.g., with respect to its frequency and severity, or that it is useless to discuss vulnerability independent of its hazard context. However, Wisner (2002) emphasizes that vulnerability is also how easy for a system or element recovers from a negative impact; therefore coping and recovery must be part of the analysis. It is important, then, to understand vulnerability as both: susceptibility—the weakness of a system or element and how much it can be affected by a stress or perturbation—on the one hand, and coping capacities—as the ability of a system or element to respond to and recover from the effects of stress or perturbations—as explained in many vulnerability approaches, e.g., Birkmann (2007); Wisner (2002) and—to a certain extent—Bohle (2001), on the other. In this sense, since cities grow reflecting socioeconomic differences, shaping in a meaningful manner risk distribution, this reality needs to be taken into account in urban planning processes, risk and emergency plans, and others (Adger et al. 2004).

As a result, an increasing number of interdisciplinary global and local proposals have been developed to measure risk and vulnerability with a wide set of indicators and indices (Birkmann 2006 in Birkmann 2007). Methods for disaster risk and vulnerability assessment and quantification are further to be developed. However, while hazard mapping and quantification are far ahead (Cardona 2005, p. 57 in Birkmann 2007), broad indicators appear repeatedly in social vulnerability analyses. For the present work, different proxies for each indicator will be used. The main objective is to propose an information display tool to inform decision making when planning the future development of the city of Quito. This tool can be developed further and could include more information than the proposed to improve its results.

2 Vulnerability, Exposure, Hazard, and Risk: Definitions

Usually, vulnerability for natural hazards means the potential for loss. However, vulnerability varies over time and space (Liverman 2001), implying that losses vary between social groups settled in different geographical locations over time (Eakin and Luers 2006). Therefore, it is natural to find differences between the concepts that define risk, vulnerability, hazard, and how one explores the relationship between them, especially when the literature on these topics is broad and extensive (Hoogeveen et al. 2004; Alwang et al. 2001). In the most basic way, vulnerability is a term used to describe susceptibility to damage or harm (J. X. Kasperson and R. E. Kasperson 2001; Liverman 2001), however, the debate has being centered around, and efforts have been devoted on how to characterize vulnerability in theory and practice (J. X. Kasperson and R. E. Kasperson 2001; Alwang et al. 2001; Kelly and Adger 2000; Turner II et al. 2003; Klein et al. 2003 cited in Liverman 2001).

For the purpose of the present work, we will use Birkmann's (2006) approach to define and correlate the terms as displayed in his book "Measuring vulnerability to natural hazards: toward disaster resilient societies." Birkmann argues that instead of looking at disasters in an isolated way, they should be understood as the complex interaction of "a potentially damaging physical event (e.g., floods, droughts, fires, earthquakes, and storms) and the vulnerability of a society, its infrastructure, economy, and environment, which are determined by human behavior." Furthermore, Birkmann quotes the final document of the World Conference of Disaster Reduction "Hyogo Framework for Acton 2005–2015," cited after in the "Sendai Framework for Disaster Risk Reduction," where its declaration points out:

> The starting point for reducing disaster risk and promoting a culture for disaster resilience lies in the knowledge of the hazards and the physical, social, economic and environmental vulnerabilities to disasters that most societies face, and of the ways in which hazards and vulnerabilities are changing in the short and long term, followed by action taken on the basis of that knowledge. (ISDR, UN 2005)

Once the complexity of the topic is understood, and to better understand the terms, Birkmann establishes that "... vulnerability and coping capacity manifest themselves once a vulnerable community is exposed to a hazardous event" (Birkmann 2006). Definitions for hazard, exposure, vulnerability, disaster risks, coping capacity, and recovery are presented[1]: Hazard is understood as: "A dangerous phenomenon, substance, human activity or condition that may cause loss of life, injury or other health impacts, property damage, loss of livelihoods and services, social and economic disruption, or environmental damage." (The United Nations for Disaster Risk Reduction 2007). Exposure is defined by "the situation of people, infrastructure, housing, production capacities and other tangible human assets located in hazard-prone areas" (The United Nations for Disaster Risk Reduction 2007). Vulnerability implies: "The characteristics and circumstances of a community, system or asset that make it susceptible to the damaging effects of a hazard" (Alwang et al. 2001). After, disaster risk is defined by the interaction between hazard, exposure and vulnerability (The United Nations for Disaster Risk Reduction 2007): "The potential disaster losses, in lives, health status,

[1]Birkmann uses the definitions acquired from the United Nations International Strategy for Disaster Reduction (2002), included in the Disaster Reduction and Sustainable Development: Understanding the Links between Vulnerability and Risk Related to Development and Environment. Available at: http://www.unisdr.org/files/3620_UNdocs21.pdf. However, for the present work updated terminology from The United Nations for Disaster Risk Reduction was used. Retrieved from: https://www.unisdr.org/we/inform/terminology.

livelihoods, assets and services, which could occur to a particular community or a society over some specified future time period." (The United Nations for Disaster Risk Reduction 2007). After, coping capacity is defined as (The United Nations for Disaster Risk Reduction 2007): "The ability of people, organizations and systems, using available skills and resources, to face and manage adverse conditions, emergencies or disasters." Finally, recovery is defined as: "The restoration, and improvement where appropriate, of facilities, livelihoods and living conditions of disaster-affected communities, including efforts to reduce disaster risk factors" (The United Nations for Disaster Risk Reduction 2007).

In other words and in a simpler manner, a hazard is any potential event that can decrease the quality of life and affect in a negative way the environment where communities or groups of people thrive; exposure is physical conditions and location of settlements in hazard-prone areas; vulnerability is related to the ability of those groups of people or communities to comeback and recover from that event and have the same quality of life as before the event, or even better. The possible impacts of the first, plus the conditions of the second and the weaknesses of the third, configure the dimension of disaster risks.

3 Measuring and Mapping Risk

As a result, dimensioning risk requires an understanding of how vulnerable people are and how can they be affected by disasters. This means not only understanding the hazards and exposure, but also recognizing that social vulnerability is much more than the likelihood of buildings to collapse or infrastructure to be damaged (Cannon et al. 2003). The analysis becomes complex since it requires a multidisciplinary approach allowing to look from a multidimensional perspective. The first hurdle is found when different disciplines use alternative meanings and concepts of vulnerability. Then, diversity of concepts leads to diverse methods of measuring it, explained by a discipline's tendency to focus on different components of vulnerability (Alwang et al. 2001), and how they are measured. The main disciplines include, considered as categories, in some cases: economics (including the poverty dynamics, asset-based, sustainable livelihoods, and food security literatures), sociology/anthropology, disaster management, environmental science, and health/nutrition. Vulnerability has been an important theoretical topic, and effort has being made to clarify its approach in global change research for around four decades (e.g., Bohle et al. 1994). Sadly, analysis on socioeconomic vulnerability has gained relevance only in the last decades (e.g., McCarthy et al. 2001). Consequently vulnerability analysis and later assessment methods need further work and development, where most thematic areas are in different advancement state and still forming.

However, it is recognized that vulnerability assessment may vary in accordance to information availability, scales within different categories and contexts (Bohle et al. 1994; Rygel et al. 2006), (e.g., In economics GDP and the Gini index were good proxy indicators in city or country scale where they were statistically significant, but not for smaller scales), therefore choosing indicators accordingly is necessary. For the purpose of the present job, indicators of "relative" vulnerability and, thereby, a context-sensitive index of risk are presented in accordance with the existing information and applicable scales. To achieve this outcome, this section will first describe the calculation method of overall risk. Second, indicators of hazard distribution, exposure, and social vulnerability will be proposed. Third, as suggested by Rygel et al. (2006), Pareto ranking is used to calculate and present the results of this index in a graphical manner, with the aim to help prioritize attention to different sectors in Quito, hence configuring a decision-making tool.

4 Calculation of Overall Risk and Component Analysis

Risk is configured by the following formula (Molin Valdez 2012):

$$\text{Risk} = \frac{\text{Hazards} \times \text{Exposure} \times \text{Vulnerability}}{\text{Coping capacity}}$$

For the present case, coping capacity would not be analyzed since it requires an understanding of the effectiveness of national and local regulations over procedures and competences when a disaster occurs. Combined into the other three components (hazards, exposure, and vulnerability), indicators are used in an orderly fashion to provide an understanding of each one of them. Maps of individual indicators and/or component scores are the starting point of the analysis. Nevertheless, by having many indicators, compiling them into the three components of the aforementioned formula enables combining them into a single measure to facilitate understanding the multidimensionality of risk (Clark et al. 1998). For the present case, every indicator was assigned with averaged score to combine them into a single measure for every component, and later for the index.

This decision was made based on Rygel et al. (2006) warning on two possible problems when averaging indicator's scores. The first problem comes when having decided to use simple averages for all indicators, because it also means that a weighting scheme could be created. This option could be valid if it is decided that the risk index should prioritize one component or indicator over another, however, and since all are of relative importance, prioritization might mean subjectivity. This creates a second problem, since when averages are used, the contributions of a component or indicator can be distorted, especially when there are extreme high and low values. However, it is acknowledged that extreme values indicate that more attention is needed. In this way, to avoid these problems, establishing a Pareto classification scheme without prioritization components or indicators has been followed, in order to organize the indicators in a series of ranges.

Pareto ranking is a method for ordering cases on multiple criteria that has become popular in different contexts, because it has the ability to depict situations that are of significant relevance in an analysis. The rationale behind the Pareto ranking method is as follows: Each case i is considered within of a set of n indicator scores, $\{is1, is2, \ldots, isn\}$. Simplifying, but without losing objectivity, it is assumed that a higher score on any individual indicator means greater contribution to a component. For example: when sectors A and B are compared (in this case census sectors are compared), sector A may represent a greater risk than sector B, but only if within the values of the scores of sector A at least one is greater than those of sector B. It is clear that when more variables are included, the analysis becomes more complex. This problem is solved by assigning relative weights only to each component within a determined range. For example, if sector A has scores higher than sector B in some components, but sector B has a higher score than sector A in other components, a simple sum of all components allows determining which sector represents the greatest risk. This methodology allows a classification of multiple components without giving relative importance to different components.

A risk calculation model by means of indicators is the most complex part of the proposed method. In the case of Quito, qualitative and quantitative variables with different scales and with different accuracies due to information quality had to be combined. A process was developed to allow standardization; the territorial unit chosen for the analysis was the census sector, because it allows the best representation of the diversity of each selected variable. The risk calculation method is divided into three phases: (1) Creation of intermediate maps for each variable; (2) Transformation of each map of internal variables to a normalized range through a linear regression, and (3) The combination of all maps by the Pareto method.

For the first phase, 14 variables were selected to be transformed into reference maps spatially. These variables are divided into two groups: The first group includes hazard variables that are provided with qualitative values in which their intensity is represented. In the proposed method, these linguistic variables are transformed into diffuse numbers between 0 and 1 using scales such as those proposed by Chen and Hwang (1992), see Fig. 1, where the area is provided with a weight in the following way; both values, intensiveness of possible event and area affected, are given a value between 0 and 1 and later combined in a weighted sum that result, again, in a value between 0 and 1 representing the degree of danger.

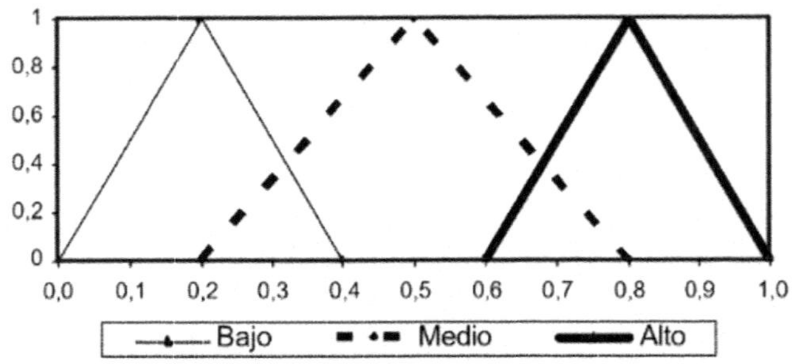

Fig. 1 Scale of three linguistic terms (Chen and Hwang 1992)

For the second variable group in the first phase, all variables coming from the Census of Population and Housing 2010 are included, which are already referred to the census sector. For this group of variables, the percentage of vulnerable population was calculated within each variable, and values between 0 and 1 were assigned. For the case of exposure and access to basic services, where there was more than one parameter for an indicator (e.g., access to water, access to electricity, access to sewage), a similar process as the one for hazards was carried out to combine each parameter in order to obtain a single value representing the indicator.

In the second phase, a process of normalization of the 14 maps in each of the census sectors is carried out through a linear regression. In the third phase, all the variables are combined through the Pareto method.

4.1 Hazards

Quito is a city that could be affected by five different hazards: (1) Earthquakes, (2) Volcanic eruptions, (3) Flooding, (4) Wildfires, and (5) Landslides (MDMQ 2016). All of them are presented spread unevenly across the city, in terms of location and intensity. At the moment, earthquakes are not included since a micro-zoning seismic plan will be developed in the future. By mapping all the different hazards and overlying the maps, we found that zones of the city could be affected by several hazards at the same time. Also, the intensity of how hazards could manifest differs along the territory. Assigning weights on the intensity and periodicity was based on existing literature describing each event. However, it is acknowledged that this could be subjective (Fig. 2).

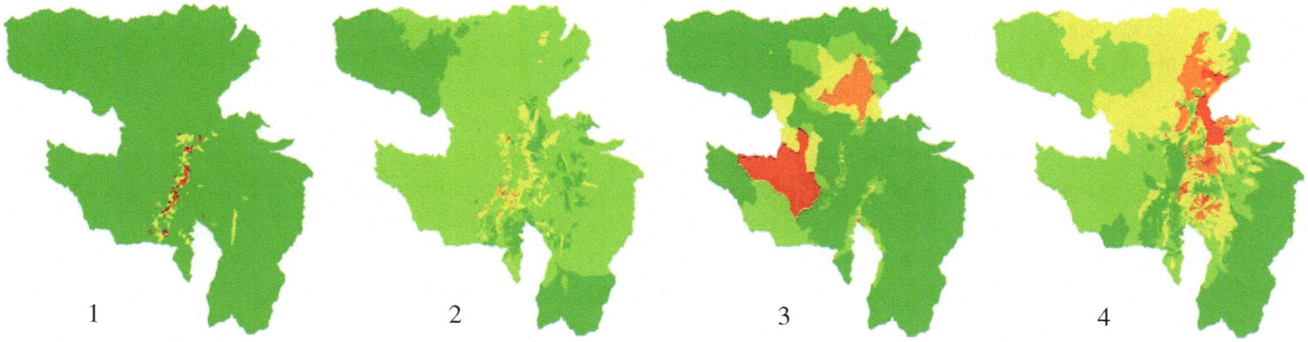

Fig. 2 Map1: flooding; map2: landslides; map3: volcanic eruptions; map4: wildfires

Indicators chosen according to information availability and scales:

1. Number of natural hazards affecting area.
2. Area possibly affected by different natural hazards and its severity.

4.2 Exposure

According to average numbers of lots and families per neighborhood, Quito will have to manage in the coming years the regularization of 45,000 lots, where approximately 180,000 people live, i.e., approximately 8% of the city's population (MDMQ 2012 cited in MDMQ 2015, v). A first diagnosis to establish the dimension of the informal land occupation in the city, conducted between November and December 2009 by each Zonal Administration (political administrative offices), revealed that there were 439 irregular neighborhoods.[2]

Furthermore, informal settlements usually occurring in adverse topographic conditions take place in a twofold manner: (1) Increases a significant factor of environmental deterioration and anthropogenic risks due to infrastructure built on fragile ecosystems or lots with immitigable risk conditions, and (2) Increases vulnerability, since many of the buildings are built with no technical assistance, hence no seismic resistance engineering. As a volatile mix, settlements in immitigable risk zones with precarious building systems spreads risk unevenly and widely along the territory (Fig. 3).

Indicators chosen according to information availability and scales:

1. Buildings with precarious construction systems.

[2]This information is based on interviews with local neighborhood leaders, might not be 100% accurate.

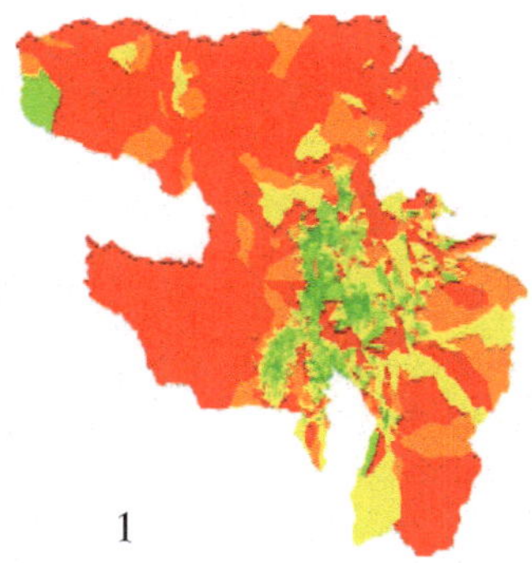

Fig. 3 Map1: precarious construction systems, exposed buildings

4.3 Socioeconomic Vulnerability

While revising vulnerability assessment literature, it was notable that some broad indicators of vulnerability appear repeatedly. However, different variables were chosen to represent each indicator, depending upon factors such as availability of data, context, and scales as explained before. Most variables were derived from the 2010 Ecuador Census and were analyzed at the census sector level. Small inconsistencies were found when comparing census sector level areas with political administrative areas, hence the first hurdle was found. The number of people or housing units in a particular category was divided by the area of a census sector. By this means, the use of percentages is an advantage in the analysis that allows reducing a set of complex variables to a smaller set of components that otherwise cannot be compared between sectors. When doing this exercise, sometimes radical differences between percentages in the analysis were identified, especially in less dense areas, for which more in-depth investigation of the data and on site is recommended.

Since the education and health systems are responsibility of the central government, this information was not included at this time. The vulnerability indicators used in this study, the ones that describe poverty, gender, age, and disabilities, are explained in the following paragraphs:

The first vulnerability indicator is poverty. Poor population faces more difficulties to overcome a catastrophic event

(Fothergill and Peek 2004). Besides being a chronic stress, since economic constraints in covering basic needs and priorities, such as food, transportation, and basic services already puts pressure on low income budgets, it also prevents investing in preparedness, emergency supplies, or other preventive measures (Clark et al. 1998). As a result, people less prepared suffer from higher mortality rates during and after a disaster (Blaikie et al. 1994) and find it harder to repair or replace their living units (Morrow 1999). While is clear that the wealthy suffer greater monetary value of material losses, this losses are relatively far more calamitous for the poor (Morrow 1999), since there is a close correlation between socioeconomic levels and the quality of housing conditions. Furthermore, during and after disasters, access to lifelines, such as communications and transportation, is limited or null for the poor (Clark et al. 1998) (Fig. 4).

Indicators chosen according to information availability and scales:

1. Share of economically inactive population.
2. Non-affiliated to the social security system.
3. Unemployed population.
4. Access to communications.
5. Access to transportation.
6. Access to basic services.

Since the country does not have a census with income per family data, it is not possible to include any indicator related.

Another variable that affects vulnerability is gender (Enarson and Morrow 1997). Single mothers, either never married or divorced, usually have lesser means to provide for their own (Bianchi and Spain 1996), thus making them less prepared to face a disaster or have harder times to recover. This condition is also of special importance during a disaster, since their ability to seek safety is severely diminished when being responsible for the very young and the very old, which usually require additional help and supervision (Fothergill 1998). Furthermore, gender-related inequalities are a constant condition in the developing world, hence the importance of placing women at the core when planning for sustainable development (Denton 2002) and when undergoing a risk analysis (Fig. 5).

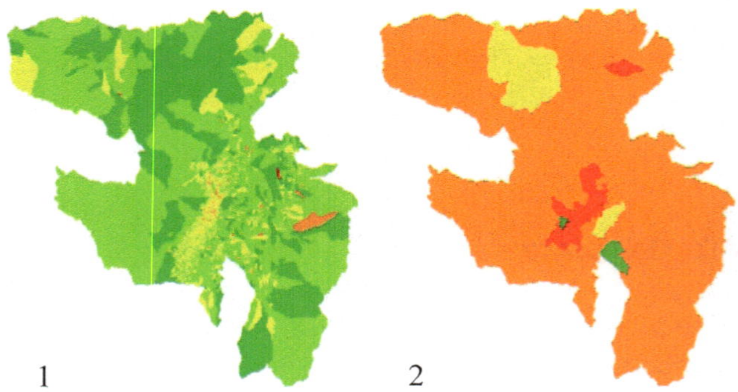

Fig. 5 Map1: caregiving women; map2: single mothers

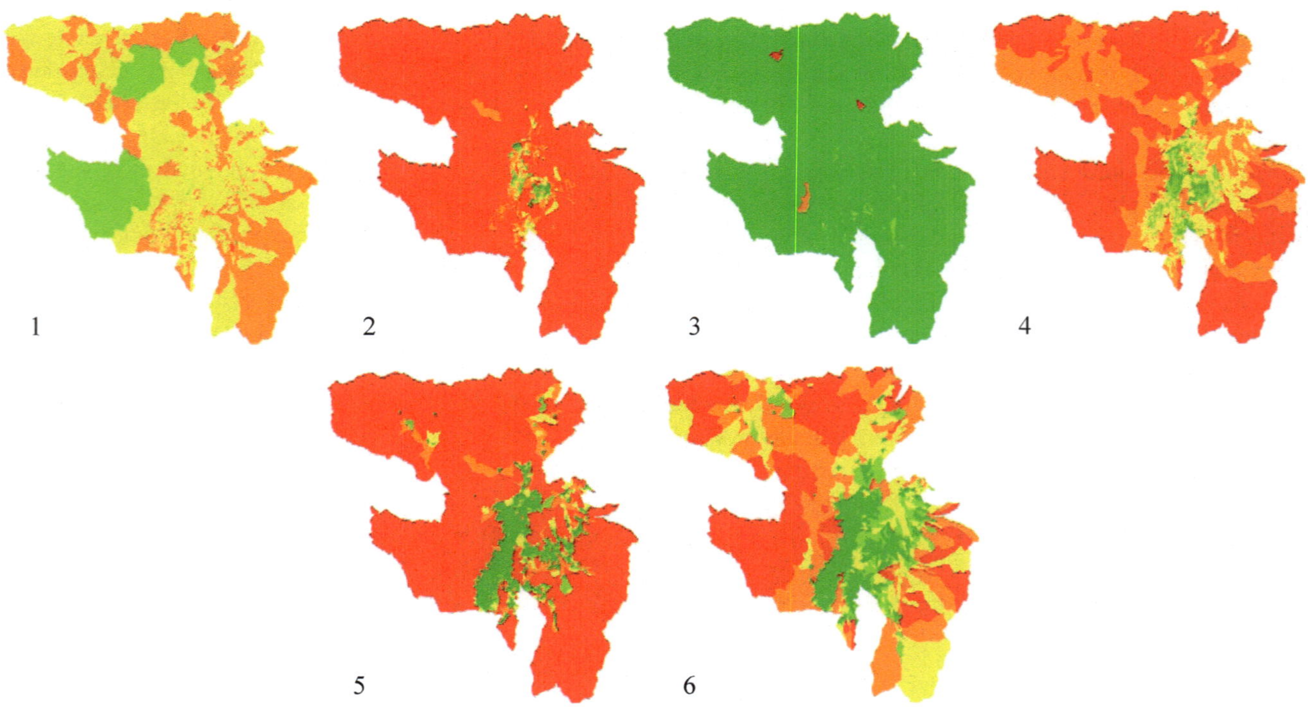

Fig. 4 Map1: economically inactive population; map2: non-affiliated to the social security system; map3: unemployed population; map4: access to communications; map5: access to transportation; map6: access to basic services

Indicators chosen according to information availability and scales:

1. Caregiving women.
2. Women with children without partner.

Another sector of the population that is less able to respond effectively to disasters is people living with mental or physical disabilities. Additional assistance is required during a disaster and specific infrastructure is needed to prepare and recover from such events. Furthermore, emergency managers are required to identify areas with high concentrations of disabled people for preparatory and early evacuation measures (Morrow 1999) (Fig. 6).

Indicators chosen according to information availability and scales:

1. Population with disabilities.

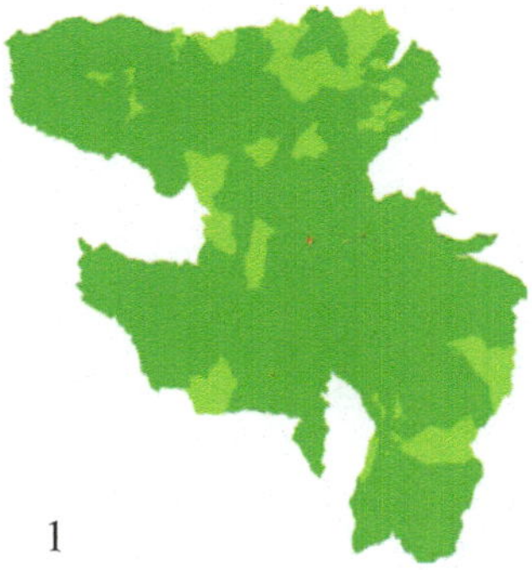

Fig. 6 Map1: population with disabilities

5 Results

Planning preventive measures and avoiding risk situations require understanding the distribution of hazards and exposure, while localizing social vulnerability can be useful to plan reactive measures, especially when prioritizing the most needed (Adger et al. 2004). As shown in Fig. 7, in a general way census sector level with higher risk values

Fig. 7 Quito's risk index graphical result

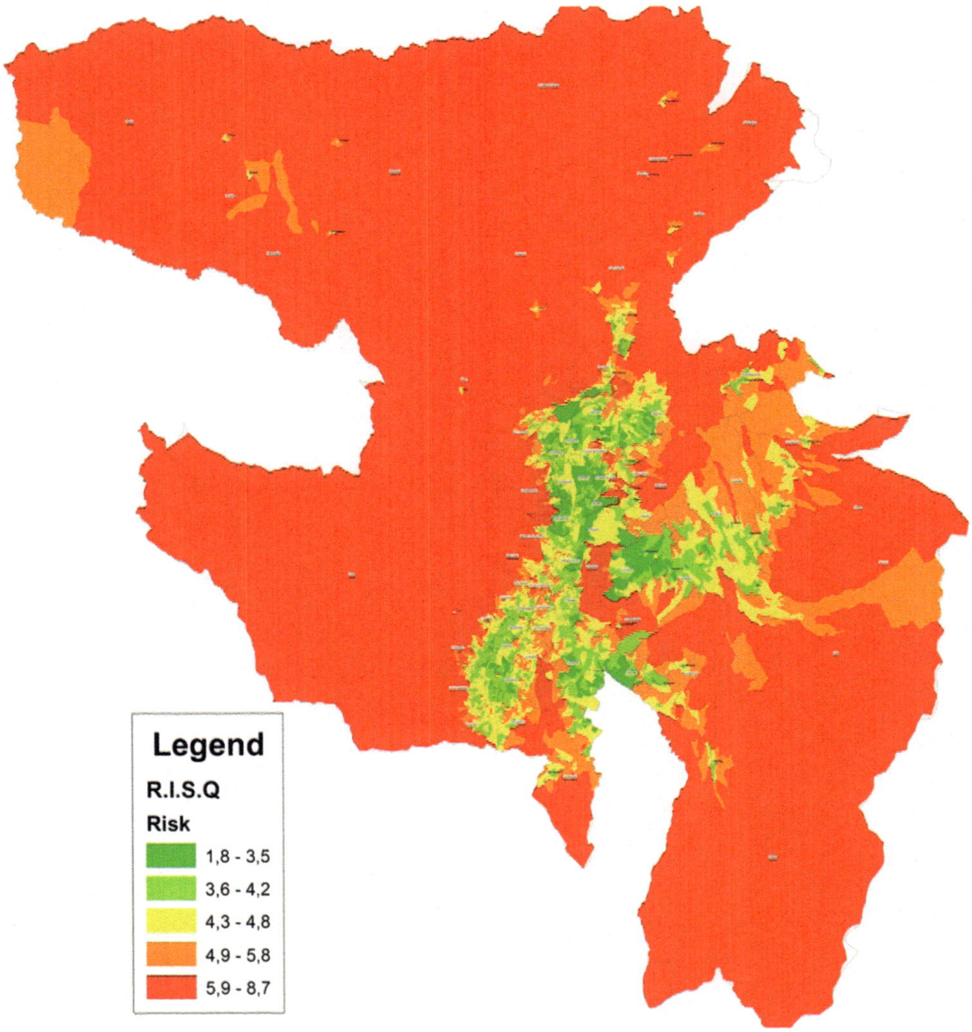

Table 1 Indicators by variable in each census sector

Census block group	Flooding	Wildfire	Landslides	Volcanic eruptions	Caregiving women	Population with disabilities	Access to communications	Single mothers	Precarious construction systems	Access to basic services	Access to transportation	Economically inactive population	Unemployed population	Access to social security system
170150154003	0.010	0.301	0.439	0.010	0.187	0.050	0.447	0.210	0.337	0.003	0.000	0.460	0.018	0.844
170150154004	0.010	0.307	0.391	0.010	0.284	0.035	0.398	0.210	0.228	0.008	0.000	0.464	0.008	0.848
170150154005	0.010	0.262	0.418	0.010	0.231	0.030	0.374	0.210	0.242	0.003	0.065	0.421	0.006	0.878
170150154006	0.010	0.291	0.597	0.010	0.152	0.072	0.551	0.210	0.600	0.012	0.688	0.419	0.007	0.930
170150154007	0.010	0.198	0.502	0.010	0.192	0.034	0.548	0.210	0.409	0.008	0.990	0.452	0.007	0.935
170150154008	0.010	0.283	0.498	0.010	0.337	0.035	0.403	0.210	0.289	0.003	1.000	0.471	0.034	0.883
170150155001	0.010	0.269	0.403	0.011	0.362	0.024	0.483	0.210	0.377	0.018	0.198	0.577	0.032	0.827
170150155002	0.012	0.276	0.428	0.250	0.221	0.000	0.172	0.191	0.137	0.000	0.040	0.400	0.033	0.424
170150155003	0.013	0.297	0.594	0.103	0.310	0.031	0.121	0.191	0.070	0.002	0.154	0.380	0.023	0.414
170150155004	0.010	0.189	0.578	0.094	0.340	0.035	0.184	0.191	0.064	0.016	0.000	0.309	0.012	0.506
170150155005	0.013	0.253	0.512	0.292	0.397	0.021	0.126	0.191	0.058	0.002	0.024	0.364	0.010	0.321
170150155006	0.010	0.155	0.423	0.219	0.326	0.051	0.113	0.191	0.100	0.004	0.000	0.400	0.020	0.472
170150156001	0.010	0.120	0.453	0.523	0.295	0.061	0.116	0.210	0.081	0.000	0.000	0.438	0.031	0.466
170150156002	0.010	0.172	0.436	0.541	0.382	0.043	0.099	0.210	0.087	0.000	0.000	0.432	0.015	0.494
170150156003	0.108	0.105	0.485	0.299	0.341	0.040	0.230	0.210	0.317	0.002	0.000	0.430	0.034	0.673
170150156004	0.010	0.183	0.438	0.520	0.315	0.038	0.114	0.210	0.102	0.000	0.000	0.417	0.016	0.569
170150156005	0.367	0.213	0.442	0.574	0.429	0.035	0.118	0.210	0.087	0.002	0.000	0.430	0.036	0.468

(color red) are located mainly at the periphery of the consolidated urban core of Quito. This is the result of the presence of four or more indicators of social vulnerability with high values, mainly to the lack of access to basic services, communications, means of transport, social security, or the high presence of economically inactive population; as well as the precarious constructive systems that prevail in rural areas.

However, since the terrain is not homogeneous in its geographical properties, intensities of hazards may occur differently, and since social vulnerability have different connotations, even within one census sector (Table 1), maps may not represent the exact location of the indicator, then smaller scales to better understand risk is advised.

Explained in a different way, in urban areas the census sectors contain one or more blocks with high urban consolidation characteristics, therefore results in the maps have higher accuracy to represent risk. In the case of rural areas, census sectors comprise areas of approximately 80 dwellings and may contain part of a settlement, a single complete settlement, a group of complete settlements, or part of one or more settlements, which are dispersed. Therefore, maps showing results do not allow a precise determination of the exact location of risk. In this sense, results could correspond to a fraction of a census sector, like a parish header, where the maps represent the same level of risk for a larger area due to the higher index numbers concentrated.

6 Conclusions

If understanding the distribution of hazards is already challenging, comparing different hazards add substantially to the challenge when analyzing a multi-hazard city with different intensities and periodicity. At the same time, socioeconomic vulnerability indicators and the construction of an overall risk index from those indicators depend on information availability. For the present case, there was no economic information per family available; hence, the analysis was adapted to indicators that could show poorness in people based on other indicators related from the 2010 census. The next hurdle was found when deciding the method of aggregation: During the exercise, a question was raised several times by the research team: Will it be desirable to give importance to certain conditions that add to risk, and if that was the case, will different indicators receive differentiated weights to modify the outcomes depending on established priorities? It was recognized that in some highly constrained circumstances, when risk is well understood, it is possible—and perhaps even desirable for planning and prioritization—to assign differentiated weights. However, due to the large number of components risk becomes so complex that it is intrinsically difficult to understand. It was concluded that it is more practical to analyze the different components and decide how to proceed from there, than to develop a weighting scheme to avoid subjective decisions. Moreover, assigning weights to different indicators might be unwise, especially when acknowledging that some components might change shifting priorities.

This paper delivers a risk index, which is graphically displayed, that enables informed decision making and effective urban planning for multirisks cities. The prioritization of interventions is possible through the understanding of the distribution of hazards, while localizing social vulnerability helps prepare preventive measures and provide additional assistance where is most needed. The exercise determines the relative existence of risk without the compromising practice of assigning weights to the various indicators. It applied the technique to a highly complex, multi-hazard threatened, with different levels of exposure and varied in socioeconomic conditions city: Quito. The exercise demonstrates the advantage of Pareto ranking over simple averaging of the indicators part of the three different components. While there are numerous other approaches just as practical, this method represents a step forward in our ability to construct useful indices of risk to prioritize actions with the final goal to reduce environmental impacts and propose effective, sustainable, and resilient urban strategies. This research also determines that results should be analyzed in greater detail in rural areas. Given the level of information available, it is not possible to establish risk location with higher accuracy.

7 Further Research

This proposal needs to be further developed. Other indicators should be added, such as economic income (if the next census exercise included such questions), or required costs for reducing physical vulnerability, or the microseismic plans of the city, among others. Furthermore, if it was possible to include other indicators, like education access, or health access, just to name a couple, this risk index could become a development index that can be graphically displayed and can inform decision making in urban and development planning, taking the risk dimension into account.

References

Adger, W.N., Brooks, N., Bentham, G., Agnew, M. and Eriksen, S.: 2004, 'New indicators of vulnerability and adaptive capacity', Tyndall Centre Technical Report 7, Tyndall Centre for Climate Change Research Norwich, UK, accessed at http://www.tyndall.ac.uk/publications/techreports/techreports.shtml.

Alwang, J, Siegel, P, Jorgensen, SL. 2001. Vulnerability: a view from different disciplines. Soc. Prot. Discuss. Pap. Ser. 115, Soc. Prot. Unit, World Bank, Washington, DC. 42 pp.

Bianchi, S.M. and Spain, D.: 1996, 'Women, work, and family in America', Population Bulletin 51(3), Population Reference Bureau.

Birkmann, J. (Ed.), 2006. Measuring Vulnerability to Natural Hazards—Towards Disaster-Resilient Societies. UNU Press, Tokyo, New York.

Birkmann, J. (2007). Risk and vulnerability indicators at different scales: Applicability, usefulness and policy implications. Environmental Hazards, 7(1), 20–31.

Blaikie, P., Cannon, T., Davis, I. and Wisner, B.: 1994, At Risk: Natural Hazards, People's Vulnerability, and Disasters, Routledge, New York.

Bohle, H.-G., 2001. Vulnerability and criticality: perspectives from social geography. IHDP Update 2/2001, Newsletter of the International Human Dimensions Programme on Global Environmental Change, pp. 1–7.

Bohle, H.-G., Downing, T.E. and Watts, M.: 1994, 'Climate change and social vulnerability: The sociology and geography of food insecurity', Global Environmental Change 4, 37–48.

Cannon, T., Twigg, J., & Rowell, J. (2003). Social vulnerability, sustainable livelihoods and disasters. Report to DFID Conflict and Humanitarian Assistance Department (CHAD) and Sustainable Livelihoods Support Office.

Cardona, O.D., 2004. The need for rethinking the concepts of vulnerability and risk from a holistic perspective: a necessary review and criticism for effective risk management. In: Bankoff, G., Frerks, G., Hilhorst, D. (Eds.), Mapping Vulnerability: Disasters, Development and People. Earthscan Publishers, London.

Cardona, O.D., 2005. Indicators of disaster risk and risk management—main technical report. IDB/IDEA Program of Indicators for Disaster Risk Management, National University of Colombia, Manizales. Available at http://idea.unalmzl.edu.coS.

Chen, S.J & Hwang, C.L. (1992). Fuzzy Multiple Attribute Decision Making. Berlin: Springer-Verlag.

Clark, G., Moser, S., Ratick, S., Dow, K., Meyer, W., Emani, S., Jin, W., Kasperson, J., Kasperson, R. and Schwarz, H. E.: 1998, 'Assessing the vulnerability of coastal communities to extreme storms: The case of Revere, MA., USA', Mitigation and Adaptation Strategies for Global Change 3, 59–82.

Denton, F., Climate change vulnerability, impacts, and adaptation: Why does gender matter?, Gender & Development Vol. 10, Iss. 2, 2002.

Eakin. H., and Luers, A. M., Assessing the Vulnerability of Social-Environmental Systems, Annual Review of Environment and Resources, Vol. 31: 365–394 (Volume publication date November 2006), First published online as a Review in Advance on July 18, 2006, https://doi.org/10.1146/annurev.energy.30.050504.144352.

Enarson, E. and Morrow, B.H.: 1997, 'A gendered perspective: The voices of women', in W.G. Peacock, B.H. Morrow and H. Gladwin, (eds.), Hurricane Andrew: Ethnicity, Gender, and the Sociology of Disasters, International Hurricane Center, Laboratory for Social and Behavioral Research, Miami, FL, 116–140.

Fothergill, A.: 1998, 'The neglect of gender in disaster work: An overview of the literature', in E. Enarson and B.H. Morrow (eds.), The Gendered Terrain of Disaster: Through Women's Eyes, Praeger Publishers, Westport, CT, 11–25.

Fothergill, A. and Peek, L.A.: 2004, 'Poverty and disasters in the United States: A review of recent sociological findings', Natural Hazards 32, 89–110.

Hoogeveen, J., Tesliuc, E., Vakis, R., & Dercon, S. (2004). A guide to the analysis of risk, vulnerability and vulnerable groups. World Bank. Washington, DC. Available on line at http://siteresources.worldbank.org/INTSRM/Publications/20316319/RVA.pdf. Processed.

Kasperson, JX, Kasperson, RE. 2001. Workshop summary. Int. Workshop Vulnerability Glob. Environ. Change, Stockholm: Stockh. Environ. Instit.

Kelly, PM, Adger, WN. 2000. Theory and practice in assessing vulnerability to climate change and facilitating adaptation. Clim. Change 47:325–52.

Klein, RJT, Nicholls, RJ, Thomalla, F. 2003. Resilience to natural hazards: How useful is this concept? Environ. Hazards 5:35–45.

Liverman, DM. 2001. Vulnerability to global environmental change. See Ref. 99, pp. 201–16.

Ludena, C.E., S.W. Yoon. 2015. Local Vulnerability Indicators and Adaptation to Climate Change: A Survey. Inter-American Development Bank, Technical Note No. 857 (IDB-TN- 857), Washington DC.

McCarthy, J.J., Canziani, O.F., Leary, N.A., Dokken, D.J. and White, K.S. (eds.): 2001, Climate Change 2001: Impacts, Adaptation & Vulnerability. Contribution of Working Group II to the Third Assessment Report of the Intergovernmental Panel on Climate Change (IPCC), Cambridge University Press, Cambridge, UK.

Molin Valdes, H., Rego, A., Scott, J., & Aguayo, J. V. (2012). How to make cities more resilient: a handbook for local government leaders. *Geneva: United Nations UNISDR.*

Morrow, B.H.: 1999, 'Identifying and mapping community vulnerability', Disasters 23, 1–18. National Hurricane Center: 2003, 'Hurricane awareness: Storm surge', accessed at http://www.nhc.noaa.gov/HAW2/english/stormsurge.shtml.

Municipio del Distrito Metropolitano de Quito, 2016, Atlas de Amenazas Naturales, Quito.

Municipio del Distrito Metropolitano de Quito. (MDMQ, 2012, March). Plan Metropolitano de Desarrollo y Ordenamiento Territorial 2012–2022.

Municipio del Distrito Metropolitano de Quito. (MDMQ, 2015, March). Plan Metropolitano de Desarrollo y Ordenamiento Territorial 2015–2025. (i) Diagnóstico Estratégico Eje Social, (ii) Diagnóstico Estratégico Eje Económico, (iii) Diagnóstico Estratégico Eje Ambiental, Diagnóstico Estratégico Eje de la Movilidad (iv), Diagnóstico Estratégico Eje Territorial (v), Diagnóstico Estratégico Centro Histórico.

Rygel, L., O'sullivan, D., & Yarnal, B. (2006). A method for constructing a social vulnerability index: an application to hurricane storm surges in a developed country. Mitigation and adaptation strategies for global change, 11(3), 741–764.

Thywissen, K., 2006. Core terminology of disaster reduction: a comparative glossary. In: Birkmann, J. (Ed.), Measuring Vulnerability to Natural Hazards—Towards Disaster Resilient Societies. UNUPress, Tokyo, New York, Paris.

Turner II BL, Kasperson RE, Matson PA, McCarthy JJ, Corell RW, et al. 2003. A framework for vulnerability analysis in sustainability science. Proc. Natl. Acad. Sci. USA 100:8074–79.

United Nations International Strategy for Disaster Reduction (UN/ISDR), 2004. Living with Risk. A Global Review of Disaster Reduction Initiatives. 2004 version. UN Publications, Geneva.

United Nations International Strategy for Disaster Reduction (UN/ISDR) (2005, March). Hyogo framework for action 2005–2015: building the resilience of nations and communities to disasters. In Extract from the final report of the World Conference on Disaster Reduction (A/CONF. 206/6) (Vol. 380).

United Nations International Strategy for Disaster Reduction (UN/ISDR) (2007, August 30). Terminology. Retrieved from: https://www.unisdr.org/we/inform/terminology.

Wisner, B., 2002. Who? What? Where? When? in an emergency: notes on possible indicators of vulnerability and resilience: by phase of the disaster management cycle and social actor. In: Plate, E. (Ed.), Environment and Human Security: Contributions to a Workshop in Bonn, 23–25 October 2002, Germany, pp. 12.

Resilient Thinking: An Important Tool to Contribute to Sustainable Development

David Jacome Polit

Abstract
This article presents a review of the most relevant literature on resilience, from a practical point of view, and proposes a synthesis to provide a fresh perspective on the concept. The objective is to position the concept of resilience and its practices, or "resilience thinking," as a multidisciplinary undertaking that can give rise to transversal initiatives to seek stability in people's well-being and to provide a continuous life quality. Resilience has the ability, then, to generate debate and analysis on various, often archetypal, issues forcing policy makers, analysts, and professionals, in general, to look at the consequences of actions from a systemic perspective. From this point of view, resilience necessarily addresses aspects of sustainability and sustainable development that can contribute in our path to a stable world. Questions will be addressed such as: What is the relationship between resilience and sustainable development? And, how do we look at resilience from an urban perspective? Notwithstanding that, perhaps, the answer lies solely in the fact that humanity looks to the future and wants and must preserve much of what it now values for the future generations. It wants sustainability, and resilience is one way to achieve it.

Keywords
Resilient thinking • Systems thinking • Sustainable development

1 Introduction

Turbulence is a known phenomenon. Characterized by rapid space–time changes of pressure and velocity, its unpredictability is the result of its randomness. The occurrence of events capable of unpredictably disrupting the status quo can have important and difficult-to-face consequences. Something similar occurs in socioeconomic systems and their interaction with the natural environment. In a publication dated February 10, 2017, scholars from two prestigious institutions presented a research paper entitled "The Anthropocene equation." In this paper, Gaffney and Steffen (2017) propose an equation capable of describing the impact of human activities on the planet and argue that these now rival the great forces of nature by driving changes in the terrestrial system. In an interview in a London newspaper,[1] Steffen says that, on the one hand, the astronomical and geophysical forces have produced a rate of change of 0.01 °C per century in the average terrestrial temperature. Meanwhile, the emissions of greenhouse gases caused by humans in the last 45 years have increased the rate of change of temperature to 1.7 °C per century, dwarfing the rate of natural origin, where "The human magnitude of climate change looks more like a meteorite strike than a gradual change."

The world is made of infinity of systems that act on it, often closely related, such as ecosystems and cities. Its complexity is evident in its many facets, e.g., globalization, or speed of change and interdependencies between these systems. Cities have often been conceived as highly adaptive and complex systems (Meerow et al. 2016) composed of small systems: critical services, buildings, climate, businesses, transport systems, financial markets, etc., but also social and economic systems: civil organizations, governments, academia, political organizations, unions,

D. Jacome Polit (✉)
Instituto Metropolitano de Planificación Urbana (IMPU), MDMQ, 100 Resilient Cities, Quito, Ecuador
e-mail: djacomepolit@gmail.com

[1]https://www.theguardian.com/environment/2017/feb/12/humans-causing-climate-to-change-170-times-faster-than-natural-forces.

C. Alalouch et al. (eds.), *Advanced Studies in Energy Efficiency and Built Environment for Developing Countries*, Advances in Science, Technology & Innovation, https://doi.org/10.1007/978-3-030-10856-4_7

unemployed population, agricultural production, financial markets, etc. This configures a multiplying factor of complexity over time, due to demographic expansion and its demand for resources, urban growth, etc. It is obvious that these problems cannot be solved independently, with regular or conventional mechanisms, which is why an equally sophisticated approach is required to generate practical solutions (Kupers et al. 2014). Resilience, on the other hand, is the ability of a system to absorb disturbances and reorganize itself while undergoing a mode change, retaining essentially the same function and identity.

2 Resilience: A Necessarily Multidisciplinary Concept

Derived from the Latin word *resilīre*, which literally means "recovering," early use of the concept of resilience in the context of threats or disruption has succeeded in positioning it as standard. But the concept is much broader, and unfortunately, there is very little consensus on the definitions of "resilience," "positive adaptation," "positive outcome," and so on, even within a single discipline (Martin-Breen and Anderies 2011). Defining resiliency, then, is not a simple task, since it must describe the properties of an isolated object or the properties of a system, defined by a collection of agents and functions and the relationships between them. This set of agents can be comprised of anything: living cells, organs, people, communities, buildings, cities, or countries. For this reason, the nature of what is studied must first be clarified before defining any property that it possesses, including resilience.

Judith Rodin in her book "The Resilience Dividend," is proposing that the concept that best defines resilience, often adopted by other disciplines, is the proposal presented by the celebrated ecologist Brian Walker, who states that "it is the ability of a system to absorb a disturbance and maintain its basic functions and structure." Walker in the early 1970s was working on a line of research at Harvard Forests in New England, whose aim was to observe how physical, biological, and human systems interact to change the planet, as the concept became more powerful. At that time, Walker argued that resilience is NOT about NOT changing and exemplified that forests and savannas are systems in constant change, absorbing disturbances generated by anthropogenic or natural factors, among which are forest fires, droughts, etc.

From this point of view, resilience is a concept reminiscent of Darwin's theory of evolution of species, where he proposes that the species that survives is NOT the strongest or most intelligent, au contraire, is the one best adapted through small variations that allow it to compete, survive, and reproduce. It should be understood, then, that resilience building is not necessarily visible, and it is a theoretical construction, a dispositional property that relates to the response of an individual or system to possible future events (Martin-Breen and Anderies 2011). This proposal can have a myriad of possible options. Carpenter and Brock (2008) have described resilience as a "broad, multifaceted, and unorganized set of concepts, each related to some aspect of the interaction of transformation and persistence."

Thus, the vagueness of the concept of resilience allows multidisciplinary scientific collaboration. The wide variety of meanings, depending on the disciplines that address it, allows several interested parties to gather around a common term without needing to be agreed upon an exact definition. Meerow et al. (2016) propose the following concept to define the urban resilience, which shows the multidimensionality of the same:

> Urban resilience refers to the ability of an urban system - and all its socio-ecological and socio-technological constitutive networks through temporal and spatial scales - to maintain or rapidly return to the desired functions in the face of a disturbance, adapt to change and rapidly transform systems that limit current or future adaptive capacity.

3 System's Resilience

We live in a highly dynamic world. Even if we do not consider global warming, social, technological, economic and ecological conditions are constantly changing, often imperceptibly outside the human scale of time (Martin-Breen and Anderies 2011). From this point of view, being resilient does not mean returning to an earlier or normal state, it is being able to adapt to these changes and continue to "function." Maintaining this function, instead of maintaining an initial state, means that there are parts that interact and that when affected by a disturbance may or may not recover, and if they do, recovery times differ from one to another. Indeed, examples of traditional ecosystem management demonstrate very clearly: In attempting to prevent any change, an ecosystem collapses. To incorporate an understanding of the change of these parts, then, we need to consider resilience in systems.

The resilience analysis is categorized into three levels: (1) structural resilience which means returning to an initial state, where it is usually understood as the resilience of a single element, (2) integrative resilience related to understanding these changes and adapting accordingly; this adaptation is the result of a group of components that often change at different rates, and (3) transformative resilience, where the capacity for self-organization or transformation operates at many levels, and where a component or sub-system can generate new relationships, configurations, or new forms of interaction, in a multidimensional and multiscale way, being able to function in a prosperous and continuous way. These systems are called complex adaptive systems and propose a new way of thinking about resilience (see Jacome Polit 2016 for further information).

As a consequence, systemic thinking is essential to understand the forces that influence resilience and sustainability, being useful to identify the multiple interdependencies between human and natural systems, and to propose policies and practices that help us move toward a resilient society besides a sustainable one. However, although it is true that the different systems that act in the biosphere are constantly changing, and whose effects until a few decades ago affected us in long time scales allowing a better adaptation, it is also true that our capacity of adaptation decreases as these systems surpass certain thresholds and their behavior begins to have consequences much more severe and unpredictable, or even lethal, e.g., if we lose the provision of essential ecosystem services like water, oxygen, or food. In other words, if the adaptive capacity depends on new configurations or forms of interaction between different systems because there are changes in the boundary conditions, it is also true that there can be a time when the capacity for adaptation is minimal or zero because it is not possible to achieve the required rhythms of change. Then, if cities are conceived as complex systems, resilience is a property of that system that makes it capable of adapting to known and unknown changes, particularly related to high-impact shocks and low probability of occurrence.

Cities are commonly considered complex systems. If we consider that populated centers (at least 50,000 residents) cover less than 3% of the Earth's surface and generate approximately 71% of global carbon emissions related to energy generation (International Panel on Climate Change, IPCC 2014), it is paradoxical that being the cities that will welcome the majority of the world's population, its operation, and its craving for resources will make us more vulnerable, since they give dimension to the problem instead of being part of the solution. This situation puts a lot at risk, the high concentration of people with a significant demand for resources, supported by complex social and economic systems, requires infrastructure and urban services to thrive, which can also mean a high concentration of risks.

4 Resilience and Sustainable Development, a Proposed Analysis

COP-21, Paris 2015, part of the United Nations Conference on Climate Change, succeeded in committing the vast majority of nations to combat this phenomenon and its effects and to contribute to sustainable development. In this and other forums, sustainability and resilience have emerged on several occasions, one as an agenda and the other as a tool. This part of the article is not intended to build an inventory of agreements and agendas with commitments to contribute to sustainable development, or to reduce greenhouse gases that produce climate change or, if not, to build resilience; the intention is rather to contrast these concepts and establish their relationship.

As a concept broader than resilience, sustainable development requires balanced economic and social growth based on a development that avoids degrading the environment in order to meet our needs without compromising the needs of the future generations (World Commission on the Environment and Development 1987). It should be emphasized that sustainable development requires comprehensive planning that adequately balances the environmental, economic, and social dimension; otherwise, there is a risk of promoting unbalanced development, allowing the strengthening of one or two dimensions to the detriment of the other, or a false development because it is unique in one dimension, or a fragile development, since it lacks the necessary means of support in all dimensions to be sustainable (Davis et al. 2016).

However, sustainable development can dispense with resilience in an explicit way by being replaced by other, less far-reaching concepts, such as risk mitigation. But an analysis of the two concepts, sustainable development and resilience, creates a clear relationship between the one and the other, since on the one hand in the short term, if we include disturbing events of important magnitude that cannot be predicted, then sustainable development, over time, requires a capacity for resilience at all times. On the other hand, in the long term, the capacity for resilience promotes sustainable development in the context of uncertainty, since it requires to contemplate strategies to combat chronic stresses that weaken different systems by making them vulnerable, such as the eradication of extreme poverty, solutions to face climate change, and its effects or the search for solutions to the growing demand for resources, among others. These challenges require strategies to change the trend and build resilience in cities through the planning and promotion of sustainable cities.

This contradicts the interpretation that sustainability is a goal that must be aspired collectively. In fact, sustainability is not the result of a manageable trajectory, but rather is the result of a continuous adaptation to new conditions. Sustainable development must, therefore, be based on a worldview where growth and transformation are inevitable. Resilience, on the other hand, is an attribute capable of dynamizing a system, which allows it to resist the disorder and to thrive in a world in constant change. Notwithstanding this, it is necessary to understand that as systems become larger and more structured, their resilience may decrease, making them more vulnerable to external shocks and internal decay. One could say, on the one hand, that resilience is defined more clearly as the property of a system capable of dynamically facing different challenges and, as such, presents a space of intellectual and practical proposals. Sustainable development, on the other hand, is a powerful idea and intuitively attractive, but at the same time difficult to apply. Button (2002), for example, suggests that this definition is particularly vague. He argues that it is very difficult to territorialize the concept of sustainable development in urban areas because of its complexity, since it only refers to temporal effects from a generational point of view without addressing the key geographical aspects and the inherent dynamism of cities.

As a clear example to solve this paradox, the Stockholm Resilience Center, part of the University of Stockholm, proposes that building resilience and addressing the concept of sustainable development are achieved by studying how this capacity can cope with unexpected events, e.g., related to climate change. This approach proposes to see people as active and influential actors within the dynamics of ecosystems and places them as agents in close interaction and with a capacity to influence the biosphere, understood as the sphere of air, water, and land in which are all manifestations of life (Applying resilience thinking: Seven principles for building resilience in social-ecological systems). The objective is to clearly establish the primary problems generated by our way of life: the depletion of resources, the deterioration of ecosystems, and the deterioration of people's health, caused mainly by the insatiable demand for goods and services, which is addressed through the use and exploitation of non-renewable resources and the generation of a high amount of waste (Kristinsson 2012).

Taeke de Jong, Professor Emeritus of Technical Ecology and Planning at the Technical University of Delft, defines the environment in which we develop as the site that meets the necessary conditions to live, and where the ecology that is part of biology, studies the relationships between organisms and their environment. This premise places us as actors in close relationship with the environment. All the elements of interaction, then, are grouped into four components: technical, biotic, abiotic, and the atmosphere. Kristinsson (2012) then gives a brief description of them, where the technical component, also defined as anthropogenic, involves everything industrially produced: buildings, roads, aqueducts, products, etc. Because this component represents everything built, including infrastructure and commodities, it requires that it can be designed and built or produced so that its parts can be reused and with the widest possible lifetime to prevent further predation. The biotic component involves all living beings, visible or invisible, which has the ability to provide, for example, building materials or fuel, such as wood. The wrong interaction with this component exacerbates its functionality and is putting the ecological system at risk. The third component, the abiotic, which resembles the lunar surface if it were stripped of the other two components, is where we find resources such as metals, fossil fuels, and even water. In fact, many of the building materials are sediments that come from rivers. Finally, there is a fourth component, being the invisible layer of the earth, which includes air, light, heat, moisture, radiation, ozone, etc. This component is much more vulnerable and thin than can be assumed, as it would be comparable to the thickness of plastic wrap around a football.

In general terms, one could say that sustainability and resilience reinforce each other. The more sustainable the world is, the less it is exposed to unpredictable shocks, while the more resilient, the less risk there is of endangering our future well-being (Fiksel et al. 2014). However, there may be an imbalance between resilience and sustainability, on the basis that although the two concepts are closely related, a system is not necessarily resilient if it is sustainable, and vice versa. To illustrate this, the energy mix in Ecuador and its generation sources is proposed as an example. By 2020, the country is expected to produce 86% of its energy mainly through hydroelectric sources. Of this 86, 32% will be generated from a single unit: the Coca Codo Sinclair hydroelectric plant (Senplades 2009 quoted in Jácome Polit et al. 2016). This undoubtedly makes Ecuador a country that contributes to sustainable development by largely avoiding the emission of greenhouse gases as a by-product of the generated electric energy. However, this also shows that we become a less-resilient country in terms of what electric power generation is concerned. Depending on a single type of energy, where a third of it is generated from a single source, it makes the country vulnerable to the effects that can be produced, for example, climate change, understood as the change of the patterns of the climate through the increase and intensity of extreme events, among them droughts (UNFCCC 2007). On the other hand, if due to these effects, the country compensates for the lack of energy with sources of easy implementation, but polluting and with important emissions of greenhouse gases, such as the generation of thermal energy based on fossil fuels, resilience is gained in the power generation system, but important contributions to climate change are made, as opposed to little or none to sustainable development.

5 Other Resilience Considerations

5.1 Resilience, as a Condition, Is Neither Good nor Bad

Then, it is necessary to clarify that resilience must be understood as a condition, which may be desired, but which is neither good nor bad. For example, in a savanna or a forest, the loss of an organism or diminishing natural areas produces important imbalances, achieving the adaptation of remaining species. This adaptation of the ecosystem, understood as a condition of resilience, becomes dangerous, since the lack of one species can mean overpopulation of another species to the detriment of a third, which triggers other problems (McKinney 2006). This means that any action that is aimed at building resilience must always be systematically examined.

This proposal is clear within the context of ecological and environmental systems, since this is where the concept has been most developed. The proposed approach should then seek to understand the relationship and interdependence between human systems and natural systems and to identify policies and practices that help build a resilient and sustainable society (Fiksel et al. 2013). Thus, studies on resilience in the field of ecology have been able to reveal cases where decision makers have been the cause of unwanted and unforeseen consequences from the implementation of inadequate policies, since they influence the use of different ecosystem services, such as water we use to cook and drink, crops that grow to nourish us, or the ability of vegetation to regulate the climate, causing changes in the biosphere in a myriad of forms, often negative, weakening these systems. In this sense, the concept of resilience and its application, or "resilience thinking," is very valuable and necessary in terms of analysis, elaboration, and discussion on aspects of sustainability and sustainable development.

5.2 Sometimes Reducing Resilience Is Necessary

A drug cartel or a nasty virus can be very resilient. To achieve change, a (temporary) reduction in resilience may be needed. So it is necessary to understand that resilience building is not always desirable, and it is often necessary to lose resilience to pursue higher goals such as sustainability, justice, or equity. An example can be found easily in socioeconomic systems. If we consider the theory of "poverty traps," which refers to the balance of poor economic systems in balance with other prosperous ones, where if an individual living in poverty tries to grow beyond his equilibrium, he will soon return to the same state, unless this growth occurs through an extraordinary jump. Barrett and Swallow (2006) explain this phenomenon by postulating stages of equilibrium with a similar duality in business, industry, and government services: Those serving these poor sectors cannot grow because they are within the aforementioned equilibrium, while the poor, in turn, have no space to grow. The system is self-reinforcing and kept within a vicious circle. In this case, then, the main objective of the strategy is primarily humanitarian, since it is not to increase the resilience of these already robust systems, but to induce transformations: to change a state of two equilibria in a single one where prosperity is attainable for everyone.

5.3 Resiliency to Face Uncertainty

A resilience approach to sustainability focuses on how to build capacity to deal with unexpected change. From the point of view of complexity and turbulence, when disturbances are unknown, risk analysis becomes obsolete and traditional risk management practices are no longer useful. There are several limitations within the classic risk management paradigm (Fiksel et al. 2013):

- Risks cannot always be identified a priori. In a turbulent world, it is often vulnerable to unprecedented or unpredictable shocks.
- Even if risks can be identified, the lack of a reliable dataset often makes it difficult to quantify the likelihood or magnitude of the major threats.
- Risk mitigation and recovery practices, such as business continuity management, are typically oriented toward returning to "normal" conditions, prior to a disruption.

It is necessary to recognize that both the frequency of different unpredictable events and their impact seem to have intensified, especially the ones related to climate. It is of common consensus that current risk management practice does not adequately protect us from these events. Known as "black swans" (PwC 2011) or "fat tails" (Taleb 2007), these are cataloged as events that deviate beyond what is normally expected in a probability analysis and, therefore, are not predictable. These events have become relevant for some time, and however, much of the risk management remains largely dependent on traditional risk or value-at-risk (VaR) values. If someone ignores unpredictable events, they underestimate their exposure to risk.

6 Conclusions

There are no-one-size-fits-all solutions when it comes to building resilience and at the same time when it comes to contributing to sustainable development. It can be concluded that sustainable development is a path that will not be easily walked if we do not embed resilience thinking in the planning processes. As we wander away from the sustainable development path, the world will become more turbulent and recovery and adaptability will become less attainable. However, before building resilience, or decreasing resilience to achieve higher goals for that matter, three main questions must be asked: On what we want to build resilience?, against what?, and what are the trade-offs of such exercise.

Resilience is a property of a complex system, with a number of agents and functions, that is interconnected with other complex systems. If one of these systems is disrupted, directly or indirectly, unexpected consequences could happen to other systems. In this sense, resilience building also means facing unpredictable and devastating events, where different systems must be able to face, overcome, and adapt to new circumstances. Change and transformation, achieved by joining strength with flexibility, and elasticity with durability are required. From an urban perspective, adaptability comes mainly from the exclusive abundant resource in cities: people. Urban resilience building is far away from individual engagement, and it requires community's engagement. Just like some cities have grown vulnerable over time due to fractioned societies, building resilience collectively is what makes it impactful, even if it takes time.

Socio-technological systems depend on socio-ecological systems to thrive. The pressure we put on life support systems on the planet makes the world more vulnerable. Our adaptive capacity is diminished as climate events become more intense and frequent. At the same time, not strengthening vulnerable social sectors makes this capacity inherently weaker and the effects of climate events, or other related to nature such as volcanic eruptions or earthquakes, cause greater havoc and prevent a more rapid recovery. This is much more evident in cities. By hosting 50% of the population today, and expecting that in the coming decades they will receive 70% of the population, it is necessary to think about the city–nature relationship, especially when one represents the demand for resources and ecosystem services, and the other represents the supply, all this within a planet of finite resources.

Sustainable development and urban resilience building usually come together when discussing global agendas. The intention is not to engage in this debate, but to call on action. This article presents concepts and reflections around the proposal for sustainable development and resilience building. Usually, new ideas and proposals can take several years or decades before having a widespread and real impact on how we do things. Collective human thought is far greater than any intent to influence it. Therefore, fostering debate on structural issues that at the moment shape the conditions in which we live on the planet, and changing them in a smart way, is necessary. Deep changes must be driven if we are to target truly resilient cities and societies where the quality of life of all is protected by policies that reinforce these conditions without weakening the systems that keep us alive.

From this point of view, resilience thinking is fundamental to understand the broad forces that influence sustainability. When analyzing, a system approach helps unveiling the interdependencies among different systems, e.g., human and natural systems, and for identifying policies and practices that will help cities progress toward a resilient society. Clearly, to achieve sustainability we must protect critical natural capital, improve resource productivity, and avoid environmental degradation, all this within a clear inclusive agenda where no one is left behind. Resilience thinking can equip us to reach such goal by evidencing trade-offs and prepare us to transition back into the sustainable development path. However, we must be willing to foster change, where the momentary loss of certain conditions that seem to make us strong today can help us reach other objectives represented by greater goals and a more secure future. This is essential.

References

Applying resilience thinking, Seven principles for building resilience in social-ecological systems; Sturle Hauge Simonsen, Reinet te (Oonsie) Biggs, Maja Schlüter, Michael Schoon, Erin Bohensky, Georgina Cundill, Vasilis Dakos, Tim Daw, Karen Kotschy, Anne Leitch, Allyson Quinlan, Garry Peterson, Fredrik Moberg; www.stockholmresilience.su.se.

"Turbulence, A Corporate Perspective on Collaborating for Resilience"; Roland Kupers (ed), Peter Voser, Michel Liès, Maike Böggemann and Norbert Both, Marco Albani and Roland Kupers, David N. Bresch, Jaap Berghuijs, Rainer Egloff and Roland Kupers, Mark Smith, Marco Albani and Kimberly Henderson, Neil C. Hawkins and Glenn Prickett, Herman van der Meyden, Thekla Teunis, Simone Arizzi, Maximilian Egger, Dawn Rittenhouse and Peter Williams, Brian Walker, Amsterdam University Press; 2014; 176 pages.

Barrett, C., & Swallow, B. (2006). Fractal poverty traps. World Development, 34(1), 1–15. http://dx.doi.org/10.1016/j.worlddev.2005.06.008.

Button, K. (2002). City management and urban environmental indicators, Ecological Economics, 40(2), 217–233.

Carpenter, S. and Brock, W. (2008). Adaptive capacity and traps. Ecology and Society, 13(2):40.

Davis, M.J.M., Jácome Polit, D., Lamour, M.; Social Urban Metabolism Strategies (SUMS) for Cities, Procedia Environmental Sciences, VL 34, 2016, Improving Sustainability Concept in Developing Countries (ISCDC), SN - 1878–0296, http://dx.doi.org/10.1016/j.proenv.2016.04.028.

Gaffney, O., & Steffen, W. (2017). The Anthropocene equation. The Anthropocene Review, 205301961668802. http://dx.doi.org/10.1177/2053019616688022.

Fiksel, J. Goodman, I. Hecht, A. 2013. Resilience: Navigating toward a Sustainable Future. Solutions. Vol 5, No. 5. pp. 38–47 http://www.thesolutionsjournal.com/node/237208.

Fiksel, J., Goodman, I., & Hecht, A. (2014) Resilience: navigating toward a sustainable future, Jour., 5 (5), 38–47.

IPCC. (2014). Urban areas. In L. L. W. Field, V. R. Barros, D. J. Dokken, K. J. Mach, M. D. Mastrandrea, T. E. Bilir, M. Chatterjee, K. L. Ebi, Y. O. Estrada, R. C. Genova, B. Girma, E. S. Kissel, A. N. Levy, S. MacCracken, & P. R. Mastrandrea (Eds.), Contribution of Working Group II to the fifth assessment report of the Intergovernmental Panel on Climate Change (pp. 1–113). Cambridge, UK: Cambridge University Press.

Jacome Polit, D. (2016). Resiliencia, innovar y reinventarse de manera sostenida. Instituto de la Ciudad - DMQ, Revista Cuestiones Urbanas, Vol. IV- Número I, page 33–61.

Jácome Polit, D., Maldonado, D., Dávalos, D., Solar Might not Always be a Green Source of Energy, Elsevier, Procedia Engineering, Volume 145, 2016, Pages 611–621.

Kristinsson, J., Integrated sustainable design; Van Den Dobbelsteen, A. (ed), Delftdigitalpress; first edition; 2012; ISBN-10: 9052694087.

Martin-Breen P, Anderies JM (2011) Resilience: A literature review. The Rockefeller Foundation. Available at: http://www.rockefellerfoundation.org/news/publications/resilience-literature-review.

Meerow, S., Newell, J., & Stults, M. (2016). Defining urban resilience: A review. Landscape And Urban Planning, 147, 38–49. http://dx.doi.org/10.1016/j.landurbplan.2015.11.011.

McKinney, M. L. (2006). Urbanization as a major cause of biotic homogenization. Biological conservation, 127(3), 247–260.

PwC (2011). "Black Swans Turn Grey: The Transformation of the Risk Landscape." In Risk Practices: Navigating the Enterprise through a World Beset by Uncertainty. Available at: http://pdf.pwc.co.uk/risk-practices-black-swans-turn-grey-the-transformation-of-the-risk-landscape.pdf.

Senplades. (Secretaría Nacional de Planificación del Desarrollo) (2009). "Plan Nacional para el Buen Vivir 2009–2013: Construyendo un Estado Plurinacional e Intercultural". Versión Resumida. Primera Edición, Quito, Senplades, 120 p.

Taleb, N.N. (2007). The Black Swan: The Impact of the Highly Improbable. New York: random House.

United Nations Framework Convention on Climate Change, 2007, Climate Change Secretariat (UNFCCC).

World Commission on Environment and Development (1987). Our Common Future. Oxford: Oxford University Press. p. 27. ISBN 019282080X.

Car Space Recycling Algorithm: A Powerful Urban Planning and Decision-Making Tool

D. Davalos, David Jacome Polit, D. Maldonado, and J. Moreira

Abstract

The impact of private automobile in the city goes beyond carbon emissions. The land used within the city for service purposes (not only roads but also other types of services such as dealership, parking lots or buildings and auto repair and mechanic services) reshapes the urban morphology by taking away enormous area of underused land, hence evidencing an important and unsustainable environmental and economic system. Taking Quito–Ecuador as a case study, a methodology to identify, quantify, and propose a better use for the urban space allocated for private automobile, aiming at the end for a more sustainable, compact, equitable, and livable city, is proposed. In this occasion, this work continues with an already proposed car space recycling algorithm in a previous article, in two specific points: First, based on an urban index, a cluster of parameters is proposed to identify the areas of intervention best suited to apply this methodology of analysis. Second, a tool is proposed to visualize the hidden potential of underused land for other services (social or environmental) and its economic potential, by comparing the benefits obtained by private automobile oriented services land to the cost in the real estate market if those lands were fully developed to achieve higher density. Finally, this tool by enabling to visualize those results could benefit urban designers, planners, and policy makers when generating strategies for the sustainable development of a city. Strategies like mixed public–private economical inversion within the urban tissue, by taking advantage of sustainable transportation systems, can become successful by fostering moving from an underused land and economically inefficient model to an efficient urban model, all based on informed decisions.

Keywords

Sustainable urban development • Real impact of cars in the city • Land use algorithm

1 Introduction

The study of land uses in a city is relevant to understand how it changes and evolves; to understand that urban settlements are all but static. The history of the city, development, and growth of the urban tissue and all the modifications within the plot or in the behavior of the inhabitants are changes that have modified the existing land uses, adapting them to new realities and environments. Urban plans serve to compile these modifications over time allowing urban planners to analyze the structure of the city from the beginning. In this scenario, the portion of land used exclusively for mobility it's one of the most predominant uses by scale, and even if it is one of the most extendedly analyzed, there are uses for private automobile which are hidden from the regular planning specters that represent a real potentiality from a new way to understand urban regeneration.

The present document aims to develop and understand the impact of the vehicle within the urban tissue and the associated land uses. Based on the document generated previously "The hidden potential behind the planned city for cars," which verifies the hypothesis that "quantifying the real land used by and for cars, the city can be more efficient in planning and implementing a safe, clean and affordable transportation system and a livable city" (Davalos et al. 2016). This paper intends to take a step further by generating a more precise proposal that could guide a quantifiable urban design process and generate public policy on land use. Thus, one of the main objectives of this study is to make a clear

D. Davalos · D. Maldonado
Universidad de las Americas, Quito, Ecuador

D. Jacome Polit (✉)
100 Resilient Citie, Inaquito, Quito, Ecuador
e-mail: djacomepolit@gmail.com

J. Moreira
Inaquito, Quito, Ecuador

C. Alalouch et al. (eds.), *Advanced Studies in Energy Efficiency and Built Environment for Developing Countries*,
Advances in Science, Technology & Innovation, https://doi.org/10.1007/978-3-030-10856-4_8

distinction between public or private car spaces uses that usually are not quantified and so does not appear in statistics or in urban or territorial plans or even worse, as the case of Quito–Ecuador, have not been registered. In Latin America, the current model of growth based on sprawl generates an enormous amount of vacant land within the city that is usually used for private car purposes, converting it into a perfect study subject for this research.

This type of studies can foresee underused land destined for the private vehicle, which, in the case of Quito, is currently used by less than 40% of the population. Therefore, land uses can be quantified according to reality and not only based on which is proposed in the urban or territorial Plans. If it is wanted to evolve toward a more sustainable, compact, complex, and diverse cities, it is pertinent to understand how much the private vehicle really costs as a city, not only in economic terms but in general resources and specifically in space manners, notably public spaces and buildable area. A city model that prioritizes compactness and void helps to control the growth of the urban tissue, conserves the natural and agricultural soils, generates a greater sense of community, mixed uses, walkable spaces, greater density and space for recreation, and reduces space for automobiles (Molini and Salgado 2010). Several studies have already been done on city's urban models showing that the models of the compact, complex, and diverse city are the most efficient ways to introduce urban metabolism and habitability and cohesion (Hermida et al. 2013).

The methodology described in the previous paper "The hidden potential behind the planned city for cars" will allow us to identify, quantify, and propose spaces that do not prioritize the spaces of the private car. It is important to realize that the private vehicle occupies the most of underused space within the urban tissue, which is why it will be the object study in this document. The same methodology as in the previous document will be used to obtain comparative data between traditional urban indicators, and the ratio of these parameters used as car spaces. The main structure was developed in different stages, each of them as flexible as possible to allow city planners to incorporate as many variables they like (Davalos et al. 2016). We will return to the concept of an urban indicator, according to the United Nations Human Settlements Program (UN-Habitat 2004), an urban indicator is a management tool that allows to identify urban reality and to become base data for generating policies, programs, and projects that will help to improve it in a sustainable way. This research continues to open an opportunity to remove the priority of the private vehicle, looking for a scenario where better decisions can be made in planning and the re-design of the city toward a more sustainable one. It has been proven with the two cases studied before that the algorithm helped to generate a digital process easier and faster for any urban planner. It is easily applicable to a context with realities similar to those of Quito.

2 Methodology

As already explained, using as starting point the methodology proposal in "The hidden potential behind the city planned for cars" (Davalos et al. 2016), the main objective of this ongoing research is to fully develop a methodology that allows to identify and use the underused spaces destined for car parking and services as catalyst point of a new urban renewal aimed for a more pedestrian experience. To achieve this goal, the original algorithm was revised in several theoretical applications. Those experimentations help to refine this methodology aiming for simplification and to specific differentiate public and private spaces and its repercussion both on the urban structure problems and in the economic and political issues that come attached. The methodology explanation should start with the selection of urban indicators that will be used.

A series of indicators have converged in multiple reflections on the mobility beyond the streets of a city. Several combinations were studied and as a result a simplified list will be used as a guideline in this research. Out of all the urban indicators found, those chosen are described ahead:

Compactness This indicator refers to the physical support in which the city is developed, and it considers spaces where social cohesion can be generated and different plots that allow cohesion such as public transportation and building density. This indicator allows us to identify the areas of most critical imbalance and those that can be used when moving the private car away from the equation.

Inhabitant Density This measures the number of people living in the study area. It shows the population density, how much land is destined for residential use, and the demand for mobility, services, and equipment. The census data of INEC-2010 are taken within the chosen study area, as well as the intervention area to obtain this relation on square meters.

$$\mathrm{ID} = \frac{\text{number of people(hab)}}{\text{study area m}^2}$$

Absolute Compactness This relates directly to the built volume with the territory. It takes into account only the building intensity.

$$AC = \frac{\text{Total constructed volume of the study area}}{\text{Total of the study urban area}}$$

Vacant Lots These are the lots or parcels of land that are not currently intended for any specific use or have no buildings. This indicator helps to make visible that buildings could be destined to another use as public space or spaces of densification.

$$VL = \frac{(\text{Total area of buildings} * 100)}{\text{total empty lots area}}$$

Underused Land This measures the soil that is not at full capacity and finds whether the soil has lack of density or is in inadequate use.

$$UL = \text{total building volume} - \text{proposed result of the regulation}$$

Pedestrian Space This calculates the relationship between pedestrian space and motor vehicle space. It takes into account sidewalks, parterres, ramblas, boulevards, etc.

$$PE = \frac{(\text{total area of spaces intended for pedestrians} * 100)}{\text{total area of space for mobility}}$$

Public Transportation Access This measures the percentage of the population that has access to the public transport networks. It is taken into account whether the house is within the area of influence of the transport network and also that the walkable ratio for a person is 300 m.

$$PTA = \text{total number of inhabitants of the study area} - \text{total area of inhabitants with access to TP/total area of study} - \text{area of influence of TP}(300\text{m}^2)$$

Mixed Use This refers to the variety of uses that generate some exchange in the urban tissue, economic activities, associations, equipment, and institutions. Mixed use evaluates the degree of maturity of the tissue as well as the balance of uses. It allows to evaluate whether the residential, economic, and leisure uses are in balance and the needs are satisfied in the urban tissue.

Urban Complexity This simultaneously measures the diversity and frequency of the people, and with this it seeks to know the mixture of uses in the urban tissue.

$$UC = \frac{\text{number of different equipment and uses}}{\text{area of study}}$$

Mixed Use/Residential Ratio This measures the relationship between non-residential activities and the amount of existing housing. It contrasts variety and urban balance.

$$MUR = \frac{\text{Non-residential activity area}}{\text{number of houses}}$$

Green Urban Space This refers to the presence of different types of green in the urban area and their accessibility. Both the green and the surface are considered, analyzing special characteristics such as permeability and proximity. It addresses the city as a living ecosystem where urban biodiversity is decisive. It is also an environmental index of quality as a result of the life quality of the people who inhabit it.

Green Space/Inhabitant Ratio This measures the relationship between green space and population and refers to the areas of the city that display vegetation to which any citizen can access freely.

$$GSI = \frac{\text{Green area}}{\text{population}}$$

Urban Space Integration This refers to the social relations between the different age groups and social condition. Urban space integration tries to measure the degree of cohesion and co-development detached from the social mixture. It measures socio-spatial integration to stabilize the system.

Once the basic components for urban analysis are described, it is time to fully deploy the car space algorithm and its possibilities. As a starting point, the algorithm will first diagnose a preselected urban area. Based on optimal ratio application (**ORUI**) of some urban indicators (**UI**), specifically selected, explained before, this (first) stage allows to find what the specific needs of a territory are.

Those urban indicators are a small but important sample of United Nations Human Settlements Programme (UN-HABITAT) indicators and could vary according to each case.

The second stage of this methodology is to obtain a new **UI** to visualize the area destined to private cars parking and services **(SCI)**. If this area exceeds 25% of the total footprint of public space (Rueda et al. 2007), then the society should reclaim those spaces to re-design the city tissue in a more inhabitant-oriented way. This SCI will permit to obtain three key data. The first one should be the total amount of area destined for private car purposes (**ACSI**) in that particular territory. Then, these data must be used to obtain an optimal ratio variation applied to expose the percent of this area over the city (**AtCSI**). Finally, a third and fundamental component emerges. The possible cost of market place for car spaces (**AcCSI**) emerges by exposing those footprints by the maximum area buildable permit by the urban code. This information could be a key evidence of how much space the city is losing by transforming it into economic data which should finally visualize the big opportunity those spaces have and possible help to create new public politics to encourage new kinds of urban uses and investments, a new guide to manage urban development.

At this point and in order to truly be able to use this area, property emerges as an important issue. **ApuCSI** and **AprCSI** appear as the next steps as the proposal to classify the **ACSI** is found within the study area into public and private properties. This division could permit to decide which kind of space should be used as a priority, for what purpose, and what kind of public policy should be generated to obtain this objective.

Finally, this methodology proposal condensed in the car space recycling algorithm suggests three possible categories of decision making about how to use those spaces. **ORhUI** indicates that spaces used for car purposes should be used for filling the gaps and deficiencies founded when a territory is analyzed by a group of urban indicators. This could be an effective and efficient way to cover a city needs, but it depends on the group of indicators preselected. On the other hand, **ORdUI** offers a way to reuse those spaces in function of a specific urban design for that area that should take some

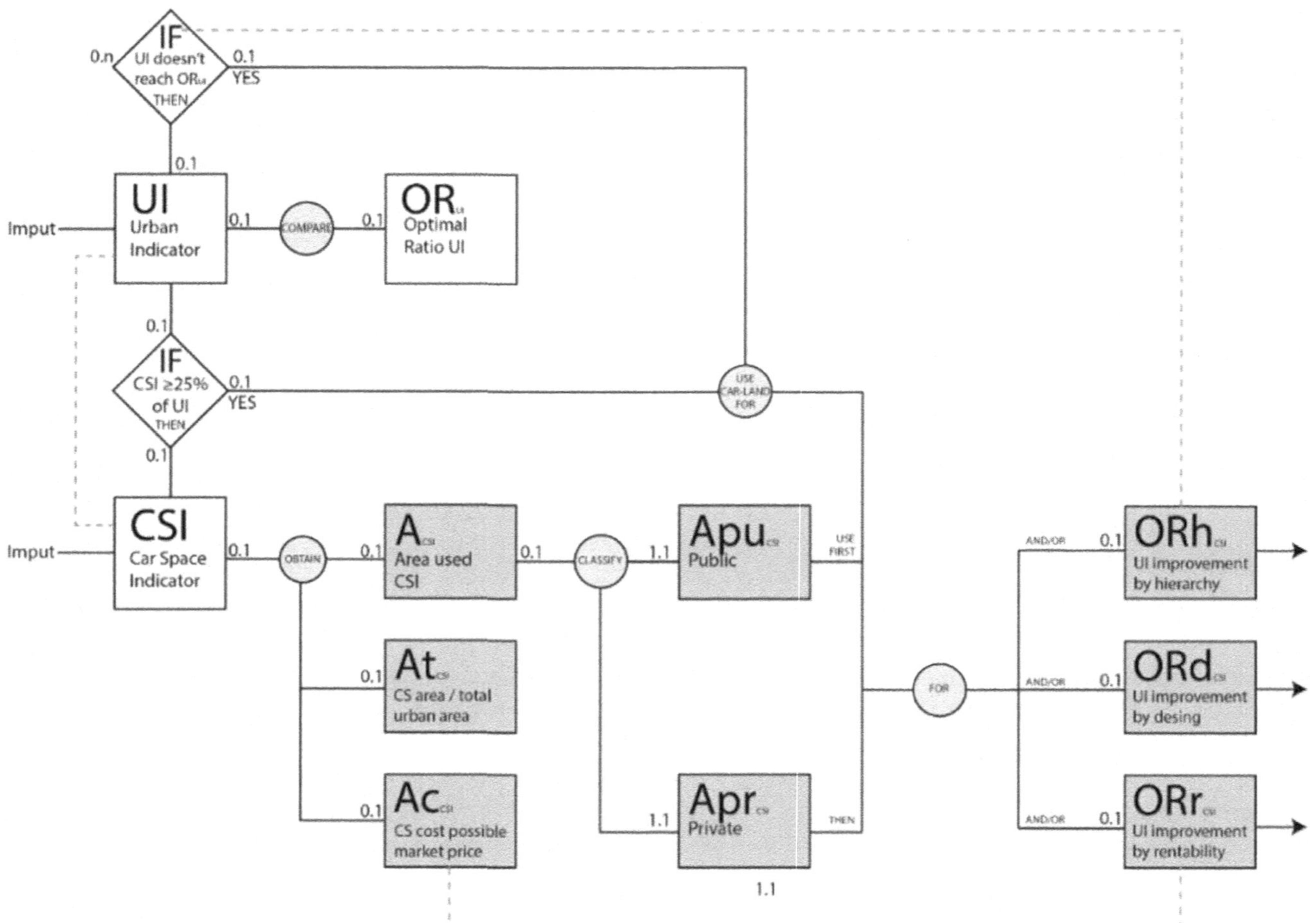

Fig. 1 Car space recycling algorithm

special considerations like city vision. Finally, **ORrUI** proposes to develop the territory in function of investment and urban profit, and even if this idea could sound as exclusionary, it is important to underline that any human settlement has an important economical role and that cities are the greatest producers of well in the world. For this methodology to become truly sustainable, efficient, and resilient, all the three possible solutions are proposed by principle that are not hierarchical so they should be used simultaneously or by iterative method approach to be decided by designers, politicians, and especially by its community.

Once again, in order to better understand this methodology, and eventually being able to digitalize it, an algorithm has been developed to graphically illustrate all the steps required (Fig. 1).

3 Application of the Methodology in the Case Study

3.1 Study Case Space Determination

The city of Quito, Ecuador (2850 m altitude), like most of the capital cities of Latin America, is the result of an ongoing process of adaptable and sprawled growth through its complex geography. As a result, the city has grown into an elongated structure from a historical center founded some 500 years ago. At the beginning of the twentieth century and as a part of the first major expansion celebrating one century of the national independence, the city expands itself toward the north and south according to the latest urban design—a new garden city was born. The sector later known as "La Mariscal" was designed and served as garden-like city for the upper class. As time passed by and the city grew, this area began to lose population as some other uses took place. By the end of the last century, La Mariscal became the main night attraction center of the city, and most of the bars, restaurants, and entertainment spaces were packed.

This transformation took with it some serious problems such as insecurity, minor robbery, and violence. But most of it, it brought some serious land use conflicts as almost consolidated urban tissue had to change form residential to mixed use. The fact of being designed as isolated "villas" present a special opportunity for the increasing fluent of private cars searching for parking and services at the point in which several houses were torn down to achieve space. Several intents had took place to regenerate and recover this major centrality and every single major have tried and only partially achieved to regain this emblematic space for its citizens. And even if this area is now a major touristic attraction, it has not been able to recover a regular population and provide real public spaces.

Mobility is also a special issue within the capital of Ecuador. While waiting for the first metro line to be constructed, the city struggles against its complex morphology (some 40 km long over 5 km wide), a disconnected urban tissue, a hypercenter that absorbs almost 70% of the city's economical activities and inefficient public transportation. All that had generated ongoing chaos to mobilize from one corner of the city to another and causing an explosion of private transportation that can be simple understood by taking official documents that project some 1,150,000

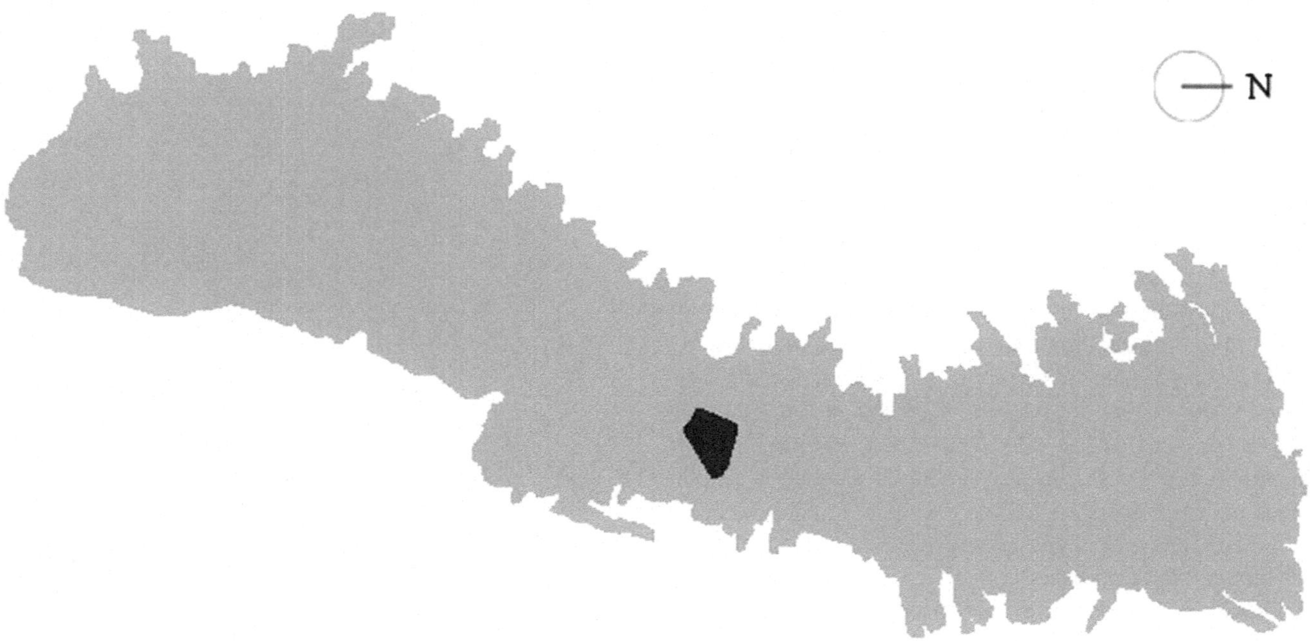

Fig. 2 Quito, Ecuador urban footprint, "La Mariscal" study case location

vehicles for the year 2030 in a city of 2.2 million of habitants (MDMQ 2014). But it seems "La Mariscal" is placed at the city's hypercenter; curious enough, this particular space appears fully covered by the public transportation system.

With all this as background, this special urban tissue was selected for testing the "Car Space recycling algorithm" with the hypotheses that if even in a so-consolidated space within the center of the city could prove the hidden potential of car spaces as catalytic regenerator, it could later be extrapolated to some other urban typologies. The results of this process will be shown below (Fig. 2).

3.2 Methodology Application

In order to validate this methodology, this ongoing research proposes its application in the study case area as previously explained. By applying the formerly urban indicator package (urban space integration, green space, mixed use, and compactness), the process allows to obtain hard measurable data and concludes that the territory selected has serious problems at almost every level, validating the selection of this sector as a possible selective candidate for an urban regeneration process.

Then, the algorithm requires to quantify the total area used for car purposes. Here, an interesting problem appears. Since the emerging of geographic information systems, urban designer and local authorities relies on these programs to gather information. In this specific case, and even if the city has a vast amount of information, the total amount of area destined for private car purposes (ACSI) cannot be identified using the provided data because a big percent of it are occupying private (vacant or underused) lots and the system will classify it as "general services" or "economical entities" proposing a major issue. Operationally, it obligates to a field quantification and data ascertainment. But most important, this vacuum of information serves to validate this research approach as it allows to expose the real amount of area that private cars are using in our cities not only in the massive road infrastructure but most important within its usable land.

Once this stage is completed, the total area generates two more important data: important area/inhabitant destined for car purposes and an impressive maximum area buildable that could become a major source of investment for the city. Also, it shows that the majority of this area (only parking and car services) are actually private spaces suggesting that in this case, the public policy should be directed to exploit those underuse spaces.

Finally, this collection of data is used to generate a series of maps and hard data to explain the potential of this methodology to visualize the car space used within an urban tissue and to effectively cover any deficiency in terms of physical urban indicators that territory might have.

3.3 Car Space Indicators. Mapping

Using GIS technology, this methodology allows to clearly visualize the impact of car space used within the selected study area. In this particular case and based on the UI previously selected, this map shows how three main problems of La Mariscal could be attacked and almost solved by using those spaces by its whole. Urban density and inhabitant density could be dramatically increased toward acceptable levels giving vitality back to the sector. Public and green space could surpass the minimum required for international standards and urban space integration could cover the hall area (Figs. 3, 4, 5, 6, 7, and 8).

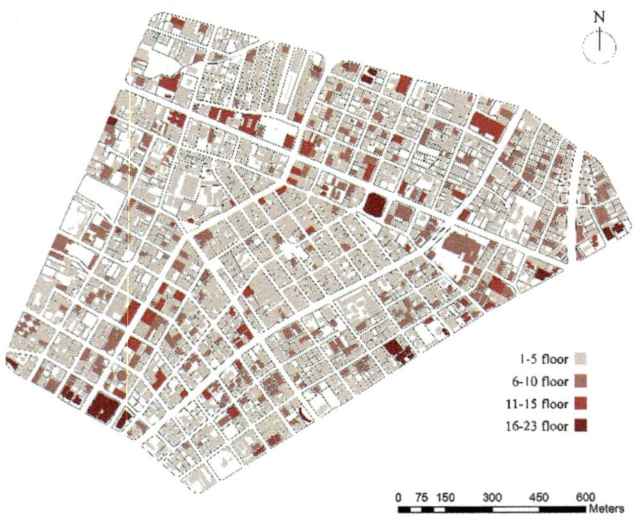

Fig. 3 Compactness in La Mariscal, actual status

Fig. 4 Compactness in La Mariscal, car space increase

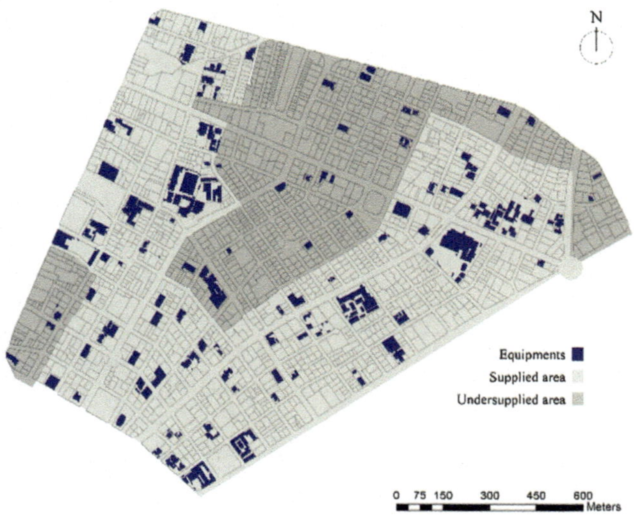

Fig. 5 Urban space integration, La Mariscal, actual status

Fig. 7 Public space, La Mariscal, actual status

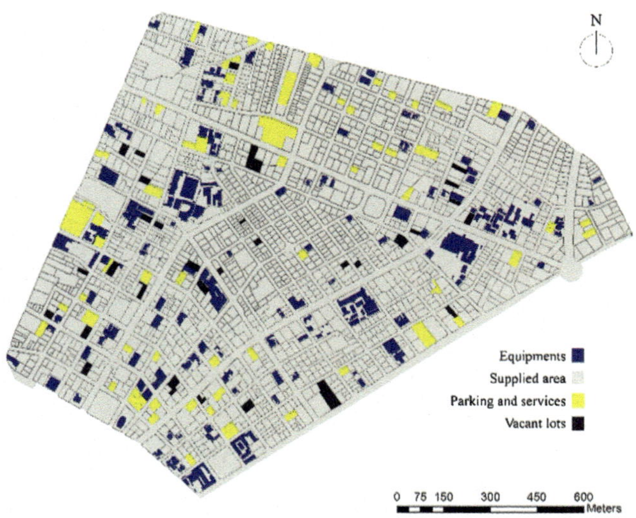

Fig. 6 Urban space integration, La Mariscal, car space increase

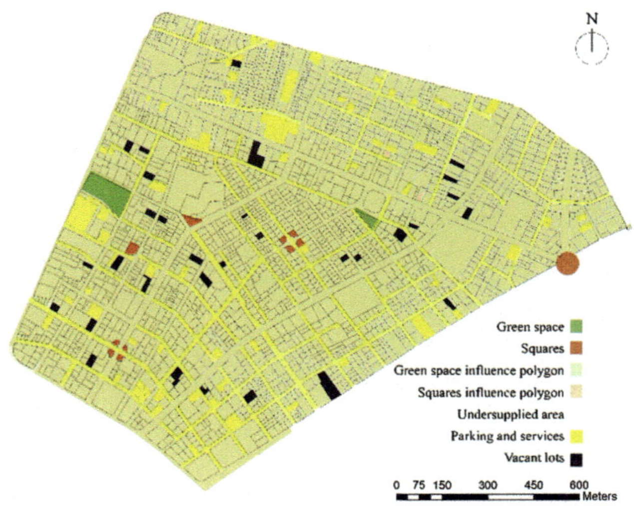

Fig. 8 Public space, La Mariscal, car space increase

3.4 Car Space Indicators. Hard Data

Following the car space integration algorithm, it becomes evident that some important data could be obtained to not only visualize the information by a GIS mapping but also to truly quantify those improvements generating a complete package of information presented as a powerful decision-making tool for urban design and public policies generation (Table 1).

By reading this information, it becomes clear how the present research could affect positively to improve a territory. In this case, by reusing the spaces used by private cars, La Mariscal could improve dramatically its performance. It could, for example, increase its inhabitant density from 37.96 to 112.80 inhab/m^2 if those spaces are fully used for dwelling. Or, if this land transforms into green space, this urban area could reach 16.06 m^2/inhab in comparison with the actual 2 m^2/inhab or it could serve to fully cover the area as an integrated urban space.

It has to be remarked that, as shown in Table 2, car space indicators can serve to obtain more complex data that could perfectly show fundamental information to rediscover a hidden potential inside the city. For instance, in the area exists an important 10.56 m^2/inhab of car space (in comparison with 2 m^2/inhab of green space) use which represents an impressive USD 670 million on market buildable space, maybe the most important data showing how much

Table 1 Urban indicators. La Mariscal

			Current condition (UI)	
Category	Formula	Optimal ratio	Data	Results
Compactness				
Inhabitant density	Inhabitants/effective surface	120 inhab/ha	7100 inhab/187 ha	37.96
Absolute compactness	Built volume, total land area	≥ 5	9,000,000 m^3/ 1,870,000 m^2	4.82
Vacant lots	Vacant lot area/total lot area × 100	$\leq 30\%$	58,000 m^2/ 1,390,000 m^2	4.17%
Underused land	Underused land area/total buildable area × 100		1,280,000 m^2/ 8,300,000 m^2	15.42%
Pedestrian/total road ratio	Pedestrian walkway area/total road area × 100	$\geq 75\%$	104,000 m^2/ 600,000 m^2	17.33%
Public transportation access	Covered by 3 different public transportation/population	100%		100.00%
Green space				
Green space/inhabitant ratio	Green space land/inhabitants	≥ 15 m^2/ inhab	14,000 m^2/ 7100 inhab	2.00
Public space/inhabitant ratio	Public space land (without roadways)/inhabitants		128,000 m^2/ 7100 inhab	18.00
Urban space integration				
Urban equipment	Area with simultaneous equipment/total land × 100 (educational, cultural 500 m/health, recreational 1000 m ratio)	$\geq 100\%$	1,215,000 m^2/ 1,870,000 m^2	65.00%

the city is losing by allowing those spaces to grow within its tissue (Table 3).

4 Conclusions

The twentieth century was, among so many things, the century of a city designed for cars. The prioritization of private cars as a key factor for mobility and the impressive amount of space and resources required for archiving its functionality has begun to be revised. Furthermore, a new global conscience of responsible sustainability that aims to a major reduction in greenhouse gas emissions while improving environmental health has evolved from optional conduct to a necessary global policy. In this global scenario, cities become the major playground to effectively achieve a better future. As part of ongoing research, this paper presents a valid way to confront an inadequate, car-oriented way of urban planning. It allows not only to identify and map some important urban indicators and the car space used but also to quantify it, extrapolate it, and effectively propose new ways to reclaim those spaces for the inhabitants.

As shown, the car space recycling algorithm provides a powerful tool to analyze a territory and proposes an interesting way of regenerating it. In the particular case of La Mariscal, Quito–Ecuador, the results presented help to visualize not only the problem that car spaces possess inside an urban tissue, but also visualize a huge potential for new development. The study shows that in the area, space destined for car purposes is oversized. Only a 17.33% of public space is meant for pedestrian, while a ratio 10.52 m^2 of car spaces (without roads) per inhabitant clearly reflects this fact. At the same time, the sector suffers from a really low ratio of green space (2 m^2/inhab) and inhabitant's density (37.96 inhab/m^2). Urban space integration and equipment accessibility are not covered either.

According to the data obtained, by using this meteorology at maximum ratio effectiveness, La Mariscal could improve its green space in more than 800% up to 16.60 m^2/inhab, increase its absolute compactness to an acceptable ratio of 5.05 or increasing its density by almost 300% at a 112.80 inhab/m^2. Any of those results clearly show the big potential that those spaces could have for a deep improvement of the territory without scarifying the original tissue,

Table 2 Car space indicators. La Mariscal

			Current condition (UI)	
Category	Formula	Optimal ratio (25%)	Data	Results
Total occupied area	Car space area (lots + roadway parking)	n/a	100,000 m^2	100,000.00
Car space	Car space area/total urban area × 100 (lots + roadway parking)	n/a	100,000 m^2/1,870,000 m^2	5.30%
Car space lots (parking and services)	Car space area/total lot area × 100	n/a	75,000 m^2/1,390,000 m^2	5.40%
Car space/inhabitant ratio	Car space area/inhabitants	≤ 4 m^2/inhab	75,000 m^2/7100 inhab	10.56
Car space lost	Car space area × total buildable area	n/a	75,000 m^2 × 600% m^2	450,000.00
Market cost	Total buildable area (in car space area) × possible market cost	n/a	450,000 m^2 × 1500 USD	675,000,000.00
Increase of inhabitants	Total buildable area (in car space area) × useful residential area/optimum housing space per person	n/a	450,000 m^2 × 0.75/24 m^2	14,000 inhab

Table 3 Optimized urban indicator by car space recycling algorithm. La Mariscal

	Maximum efficiency (OR CSI)		
Category	CSI data usage	Maximum improvement	New total
Compactness			
Inhabitant density	Increase of inhabitants	Plus 14,000 inhab	112.80
Absolute compactness	Car space area × total buildable area	Plus 450,000 m^2	5.05
Vacant lots			
Underused land	Car space area	Minus 75,000 m^2	14.52%
Pedestrian/total road ratio	Car space area	Plus 100,000 m^2	19.00%
Public transportation access	n/a		
Green space			
Green space/inhabitant ratio	Car space area	Plus 100,000 m^2	16.06
Public space inhabitant ratio	Car space area	Plus 100,000 m^2	32.12
Urban space integration			
Urban equipment	Car space area	Plus 100,000 m^2	100%

structure, architecture, or social dynamics. Moreover, by taking additional data in account, it becomes relevant that those spaces represent a 450,000 m^2 of lost built area at a market price of USD 670 million. Numbers that could become fundamental to engage new investment projects that would render viable this urban transformation while reducing urban sprawl, recovering city density and reduction greenhouse gas emission by eliminating forced mobility within the city. In the end, a new tool aims for a more compact, sustainable, and livable urban settlement for its citizens.

5 Recommendations for Further Research

As a preliminary result of ongoing research, this paper shows a major opportunity to develop a powerful urban planning and decision-making tool. The basic objectives of truly visualizing the amount of land used for car purposes, effectively quantify it and reformulate its uses to regenerate urban spaces have been covered. In this path, the next step should be to fully digitalize as exploring a deeper 3d approach while introducing carbon emission measures for appliance purposes in a real scenario, an ongoing work proses.

References

D. Davalos, D. Jacome, D. Maldonado. The hidden potential behind the planned city for cars. Elsevier. Quito' Ecuador. 2016.

F. Molini, M. Salgado. Superficie Artificial y viviendas unifamiliares en España, dentro del debate de ciudad compacta y dispersa. Boletín de la Asociacion de Geografos Españoles, 125–147. Madrid-España. 2010.

A. Hermida, D. Orellana, N. Cabrera, P. Osorio, C. Calle, La ciudad es esto. Medición y representación espacial para ciudades compactas y sustentables, Universidad de Cuenca, Cuenca, 2013.

U. N. Habitat, Urban indicator guidelines, 2004.

S. Rueda, A. Cuchí, J. Subirats, M. de Mingo, A. Díaz, Libro verde del medio ambiente urbano, Agència de ecología Urbana de Barcelona Barcelona (2007).

Municipio del Distrito Metropolitano de Quito, Diagnostico de la movilidad en el distrito metropolitano de Quito para el plan metropolitano de desarrollo territorial (PMOT), 2014.

Part II
Design for Energy-Efficient Architecture

The second part of this book provides a highly technical review of the challenges, obstacles, and opportunities associated with energy use in the developing countries as well as potential solutions needed to improve the sustainability situation in these countries. In their bid to present various ways to improve the concept of sustainability in developing countries including Jordan, Egypt, and different locations in Asia and Africa, the authors of the chapters in this part present critical assessments of energy performance.

In the chapter titled “Learning from Informal Settlements’ Architectural and Urban Features: A Study of Comfort Levels in Both Informal and Public Housing Settlements”, the author addresses an evident gap in understanding the level of efficiency informal housing in Egypt provides and its relation to sustainability. The author also addresses public housing provided by the government that is marketed as a better solution for those who are unable to afford private housing. This chapter explores the existing comfort levels in both public housing settlements and informal housing settlements and compares this in terms of comfort and sustainability.

The chapter titled “Monitoring Thermal Bridges by Infrared Thermography” focuses on thermal performance assessment of an existing building by in situ measurements of its external envelope using infrared thermography (IRT). It demonstrates how IRT can be used to improve the thermal performance of existing buildings. Whereas the chapter “Net Zero Energy Buildings (NZEBs) Potential in MENA Region: Critical Review on Egypt Case” discusses how the concept of net-zero energy building could contribute to the sustainability in the Middle East and North Africa region with particular focus on the Egyptian context.

Finally, in “Energy Retrofit of Existing Building Stock in Amman: State of the Art, Obstacles and Opportunities”, the authors analyze energy security in Jordan. Due to the fact that the majority of the resources needed to produce energy in Jordan are imported from foreign countries, the authors address issues related to energy retrofitting in order to assist the country in developing effective plans to reduce the energy consumption of buildings and therefore decrease associated energy costs leading to a more sustainable environment within cities.

Learning from Informal Settlements' Architectural and Urban Features: A Study of Comfort Levels in Both Informal and Public Housing Settlements

F. Yasser

Abstract

The city of Cairo has witnessed novel housing transformations by the government since the introduction of informal settlements in the 1960s. Despite the heavy subsidisation of these public housing settlements, otherwise known as social housing projects, which are partly payed for by the government yet still are regarded as expensive compared to the informal settlements housing unit counterpart. Informal settlement housing units have succeeded over public housing units basically by imposing an entirely organised city without the surveillance of formal institutes whether economically, socially or through mobile networks. Despite the successful functioning areas of informal settlements, they receive much negativity for their existence as being centres of energy consumption, social and environmental degradation. Currently, there is a gap in understanding how the highly dense urban clusterings of informal settlement buildings and their architectural features affect sustainable development. Public housing settlements are portrayed and marketed as the solution to informal housing complications despite their unsubstantiated design methodologies which do not coincide with many of the design considerations needed. This also raises the question of whether they can contribute to the overall and particular level of contribution to Cairo's sustainable development through comfort levels such as thermal gains and losses. The objective set were to discover the existing comfort levels of thermal gains and losses in both public housing settlements and informal housing settlements. The second objective was to gain information from professionals and residents which have experience with informal and public housing settlements on which settlement type is more comfortable in terms of thermal gains and losses. A mixed method approach was used to carry out the research, the quantitative data represented the existing thermal gains and losses using Ecotect simulation models of a selected public housing and an informal settlement sample in South-West Cairo. The qualitative data were collected using semi-structured interviews with informal settlement residents and professionals of various professional backgrounds with experience in informal settlements. It is important to mention that the conclusions found that informal settlements have poor, mediocre and good levels of sustainable development in terms of thermal gains and losses in various subdivisions due to their architectural and urban uniqueness which public housing settlements lack. These are not solely a result of the residents control over the many architectural or urban variables involved despite their awareness of them, yet it is also—to a large extent—affected by the obligatory situations and dispositions they face such as economic difficulties, land plot size, lack of professional intervention and the absence of building regulations.

Keywords

Architecture • Urban fabric • Clustering • Planning • Elevations • Sustainable development • Gains • Losses • Thermal • Ventilation • Solar light • Day light and discomfort hours

1 Introduction

Informal settlements are defined as housing built on prohibited land, where they are in violation of the law and are refused provision of services by the government. Residents of informal settlements are likely to experience poverty and disease due to the effects of quickly deteriorating housing and lack of infrastructure (Afify 2004). Informal settlements were a manifestation of the Egyptian Industrial revolution in the 1960s, where rural–urban migration lead to the urbanism

F. Yasser (✉)
Department of Architectural Systems Engineering, Modern Sciences and Arts University, 26 July Mehwar Road Intersection with Wahat Road, 6th of October, Cairo, Egypt
e-mail: farresyasser@gmail.com

C. Alalouch et al. (eds.), *Advanced Studies in Energy Efficiency and Built Environment for Developing Countries*, Advances in Science, Technology & Innovation, https://doi.org/10.1007/978-3-030-10856-4_10

of agricultural land peripheries due to the prioritization of heavy military fundings over government housing construction plans up until the mid-1970s (Al-Malky 2009, cited in Khalifa 2011). Provision for housing construction was scarce since the private sector only provided for the middle-upper and upper classes; thus, the remaining larger portions of the social scope took refuge in informal settlements (Rydin et al. 2012). The oil boom in the Gulf region caused the Egyptian expatriates to generate capital in order to ultimately invest in housing in Cairo. Since the private sector was still set for the elite and since the government did not provide sufficient or affordable housing alternatives to the illicit construction of agricultural lands, a large portion of the Egyptian expatriates thus invested their capital into the development and growth of housing in the—once agricultural—informal areas (Sims 2010) (Fig. 1).

In the recent public housing units, various inadequacies are still present which leaves the public housing (social housing) sector vulnerable to modifications by the former informal settlement residents for primary reasons such as the need for additional rooms, privacy needs and inflexibility or inefficiency of internal space (AbdelGelil 2011). Examples of these modifications would be the conversion of ground floor apartment front rooms into shops and hair salons for income production (these are unregistered and illegal), adding extensions to houses illegally to run informal businesses, the illegitimate construction of shacks to accommodate extended family members or to let and utilising pavements in front of households for small shops. Illegal amenities tap in connections between the households (Foucault 2007a cited in Massey 2014).

As explained by Serageldin (2017), the urban and architectural designs that emerged where a direct result of the size, orientation and location of the existing agricultural land plots which were transformed by their owners into family households and buildings with a hybrid of commercial bases, apartments for sale or rent as well as apartments for the land owners' family. At first, the urban and architectural expressions were pleasing and untroubling; this was ascribable to their low-rise skyline and the use of vernacular traditions in architecture despite the defects of unplanned layouts. Eventually, these positive architectural and urban features were eliminated piecemeal through the introduction of excessive heights, haphazard repetition, conflicting juxtapositions of styles and oppressive intensity of use. These all lead to the degrading levels of sunlight, daylight, ventilation, privacy and ultimately heat gains and losses. The use of reinforced concrete construction made spiralling escalations of 100–600 persons per hectare possible (Serageldin 2017). Table 1 below represents the differences between the Informal settlement and public housing sectors (Yasser 2017).

(a)

(b)

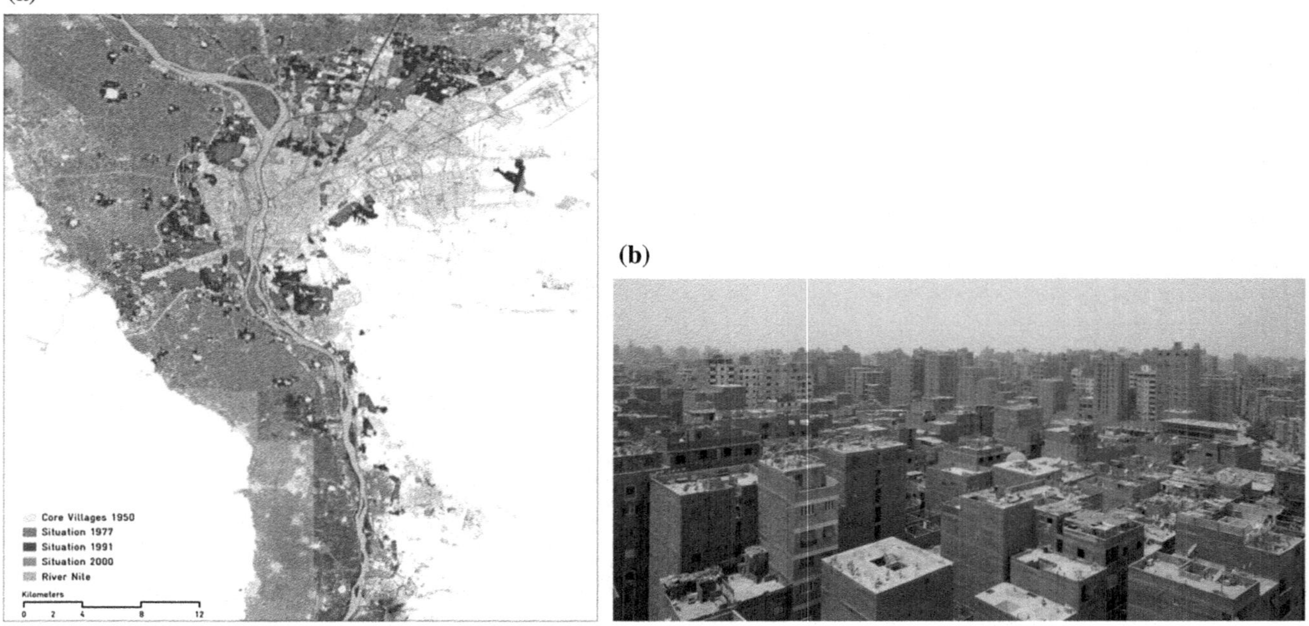

Fig. 1 **a** Figure demonstrates the development of informal settlements which took place in the late twentieth century. It is clear that South-West Cairo went through strong densification of agricultural urban centres in the late 1970s, less in the 1980s but is still ongoing in the twenty-first century. These urban developments mostly affected the land plots on the fringes of formal areas, the peri-urban centres and the agricultural lands in between (Kipper et al. 2009). **b** Figure is an image of the once agricultural land plots which have become informal settlements with red brick and concrete skeleton construction

Table 1 Differences between the Informal settlement and public housing sectors

	Informal settlements	Public housing
Floor area (m^2)	70–80, 80–100, 100–120	60, 60–80, 80–100, 100–120, 120–160
Number of floors	4–20 floors	4–6 floors
Site types	Agricultural plots In cemetery housing, Dessert land (near agricultural land) Clustering of in formal areas	Dessert land (further away but still away from the Greater Cairo region)
Tenant options	Rent, Long lease rent, Ownership	Ownership
Tenant pricing options	200–900 EGP	700–1500 EGP

2 Literature Review

2.1 Informal Settlement Urban and Architectural Features in Context

2.1.1 Density and Narrow Streets

The main characteristic of informal settlements' urban fabric is their extreme dense urban forms and narrow streets which traditionally were regarded as the only solutions for the two main housing complications which were the prevention of intense solar radiation and the provision of adequate cooling causing less exposed surfaces to the sun; thus, less fabric gains through external walls (Koenigsberger et al. 1974 cited in Almaiyah et al. 2010). These characteristics are paradoxically the same reason why they are regarded as inadequate as a proper housing solution.

However, the recommendation of these compact designs is not always the case, (Abouelmagd et al. 2013) explains that informal settlements have been given two novel institutional terminologies by the informal settlement development fund (ISDF), "Unplanned Areas" and "Unsafe Areas". The mentioned unplanned areas have been broadened to include residential areas which were not planned officially into subdivisions for building construction nor do they cohere with building and planning laws and regulations; these are predominantly previously agricultural lands—now informal settlements—which were subdivided with the intention of agricultural investment (MTI and GOGPO 2009, cited in Khalifa 2011). The nature of these mentioned areas in South-West Cairo is characterised by very narrow and extremely long streets which have many complications such as problems with the entry of fire trucks during emergencies and also environmental issues such as elevated humidity levels due to little to no sunlight penetration in each street and poor air quality for households attributed to the lack of open green spaces and over-crowdedness (Gerlach et al. 2009).

2.1.2 Architectural Features and Incompetent Tenure Ship Laws

Informal settlements not only experience infrastructure complexations but also fundamental safety problems on single buildings in specific areas caused by poorly maintained buildings, damaged sewers systems and alleyways as wide as an arm stretch (Ahmed 2013). According to Shehayeb (2009), health issues thrive rapidly by virtue of law regulations, such as the stagnation of rent increase in the 1980s, which caused landlords to neglect building maintenance. The results of these regulations were water and sewage leaks, building structure depletion, inadequate lighting, the appearance of makeshift materials and overall poor maintenance leading to health hazards. The previous statement proves how the architecture elements in informal settlements are not the main cause for the many complications they face but rather their amalgam with their specific dense urban fabric attributes and the municipal decisions made by officials as being the reasons for the health hazards faced by informal settlements residents. However, the usefulness of these urban fabric formations is not mentioned which makes a large portion of the literature biased.

2.2 The Public Housing Projects Alternative

Public housing settlements are portrayed as the main solution for the shortcomings of informal settlements especially in their building design, orientation and site urban plan or layout. However, many of the public housing settlements that are built or currently under construction do not fulfil the architectural or urban design requirements that are needed to provide any change or development in the resulting comfort levels they provide for their residents. A study by Khalil et al. (2014) on a government-designed and -planned settlement was conducted to test the thermal comfort levels present using Envi-met simulation software. The selected government housing project was "Ibni Beytak" translated as "build your home" and was subdivided into small 150 m^2 residential land plots for middle-income citizens to build their own homes with strict regulations limiting the total building construction to 63 m^2 per land plot and a maximum of four-storey heights. The results showed the buildings having a lower than average comfort rate of 40% throughout the year between 9 a.m. and 3 p.m. Sayed et al. (2012) states that the reasons for this type of thermal discomfort and heat gain is the government's lack of adequate design tools in the construction of new communities and the fact that public housing projects are fixed prototype housing projects that are built using cookie cutter urban planning strategies regardless of the topographical and climatic conditions specific to each site. Furthermore, Shehayeb et al. (2011) claimed that the current informal housing strategy is a solution, imperfect because of the state's rejection towards intervention with architects and urban planners so as to develop this already built available solution which are informal settlements instead of imposing an imported blueprint for a public housing utopia. Her suggestions of intervention where experimented through the research conducted by Khalil et al. (2014) which simulated the possible outcomes for the government-planned housing project Ibni Beytak by implementing drastic changes to the simulation model through increasing the levels of greenery in many locations throughout the site as commended by Sayed et al. (2012) by using urban design essential strategies for their positioning and level of density to examine the air temperature changes before and after the modifications were made (see Fig. 2). Also, a list of design techniques are provided by Rehan (2014) in Table 2 which demonstrate a set of templates for designing new cities more efficiently reducing energy use whilst taking into consideration the residents comfort.

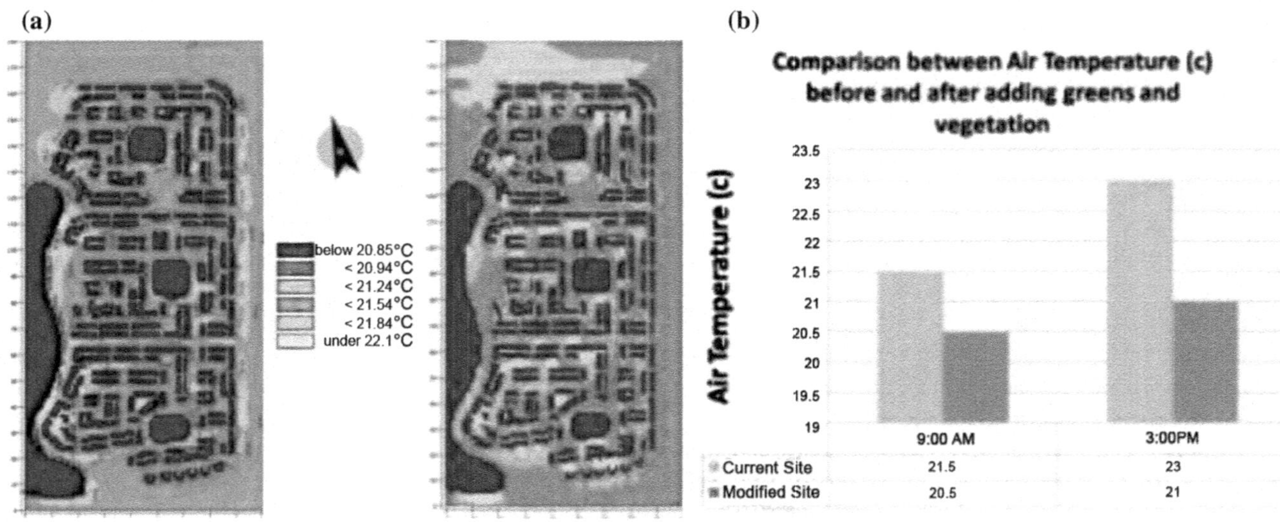

Fig. 2 **a** Air temperature changes are demonstrated visually before the modifications made which were the densification of greenery. **b** Air temperature changes after the modifications. **c** The graph represents the outcomes with exact temperatures (Khalil et al. 2014)

Table 2 A set of design techniques targeting the development of public housing plans in order for them to be designed using a larger number of passive design methods (Rehan 2014)

Passive design development in buildings	Enhance applications of passive ventilation to reduce the dependence on air conditioning and fans
Reflective materials	Encourage the use of high-albedo pavements, change the colour of pavement, roofs, and facades to light colours in order to reduce the ambient temperature of the city
Vegetation	Large trees should be planted to provide shading, plus they are important in mitigating the UHI effect
Green roofs	Promote green roofs to improve the air quality
Water installations	Integration of water installations into the landscape to reduce UHI effects and conserve water resources through reuse and system efficiency
Green corridors	Improve air quality through using green ventilation corridors and to minimize heat gains in urban environments
Street orientation	Promote the thermal comfort of city dwellers through promoting good wind circulation
Environmental management	Focus on good transportation planning using of sustainable modes of transportation, which is essential to minimize heat gains in urban environments

2.3 Informal Settlement Urban and Architectural Features in Context

The individualism that informal settlements represent out of obligation and despite repression comes with its own set of architectural characteristics. These characteristics allow for certain comfort level alterations such as protrusions for shading and wind capturing, staircase cores and service shafts for wind current generation. However, their individualism has also led to a regression in their living standards due to the lack of professional involvement and intervention. A project by Grigorios Dimitriades, Bernardo Menesez Falcao and Shinji Terada was conducted which had a core aim to set a new familiar archetype for informal settlements which would use the existing architectural elements in buildings with innovative rearrangements and standards application to ensure better living conditions. Juxtapositions of the current elements such as shafts, circulations patterns, protrusions and cantilevers proposed improved urban patterns through the introduction of minor spaces between buildings allowing for more light and providing ventilation whilst still using the whole land plot for construction (see Fig. 3). The proposal suggests two typical housing units to be combined with mirrored staircases on the outskirts of the building, with a few architectural and structural modifications. This proposal aims to enhance the single unit living conditions as well as those for the entire neighbourhood if replicated correctly (Elshahed 2015).

The former acknowledgement of informal settlements' potentials through their architectural and urban features is not an attempt to cleanse their existing imperfections such as infrastructure, hygiene, sunlight and living standards, yet it is to realise the possible and existing sustainable features within informal settlements. One of these features is the urban compactness they have which—aside of its provision of walkable districts—provides low energy demands depending on the orientation and street width in which the building lies in, leading to more shading and wind currents thus causing the residents to be in no need of cooling appliances. Once the acknowledgment of the potentials are at hand, a collective symbiotic relationship which involves the residents and professionals' opinions can introduce the missing needed urban or architectural elements for development (Mostafa 2013).

The loss of traditional architectural elements which would improve comfort level heat gains and losses drastically are symptoms of limited space to build them as well as their replacement by subordinate economical and well-being technological innovations such as curtains, air conditioners and fans. Finally, changing lifestyles with less focus and dedication to details, craftsmanship and authenticity are reasons for the loss of traditionally effective architectural elements which provided satisfactory comfort levels. Fences, doorways staircases, landings, windows, balconies and parapets create varying degrees of enclosure and openness which in turn provide various levels of ventilation, fabrics, direct and indirect lighting heat gains and losses. These elements such as curtains, shades and shutters amalgamate between the need for privacy through the use of opaque shields and the need to control the amount of heat gained or

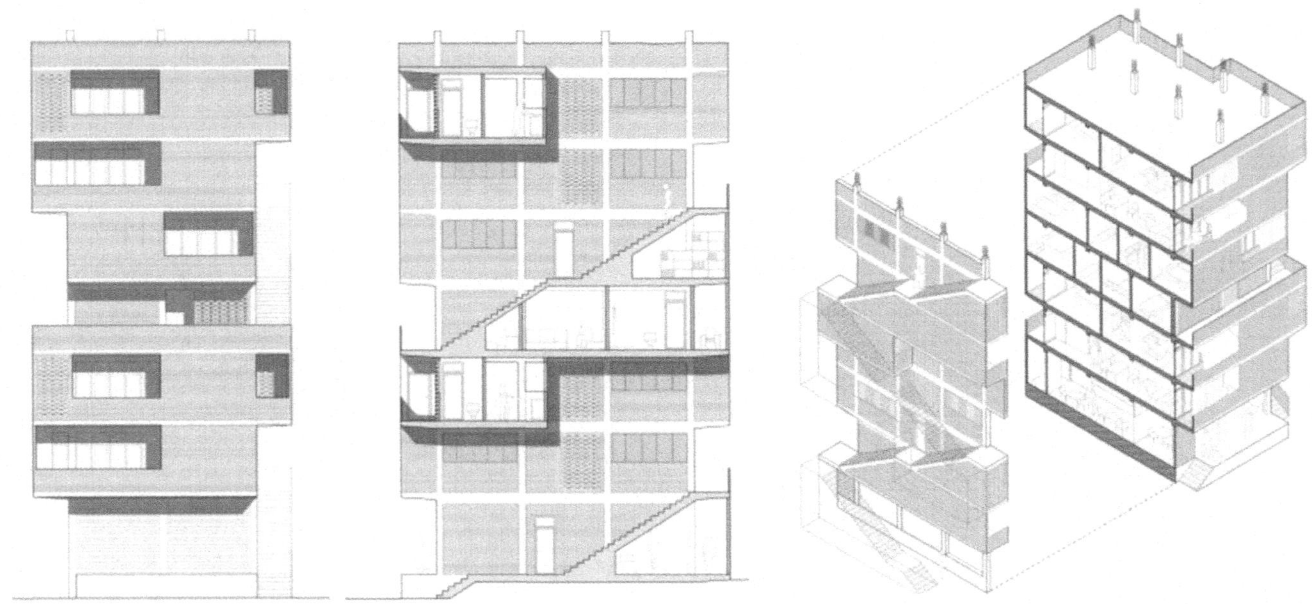

Fig. 3 Shows the proposal that Grigorios Dimitriades, Bernardo Menesez Falcao and Shinji Terada to help propose a new housing typology for informal settlements which enhance the single unit living conditions (Elshahed 2015)

lost through various mediums despite their low levels of efficiency compared to traditional architectural elements (Serageldin 2017).

3 Methodology and Analysis

3.1 Methodology

Positivist and interpretive approaches were utilised since the objective needed both quantitative and qualitative data. The positivist approach portrayed results as fixed truths and objective facts (Crotty 1998, cited in Gray 2004), this allowed for a clearer understanding of the existing comfort levels through the building performance simulation analysis which provided large quantities of quantitative data. This first methodology technique was implemented using Ecotect, an environmental analysis software, to accurately understand the comfort levels that exist within both of the existing informal and public settlement sites in South-West Cairo chosen for modelling. The informal settlement area selected was in the Konayyese area south of the Ring road and the public settlement area chosen was the "Masakin Osman" or Othmon Housing projects in October city east of the Wahat Albaharia road (see Fig. 4). Accurate weather data files were used in the process. Ecotect was used as it can perform in essence all types of environmental analysis including reliable psychometric charts, in-depth thermal analysis, shading studies, lighting and wind analysis. Its recent development as an extension within Revit only promotes the use of Revit but does not facilitate the interface nor add any new analysis features.

The interpretive approach was introduced via the qualitative data collected by conducting semi-structured interviews with informal settlement residents and professionals. This provided new levels of clarifications and definitions for both the literature and model results presenting a wide spectrum between the dichotomy of opinions on the actual

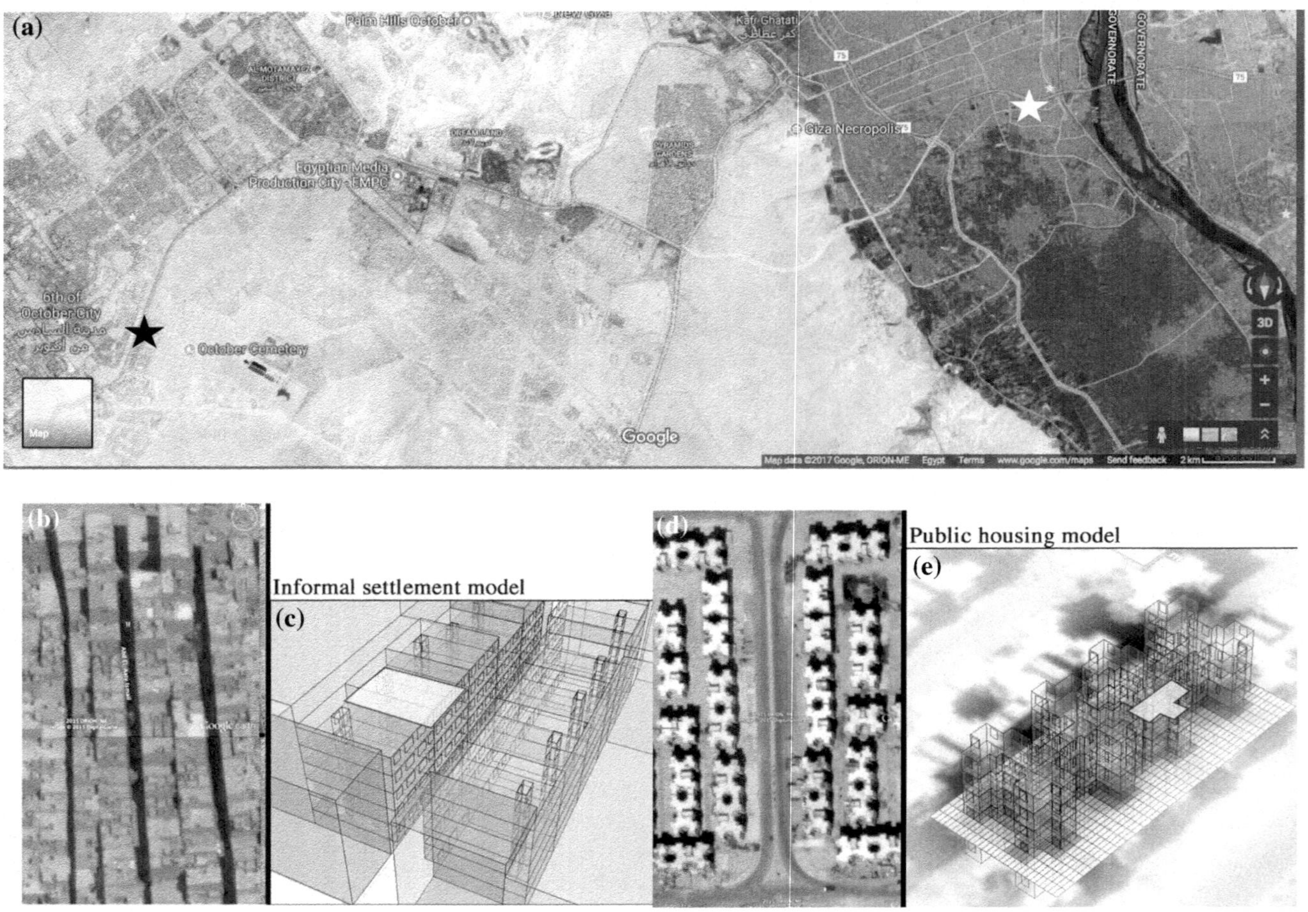

Fig. 4 **a** Figure shows the location of the informal settlement model sample (black star on the right), and the location of the selected public housing settlement model sample (white star on the left). (Google Maps 2017). **b** The first image in the figure represents the satellite image of the selected informal settlement in South-West Cairo (Google Earth 2017). **c** The second image represents the informal settlement Ecotect model (Author 2017). **d** The third image in the figure represents the satellite image of the selected public housing settlement in South-West Cairo (Google Earth 2017). **e** The fourth image represents the public housing Ecotect model (Author 2017)

levels of thermal heat gains and losses in informal settlements which result from their urban and architectural forms.

The most common type of sampling, random sampling, could not be used in its full definition to determine the essence of the specified population and to provide an equal chance of selection for all members as stated by Marshall (1996), but rather a pilot sample which was randomly selected from the selected informal settlement area to determine—as much as possible—the nature of the population at hand. This was done by asking residents of the area if they wished to participate in the research and if they knew any of their neighbours who wished to participate as well, those who came forth were given an application with general information questions and specific questions of whether they had any previous experiences of living in a public housing settlement or currently visit a relative there regularly. Those who replied in the affirmative were selected as part of the sample.

The main focus on the residents sample range was on the variety of respondent backgrounds within the cluster selected rather than on the significance of the number selected. The qualitative data collected was from two respondent groups, the first group were eight informal settlements residents of whom which all had experience with public housing schemes either directly, through friends or through relatives which in turn caused for consistent visits and tangible experiences with the settlements. The residents were of different ages, educational backgrounds, professions, genders, income levels and social backgrounds. The prioritisation of semi-structured interviews came from their capacity to provide a clear set of instructions for the interviewees allowing for reliable qualitative data through a combination of mostly open-ended and closed questions. Attitudes and subjective norms regarding comfort level responses were measured on a five point Likert-type scale ranging from "very good" to "very poor" and "very satisfactory" to "very unsatisfactory" through questions with normative statements which could not have been obtained by simple straightforward questions in a questionnaire as they needed to be asked many probe questions till they could come to a conclusion for appropriate ratings. The scale was used to measure how they see direct and indirect lighting, indoor ambient temperature and ventilation. The second group of respondents were six professionals from different professional backgrounds. They have all worked on developing and monitoring projects in informal settlements. This provided nuanced explanations for many areas that the simulation's quantitative data could not have addressed in entirety if it was done solely (see Tables 3 and 4 for the respondents' details).

Models for each of the informal settlements and public housing settlements were made. Only one flat was simulated for each model which was the flat on the third floor (the floor before the last) which had its main elevation with balconies and windows facing the east in each model (the flats are highlighted in white in both Fig. 4c, e). The models were created in full to provide accurate simulation results based on the surrounding dimensions, spaces and locations of solids or voids. Both of the samples selected had the main street between the buildings directed towards the north orientation. The length of the street sampled was 50 m in both models. The informal settlement model had blocks adhering to the simulated buildings, these blocks were not rendered but only acted as solid blocks adhering to the targeted

Table 3 Following table lists and describes the informal settlement residents that were interviewed with information on each individual's specific background

Respondent number	Gender	Age	Profession	Additional comments
1	Female	35	Garments retailer—home based	3 children, pregnant, husband is handicapped
2	Female	60	Home cleaner—part time	3 grown children, 1 teenager deceased husband
3	Female	62	Home cleaner—part time	lives with husband, sister in law and only her husband receives pension
4	Male	32	Owns a building materials hardware store in downtown	His father owns the informal settlement building which his father brother and himself live in and also let
5	Male	42	Waiter in a downtown prestigious sporting club	Is preparing his daughter for marriage and has a young son
6	Male	21	Student at university	Lives with family
7	Male	19	Unemployed	Lives with family
8	Male	52	Carpenter and sub-contractor	Father of 3 and owner of a workshop

Table 4 Following table lists and describes the professionals that were interviewed from different fields and have had previous work experiences with informal settlements

Professional respondent number	Gender	Age	Profession	Year of experience in informal settlements
1	Female	52	NGO executive coordinator	8
2	Female	49	Public housing researcher and Senior lecturer at MSA University, in the Department of Architecture	8
3	Male	42	GIS researcher and lecturer at MSA University	3
4	Female	32	Lecture assistant and Ph.D. candidate on historic slims sustainable development	6
5	Male	25	Village contractor	2
6	Male	41	Lawyer, works mostly with informal settlement residents	10

informal building structures so as to provide accurate data on the comfort level outcomes since these blocks have an effect on fabric, solar and ventilation heat gains and losses but cannot be added to the total values obtained. The openings were designed based on an on-site visit of the sample areas. After the completion of the models, the software analysis provided data with the existing fabric, solar and ventilation heat gains and losses as well as the total discomfort hours. The mentioned results were analysed and dissected showing clearer figures of the obtained data results.

3.2 Results Analysis

See Fig. 5.

3.2.1 Direct Solar Gains

The direct solar gains represent the thermal gains through the building openings (Stoios et al. 2006). The solar gains from direct penetration of sunlight rays into the informal settlement model and public housing model were 41.50 and 10.50%, respectively (Fig. 6).

In the literature review, Ahmed (2013) stated that the narrow alleyways of informal settlements provide poor environmental conditions of which inadequate sunlight is a result. Surprisingly, the results do not coincide with this statement. In contrast, the residents stated that they receive enough sunlight. Two of the residents said they depend on the sunlight they receive in the street during their walks for their daily intake of direct sunlight exposure. The professionals denied these areas receiving even the smallest

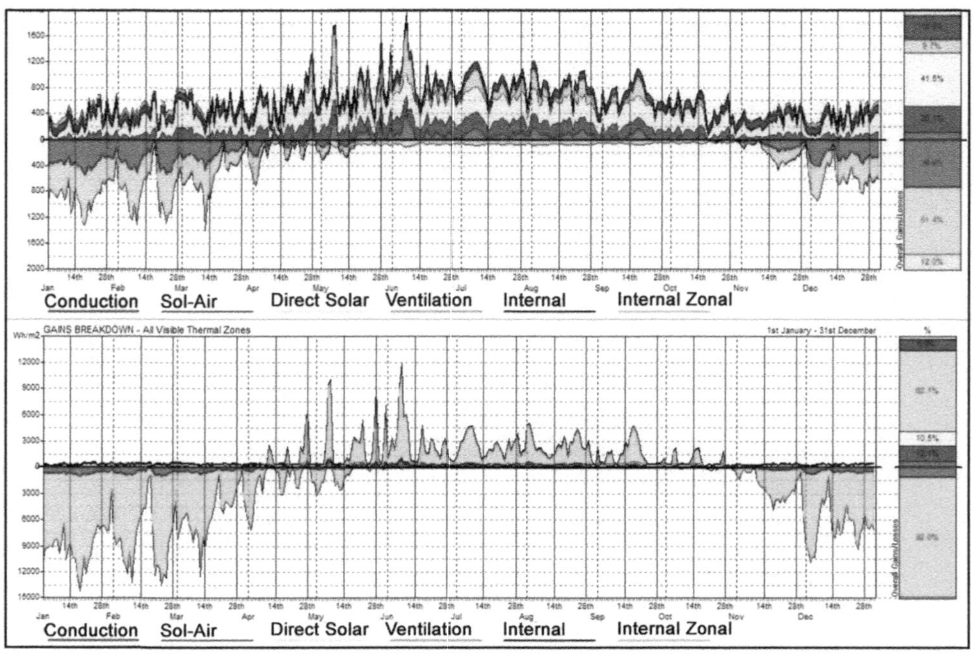

Fig. 5 Shows the outcome of all comfort level data in one graph as rendered by Ecotect for both models (informal settlement model at the top, public housing model at the bottom)

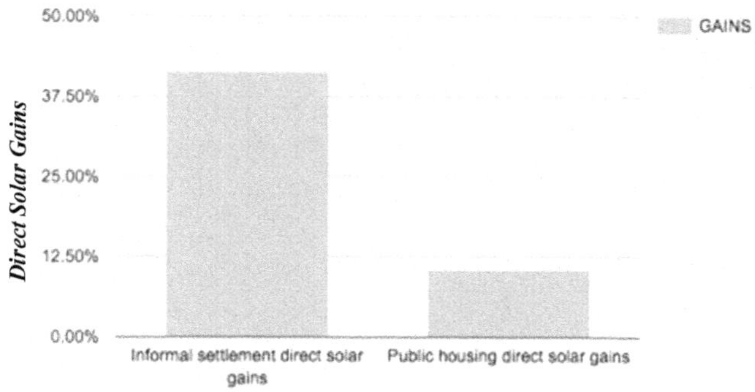

Fig. 6 Graph represents the direct solar gains in both the informal settlement model and public housing settlement model

amounts of adequate sunlight exposure and specifically mentioned this as one of informal settlements' main drawbacks. The interviewed NGO executive mentioned that, "One could not distinguish between noon and sunset times in many informal settlement regions due to the sparse amounts of direct and indirect sunlight". Although the respondents still did not strongly stress that the sunlight which penetrates and reaches their apartments directly is enough or better than the light received in public housing, it seems that the flat surfaces in the informal settlement model allowed for much more direct sunlight exposure in comparison to the public housing model which had many protruded elevations which most probably caused for the elevations to overshadow the building itself causing less chance for direct solar gain within public settlements (May 2008).

3.2.2 Fabric Gains and Sol-Air Gains

Fabric gains are the heat transfer through the building fabric. It helps to determine the heating and cooling loads and internal temperature variations of a building (Barry 2010). The informal settlement fabric gains, which were 5.60%, were almost at odds with the public housing's fabric gains of 4.80%. Unsurprisingly though, the losses in the informal model were of a much higher percentage, 36.5% compared to 7.20% in the public housing model. This brings the total losses of the building fabric to 30.90 and 2.40%, respectively (Fig. 7).

The literature did not mention the level of comfort within informal settlements in terms of the fabric gains which take place, however it did verify that the design of green corridors, relevant street orientation and good transportation planning are all essential for minimising the current heat gains in public housing projects (Rehan 2014). The residents that were interviewed on the other hand did not complain of the temperature of informal settlements and explained that their areas of residence had shops using water constantly thrice on a daily basis to cool the air temperature. Also, the constant shading effect due to the narrow street widths caused the air to be very cool and humid in informal settlements. This was confirmed by both the residents and professionals. The residents did not mention the heat of public housing settlements whatsoever. However, the professionals suggested that these places are better ventilated and are not humid due to their exposure to sunlight. The public housing researcher explained that, "The moist agricultural lands of informal settlements and there tightly woven urban fabric causes unhealthy humid climates when compared to public housing settlements which are full of sunlight and ventilation".

The reason fabric heat losses in the informal settlement model exist may be due to the high temperature differences between the alleyway and the buildings themselves due to high humidity levels, shading degrees and also due to the adhering structures next to informal settlements which absorb heat through conduction. Furthermore, public housing projects are characterised as areas of high temperatures, wide streets and large spaces between buildings which would make the existing clustering trends in informal settlements such as street water sprinkling and shading devices unproductive strategies for heat loss and cooling effect (Sham et al. 2012).

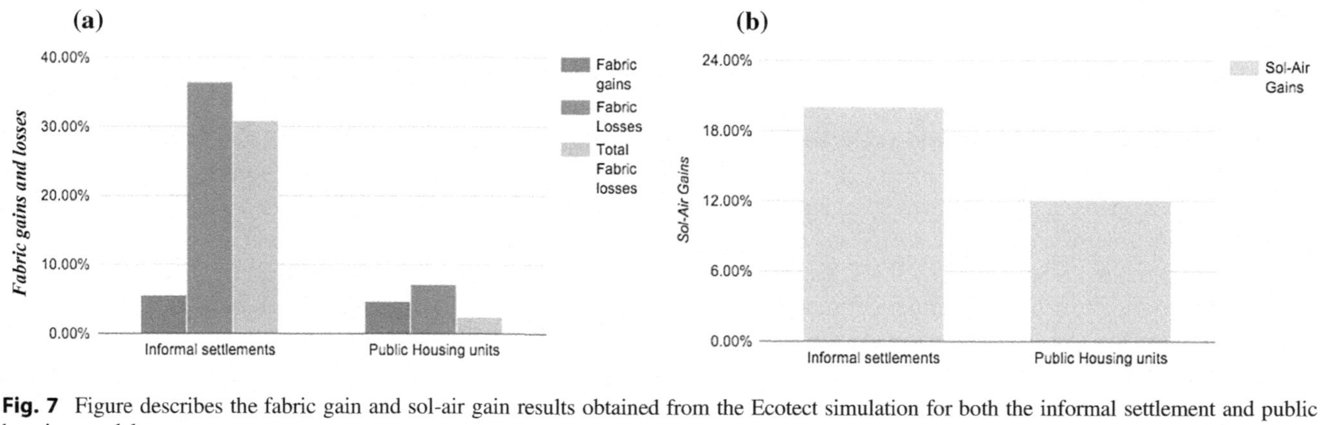

Fig. 7 Figure describes the fabric gain and sol-air gain results obtained from the Ecotect simulation for both the informal settlement and public housing models

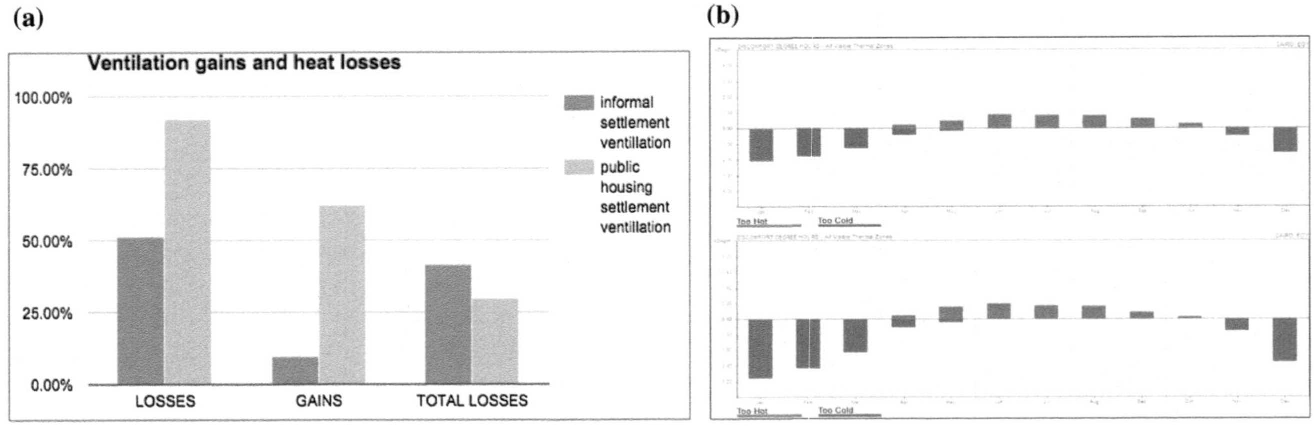

Fig. 8 **a** Figure represents the ventilation heat gains and losses in both of the models. **b** The monthly heating and cooling discomfort hours of both models, informal settlement model at the top and the public housing model on the bottom (*Note* The Ecotect software rendered the two graphs with different values on the *y*-axis, this was taken into consideration during the analysis by using the accurate values presented on the spreadsheet file exported from Ecotect)

3.2.3 Sol-Air Gains

Sol-air gains are temperature gains produced by indirect solar exposure of the building's exterior surfaces when exposed to solar radiation (Lin 2007). The gains found were almost double in informal settlements than those in public housing with percentages of 20.10 and 12.10%, respectively.

The results showing lower sol-air gains in public housing settlements denies the fact stated by Sayed et al. (2012) that these prototypes are designed without any regard to the comfort needs of their residents. They also challenge the statement that public housing structures provide feeble thermal comfort levels (Khalil et al. 2014). The respondents' answers reaffirmed the simulation's data outcome by stating that they face extreme heat in informal settlements especially during the day in the summer. However, they did not mention heat as a drawback of the apartments in public housing. These results suggest that the single flat elevation of buildings in informal settlements may be the cause of more heat induction despite less exposure to the sun which is in contrast to the public housing settlements' design of multiple elevations and outer wall protrusions being exposed to the sun. This is the main reason why public housing settlements prevent much of direct sun radiation incidence despite their multiple surfaces exposure (Fig. 8).

3.2.4 Ventilation Gains and Monthly Discomfort Loads

Ventilation heat gains and losses are the gains which take place through windows, doors as well as cracks in the building structure and vents (Taylor 2002). The total ventilation heat losses were higher in the informal settlement model compared to the public housing model with values of 41.70 and 29.90%, respectively.

These observed values justify the literature where Rehan (2014) suggests that South-West Cairo's public building prototypes need design development strategies in buildings to enhance the applications of passive design functions such as ventilation to reduce the dependence on air conditioning amongst the many other design guidelines she proposed. The respondents authenticated that the level of ventilation in informal settlements as either good or neutral in the past, although they rated it as poor for the current situation especially after the increase of average building heights which took place after the revolution of January 2011 in the absence of governmental surveillance. The respondents claimed abundant ventilation in public housing blocks due to the large spaces in front of the buildings; however, no claims of cooler temperatures in public housing were made. All of the professionals claimed these areas to be extremely humid and damp which further authenticates the public housing model's initial higher ventilation heat loss value but not its final value which gives informal settlements the greater level of heat loss through ventilation. The interpreted explanation for these results may be the lack of service courts in public housing settlements which act as a main reason for heat loss through cross ventilation in informal settlements. The single sided orientation of apartments in public housing settlements further prevents any type of adequate ventilation for heat loss to occur.

The following Monthly discomfort hour loads demonstrates the total over heating and over cooling hours for each zone in the form kdeg per hour within the building throughout the year which would in turn have an effect on the cooling and heating loads (Young 2015). The highest heat discomfort hour values in the informal settlement and public housing models were almost equal at a value of 0.9

and 1.0 kdeg/h in June, respectively. The highest cooling discomfort hours were 2.1 and 3.0 kdeg/h in January, respectively. The monthly heat discomfort hours were almost the same in both the informal settlement and public housing models throughout the year yet slightly higher in the informal settlement model with a minute monthly difference ranging between 0.05 to 0.1 kdeg/h.

The similarity between the heat degree hours in the informal settlement model and the public housing model fortifies the research by Khalil et al. (2014) and Rehan (2014) mentioned in the literature review which proved and stated that passive design is needed to enhance applications of passive ventilation to reduce the dependence on air conditioning. It was also mentioned by Sayed et al. (2012) that heat gain and thermal discomfort in public housing are the result of the government's lack of adequate design tools in the construction of new communities as was reaffirmed by the Ph.D. candidate working on historic slums sustainable development whom stated that, "Public housing projects are fixed prototype housing projects which are built repetitively regardless of the climatic conditions specific to each site". Contrary to expectations, these specific results did correlate with the residents' comparisons of public housing and informal settlement temperature difference descriptions. Four of the interviewed residents mentioned that their apartments were extremely hot and that numerous amounts of fans were needed to cool the temperature. On the other hand, nothing was mentioned in regards to public housing settlement flats other than respondent 8 stating from a health well being perspective that, "The strong airflow and heat of the sites where public housing units lie is useful for purifying the air and any contaminations that may take place in streets and apartments". The Public housing researcher, however, made clear that the overheating—and in some cases—over cooling of apartments is a result of the inadequacy of the building orientation in public housing blocks. Her research findings portray many housing projects as cookie-cut urban plans which are repeated constantly without consideration to the site's orientation or location. The rest of the professionals were concerned with the strong humidity and lack of sunlight in informal settlements and regarded public housing as more advantageous housing sites —regardless of their lack of proper orientation—with much less humidity levels as previously mentioned; however, they did not focus on the cooling and heating discomfort outcomes of each settlement type.

These results of higher heat discomfort hours in informal settlements and higher cooling discomfort hours in the winter might partly be explained by the previous findings that the overshadowing elevations of the public housing units allow for less direct heat gain resulting in less overall heat discomfort hours. However, public settlements being in the desert regions of Cairo where the temperature is relatively lower than in the city of Cairo itself or its fringes, causes much higher levels of radiant heat loss and in turn more cooling discomfort hours (Mohamed 2010). Moreover, the public housing units being exposed to the outdoors from all four sides cause much more heat loss through convection; this could explain the much higher overall cooling discomfort hours in the public housing model (Ministry of New and Renewable Energy 2015).

A depiction of the interpretations explaining why both the informal and public housing settlements had different outcomes in terms of thermal gains and losses is evident in Fig. 9.

4 Conclusions

This paper focused on the level of sustainability of informal settlements through studying comfort levels which exist due to their urban and architectural features with mention of their counter product public housing schemes. It is important to begin this conclusion by mentioning that the level of sustainability in informal settlements through comfort levels is not solely controlled by the residents over the variables they have control over, yet it is—to a large extent—affected by the obligatory situations they face such as enforced orientations for streets and main elevations, wall to wall construction preventing the provision of various ventilation strategies and the lack of proper intervention from professionals which has inhibited the possible improvement of these elements for better comfort levels. The Government's public housing schemes are much dreaded only for their outdoor spaces, clean environments and well prepared services as they do not offer the targeted informal residents the desired comfort levels that one would expect them to have. The results are still in an experimental stage which used a pilot survey and simulation analysis to understand the possible outcomes of the experiment.

The respondents varied substantially in their opinions on the architectural and urban features of both informal and public settlements presenting opposing views. The residents did not complain of cold winters nor did they complain much about the heat in the summer, they also testified to having adequate amounts of day light; however, they stressed that there are more complications when it comes to sunlight. Also, they made it clear that ventilation was not in abundance but also not absent, allowing for a placid breeze and for heat loss. In contrast, most of the professionals did not see any potential in informal settlements and viewed the relocation of informal settlement residents to public housing projects as "the best of whats available", for those who can afford it, but not the ideal solution. They explained that this is their opinion because they see the current imperfect public housing units as the only substitute

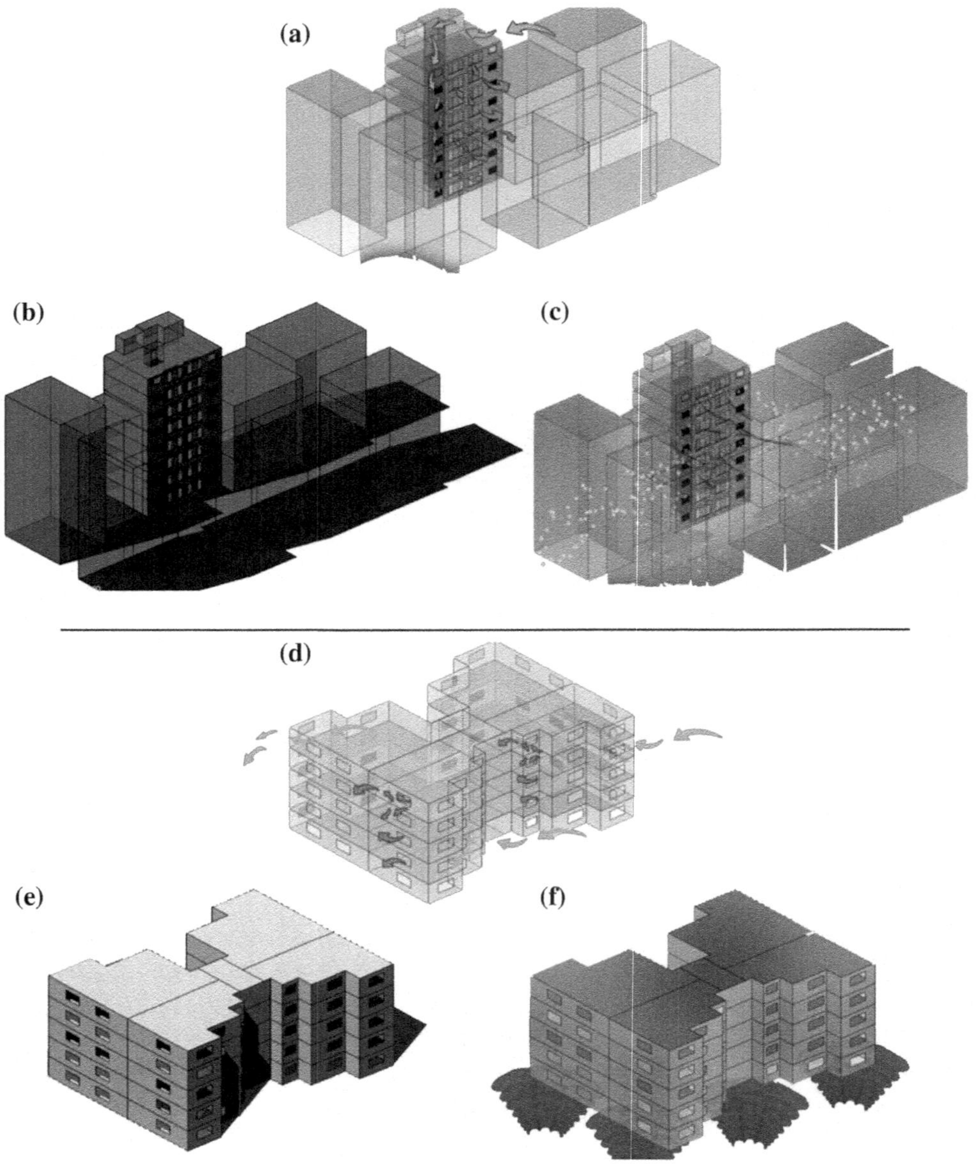

Fig. 9 **a** Model shows how the elevated shafts in high-rise informal settlement buildings can cause ventilation losses. **b** The model shows how the building has been overshadowed by neighbouring buildings and also how is has less exposure to solar gains due to it having one side of sun exposure instead of all four. **c** The rendering shows that the humidity and difference in temperature within the alleyways causes for heat loss through barometric pressure. **d** The model shows how public housing settlements lose or do not lose heat easily through ventilation depending on the orientation of the apartment. **e** The open spaces between housing blocks and protrusions of the block itself cause for over exposure to the sun as well as some shading effect. **f** The rendering shows how the heat radiance from the surrounding environment due to the lack of shading and improper design techniques causes the public housing settlement to gain more heat

to provide informal settlement residents with adequate thermal comfort and protection from health hazards. The evidence for the professionals' comments on the unsatisfactory status of public housing settlements arose from the lack of professional urban plans and layouts for proper fabric, solar and ventilation heat gains and losses as well as units which are not well designed for the best ventilation or sunlight exposure.

Surprisingly, the simulation tests performed by the Ecotect software have suggested informal settlements to be more efficient than public housing units in the level of thermal gains and losses. It is characterised with much higher direct and indirect solar thermal gains and better overall ventilation and fabric losses than public housing projects. The direct and indirect solar gains were extremely high in informal settlements almost doubling the public housing solar gain values due to the flat surfaces of the elevations which lack any protrusions, sun breakers or shaders. The final ventilation and fabric losses after deducting the initial gains were higher in informal settlements despite the higher initial ventilation losses in public housing settlements. This was due to many architectural factors such as the service shafts, unroofed staircases—in some cases—and of course the narrow and humid alleyways which allow for heat loss through convection as mentioned earlier. The presence of a very trivial but higher number of heat discomfort hours in the informal settlement model is a testimonial to their unintended efficiency in terms of their compact formations which prevent heat gains from multiple sides.

The results have also suggested public housing units to be of some degree of efficiency in their design and planning despite their distant locations and hot environments which is not at an optimal or satisfactory level of efficiency but present. Their protrusions have much to do with reducing their level of heat gains despite the fact that they are designed in a manner that obliges them to be surrounded by sunlight all day long from almost all four orientations increasing the likelihood of overheating. The inadequate loss of heat through ventilation is obvious in the lack of service shafts and units which lack any kind of exposure to the wind due to poor orientation. The overall cooling discomfort hours is a result of their imported urban plans which do not consider the site they are situated in which in this study was at a high altitude with cooler temperatures, exceptional sunlight and clearer skys. Despite these results, this does not negate the existence of some comfort level advantages in public housing blocks such as their wide streets and open areas plan typology which causes them to be clear of over humid, damp and shaded streets and instead are characterised with sufficient direct sun exposure through the wide spaces between the buildings preventing the strong probability of disease and providing for sufficient heat loss in the summer nights through the many exposed elevations.

There was not much information form the literature review in terms of the existing comfort levels within informal settlements which are produced form the architectural and urban elements in order to truly access the level of benefit which can be derived from informal settlements to be implemented in public housing settlements. Furthermore, proposals for devices that enhance or hinder certain comfort parameters were found, these researches study the enhancement of comfort levels that have to deal with areas such as lighting levels or ventilation gains and losses through refractive surfaces and modernised traditional wind catchers which is a great leap for research in informal settlements but does not provide information on the advantages of the current standing architectural and urban features which we can benefit from immediately in order to understand how can our public housing settlements be better tailored to the needs of Cairo's hot arid environment and as well to the resident's life styles. Moreover, this study was limited by the absence of more cross locational studies with various street orientations and similar building samples which would have provided greater details in the quantitative values obtained through the software renderings.

References

AbdelGelil, N. (2011). LESS SPACE, MORE SPATIALITY FOR LOW-INCOME HOUSING UNITS IN EGYPT: IDEAS FROM JAPAN. Archnet-IJAR, International Journal of Architectural Research, 5(2), pp. (24–48).

Abouelmagd, D., Kesteloot, C., & Corijn, E. (2013) Housing Projects for Low-Income Groups and Modes of Economic Integration: A Comparative Study in Greater Cairo. Tijdschrift voor economische en sociale geografie 104 (4):456–477.

Afify, A. (2004) Towards Stimulating Modern Urban Upgrading Policies For Informal Settlements In Egypt, In CIB World Building Congress.

Ahmed, (2013) Egypt's informal settlements suffer due to inadequate budget allocations, [online] Available at: https://www.youtube.com/watch?v=mOD9TGH74JQ (Accessed 7 May 2015).

Almaiyah, S., Elkadi, H., & Cook, M. (2010). Study on the Visual Performance of a Vascular Dwelling IN Egypt. First International Conference On Sustainability And The Future. Retrieved from http://csfs.bue.edu.eg/files/Library/Papers/Sustainability%20and%20the%20Future/106.pdf.

Barry, R. (2010) Sustainable Building Design with Autodesk Ecotect, Paris, Le SOMMER Environment, [online] Available at: https://www.divaportal.org/smash/get/diva2:433382/FULLTEXT01.pdf.

Elshahed, M. (2015) New Architectural Forms for the 'Informal', Cairobserver, [online] Available at: http://cairobserver.com/post/117458095714/new-architectural-forms-for-the-informal#.WGyEPrZ96Rs (Accessed 4 January 2017).

Gerlach, J., el-Jesri, M., Fink, M. and el-Sebai, N. (2009). Cairo's informal areas—between urban challenges and hidden potentials. Cairo: GTZ.

Gray, D. (2004) Doing research in the real world, London, Sage Publications.

Khalifa, M. (2011) Redefining slums in Egypt: Unplanned versus unsafe areas, Habitat International, 35(1), pp. 40–49.

Khalil, R., Hassan, A. and Saleh, H. (2014) Middle-income Public Housing in Egypt, Evaluate and Improve to Reach a Sustainable Design, ALAZHAR University, 9(33).

Kipper, R., Howeidy, A. and Wiens, C. (2009) Cairo's informal areas - between urban challenges and hidden potentials, Cairo, GTZ.

Lin, J. (2007) Introduction to Ecotect-Thermal Analysis, California, Cal Poly Pomona Department of Architecture.

Marshall, M. (1996) Sampling for qualitative research, Family Practice, 13(6).

Mohamed, M. (2010) Traditional Ways of Dealing with Climate in Egypt, In The Seventh International Conference of Sustainable Architecture and Urban Development, The Centre for the Study of Architecture in Arab Region, pp. 247–266, [online] Available at: https://www.academia.edu/582673/Traditional_Ways_of_Dealing_with_Climate_in_Egypt.

Massey, R. (2014) Exploring counter-conduct in upgraded informal settlements: The case of women residents in Makhaza and New Rest (Cape Town), South Africa, Habitat International, 44, pp. 290–296.

May, N. (2008) OVERSHADOWING ANALYSIS, 1st ed, Bedforshire, Flow Analysis Limited, p. 8,[online]Available at: http://www.crawley.gov.uk/pub_livx/groups/operational/.

Ministry of New and Renewable Energy, (2015) Thermal performance of buildings, India, Ministry of New and Renewable Energy.

Mostafa, M. (2013) In Nagati, O. and Stryker, B. (ed.), Learning from Cairo: Global Perspectives and Future Visions, 1st ed, Cairo, CLUSTER.

Rehan, R. (2014). Cool city as a sustainable example of heat island management case study of the coolest city in the world. HBRC Journal, 12(2), pp.191–204.

Rydin, Y., Bleahu, A., Davies, M., Dávila, J., Friel, S., De Grandis, G., Groce, N., Hallal, P., Hamilton, I., Howden-Chapman, P., Lai, K., Lim, C., Martins, J., Osrin, D., Ridley, I., Scott, I., Taylor, M., Wilkinson, P. and Wilson, J. (2012). Shaping cities for health: complexity and the planning of urban environments in the 21st century. The Lancet, 379(9831), pp. 2079–2108.

Sayed H., A., Hiroshi, Y., Abdelsamei Eid, M. and M. Radwan, M. (2012) Indoor Natural Ventilation Using Evaporating Cooling Strategies in the Egyptian Housing: A Review and New Approach, IJET, 4(3), pp. 229–233.

Serageldin, M. (2017) The Architecture of Housing: The Development and Morphology of Informal Housing, 1st ed, p. 2.

Sham, J., Lo, T. and Memon, S. (2012) Verification and application of continuous surface temperature monitoring technique for investigation of nocturnal sensible heat release characteristics by building fabrics, Energy and Buildings, 53, pp. 108–116.

Shehayeb (2009) Kipper, ed. (2009) Cairo's informal areas - between urban challenges and hidden potentials, Cairo, GTZ. pp 44–49.

Shehayeb, D., Sabry S., Hafez, Y. (2011). Community Oriented Activity Patterns—Background Study for the Urban Regeneration of Historic Cairo Project - UNESCO. https://doi.org/10.13140/RG.2.1.3114.3285.

Sims, D. and Abu-Lughod, J. (2010). Understanding Cairo. Cairo: The American University in Cairo Press.

Stoios, A., Bougiatioti, F. and Oikonomou, A. (2006) Thermal performance of a passive solar house for continental climate, in Florina, north-western Greece, In The 23rd Conference on Passive and Low Energy Architecture, Geneva, Department of Architectural Technology, School of Architecture.

Taylor, N. (2002) Energy Efficiency for Everyone: Analysis and development of an Energy Home efficient project, BE, The University of Western Australia.

Yasser, F. (2017) Renderings of the informal and public settlement models made for simulation in Ecotect to test the comfort levels in each area.

Young, S. (2015) Ecotect, 1st ed, p. 1.

Monitoring Thermal Bridges by Infrared Thermography

Egemen Kaymaz

Abstract

In some sections of the building envelope, either structural or formal reasons, or different thermal characteristics of components, higher heat transfer occurs lowering the overall thermal resistance of the building envelope and increasing the heating and cooling energy expenditure. Depending on the location and size of thermal bridges, the overall heat transfer coefficient of the building envelope increases while energy efficiency, user comfort and indoor air quality are negatively affected. This paper focuses on thermal performance assessment of an existing building by in situ measurements of its external envelope using passive IR thermography. A series of facade inspections were conducted particularly considering the interrelations of external walls, floor slabs, roof, RC framework and wall openings which is partly presented here. IR thermography imaging results including thermal bridges are briefly explained and discussed qualitatively through gathered information based on architectural features of the case building and conditions of the surrounding environment together with the technical data collected during the construction phase of the building.

Keywords

Building envelope • Thermal bridge • Insulation • Infrared thermography

1 Introduction

Today, the excessive energy consumption, resource depletion, environmental pollution and climate change put pressure on construction sector to consider energy efficiency throughout the lifecycle of buildings. Yet, advances in building science and technology have introduced sustainable design approaches and solutions that offer a significant potential to face the global challenges. This is also supported by energy policies and binding standards. According to the EU's Construction Products Directive (1988) and the Energy Performance of Buildings Directive (2010), building and its parts must be energy efficient and fulfil performance requirements over an economically reasonable working life. Thermal performance of building envelope is mainly associated with energy efficiency in terms of long-term performance, service life prediction and durability of buildings and its parts, in addition to user comfort and indoor air quality issues.

The user comfort depends on thermal performance of building regarding heat losses and gains through building envelope. According to Turkish Guidelines for heat insulation in buildings standard, thermal comfort is defined as the least energy of the human body uses to adapt himself to the environmental conditions (TS825 2013). Indoor temperature of 19–22 °C and relative humidity of 55–65% are defined as necessary conditions to maintain optimal comfort in buildings. In addition to constant indoor ambient temperature, the difference between the minimum interior surface temperature and indoor air temperature is required to be no greater than 3 °C and maintain homogeneous throughout the year. The standard also specifies limited energy consumption and the acceptable U-values of building assemblies for different climate zones in Turkey.

As energy conservation is the leading issue of our time, heating and cooling energy consumption is required to be low in almost all types of buildings. Heat energy flows through the building enclosure via opaque and transparent components by conduction, convection and radiation. Researches maintain that the main energy losses through building envelope occur at external walls, roofs and windows. To raise overall thermal resistance of the entire building envelope, heat transfer and temperature distribution through the building assemblies should be controlled by applying insulation as well as preventing thermal bridges,

E. Kaymaz (✉)
Faculty of Architecture, Uludag University, Bursa, Turkey
e-mail: ekaymaz@uludag.edu.tr

C. Alalouch et al. (eds.), *Advanced Studies in Energy Efficiency and Built Environment for Developing Countries*,
Advances in Science, Technology & Innovation, https://doi.org/10.1007/978-3-030-10856-4_11

unconditioned air leaks, moisture and dampness. Thermal performance properties expected from a building envelope can be regarded as: (i) lower heat loss during heating period, (ii) lower heat gain during cooling period, (iii) capability of heat storage, (iv) lack of air infiltration, (v) keeping inner surface temperatures of required levels and last but not least (vi) lack of thermal bridges.

1.1 Thermal Bridge

In some sections of the building envelope, either structural (mainly in beams, columns, shear walls and lintels) or geometric reasons (corners and junctions of walls and floors), or different thermal characteristics of components (window frames, door sills, etc.), higher heat transfer occurs lowering the overall thermal resistance of the envelope. These types of thermally critical elements are referred as "thermal bridges". High heat flow takes place across the thermal bridges and it increases the overall heat transfer coefficient (U-value) of components and creates condensation risk indoors (Trechsel 2001). Depending on the location and size of thermal bridges, air leakage through the building assemblies may negatively affect the indoor conditions. Air flow transports moisture from either indoor or outdoor environment into the structure. This may cause a significant amount of heat loss, mould growth, deterioration of indoor air quality and defects in the building itself (Wrobel and Kisilewicz 2008). During the heating season, the impact of thermal bridges generally results in high surface temperatures from outside and low surface temperatures from inside in IR images (Fouad and Richter 2006).

1.2 Literature Review

IR thermography is a non-destructive testing method that can measure radiated heat from a material's surface. Infrared radiation emitted from a surface of an object is directly related to its temperature. Thermal patterns appear in colour hues or grey shades on thermal images generally correspond to heat energy transfer. Infrared radiation detected by the camera's micro-detectors is rendered by the camera's image processor and converted into a pixilated image called "thermogram". Each colour-based pixel on camera display is associated with a temperature level. Heat radiated from the target surface is graphically displayed and a thermal profile is obtained by rating the contrasts of the image.

The origin of IR thermography goes back to the early nineteenth century when William Herschel discovered the invisible light rays later called infrared. Later in the mid-1960s, IR thermography became a method of surface temperature mapping (Schneider 2007). Today, IR technology in building monitoring is increasingly utilized for various engineering and architectural applications. It is commonly used for the detection of building defects, such as thermal bridging, air infiltration, hidden condensation, potential moisture ingress, lack of insulation and unbalanced operation of HVAC systems.

As part of literature review, studies particularly consisting of IR thermography method and external envelope inspections are reviewed. Balaras and Argiriou (2002) gave an overview of infrared imaging technology for inspecting building envelope. They provided information on performing non-destructive testing for building diagnostics related to moisture accumulation, air leakage and structural thermal bridging along with HVAC installations problems and electrical and mechanical service systems. Barreira and Freitas (2005) conducted both the laboratory and in situ works on sensibility analysis to investigate the influence of emissivity, environmental conditions, colour and reflectivity along with wetting and drying processes on thermographic measurements. Avdelidis and Moropoulou (2002) studied the emissivity measurement techniques on various building materials, such as plasters, marbles, and porous stones in laboratory at a variety of temperatures, wavelength and surface conditions. Meola et al. (2004) aimed to evaluate building degradations with three different techniques: IR thermography, ultrasonics and electric-type geophysical methods. Investigations were carried out in laboratory on intentionally defected marble, brick and tuff specimens. In situ IR thermography was safely used to inspect the status of the mosaic tiles covering external wall of the building in concern and to detect the presence of detachments. Li et al. (2000) studied the thermographic principle of debonded building-finish system based on laboratory calibrations using test specimens contained some simulated defects. Field study was performed early in the morning after it had rained for several days and moisture was trapped in the defective wall parts. Surface temperature of the defective areas was found to be lower than that in the sound areas. The higher temperature regions were suspected as debonded areas. Choi et al. (2006) developed a quantitative evaluation method for facade insulation based on mathematical models in accordance with the IR thermal images. Temperature and heat flow pattern analysis were conducted in situ for different buildings. An insulation performance evaluation index was proposed a long with laboratory experiments conducted using IR camera, heat conduction sensor and a simulation program prior to field study. Insulation performance for the side walls of apartments were then evaluated by temperature difference ratio between defected and defect-free sections. Ribaric et al. (2009) described a knowledge-based system for the non-destructive diagnostics of facade insulation. Two images of the same building facade were taken at the same time, from nearly the same viewpoints by IR and digital

Table 1 Factors affecting data collection and interpretation by IR camera

Surface-related factors	Optical properties of surface such as texture, smoothness, absorptivity, reflectivity and permeability, emissivity (ε), heat capacity, moisture content of the material and façade orientation
Environmental factors	Solar radiation, cloudiness, precipitation, ambient temperature, pressure, relative humidity of the indoor and outdoor environment, wind speed and direction, atmospheric reflections and extraneous heat sources
Equipment related factors	Distance and angle between camera and target surface, camera temperature and colour scale settings, resolution and image recording speed

cameras, respectively. Gathered data on the temperature and the type of structural components were then transformed into a histogram using the graphical user interface system. Other studies including Martín Ocaña et al. (2004), Molin et al. (2011), Cerdeira et al. (2011), Fox et al. (2014), Kylili et al. (2014) and Cho et al. (2015) have presented the benefits of thermography to facilitate the existing buildings energy diagnostic practices and support improvements in building energy efficiency.

Thermal distributions captured from building surfaces both from inside and outside are directly influenced by thermal bridges caused by faulty constructions or deteriorations. These abnormal temperature variations adversely affect the durability of construction materials and building's energy performance as well as comfort level of occupants. Considering that the primary source of energy used in buildings in Turkey still depends on fossil fuels, analysing thermal bridges have significant potential to reduce excessive energy consumptions for space heating and cooling purposes. The main concern of this study is the thermal performance assessment of an existing building envelope by qualitative analyses of temperature data obtained by passive IR thermography. A series of facade inspections were conducted considering the interrelations of external walls with roof, floors, reinforced concrete (RC) framework and wall openings. The instant surface temperature differences for non-insulated and insulated building envelope configurations were determined in situ and identified thermal bridges in terms of IR monitoring results are partly presented here.

2 Material and Method

Building thermography is an effective research and evaluation method to assess thermal performance of building envelope. In this study, an IR camera is used to visualize the thermal bridges and generate thermal patterns of exposed surfaces. Thermal images are recorded using a Flir Systems InfraCAM model IR camera and photographs are recorded with a digital camera regarding the same capturing angle. Indoor and outdoor ambient temperature and relative humidity are measured with a simple handheld device. A laser metre is used to measure the distance between camera and the target surface.

There are a number of factors that complicate the collection and interpretation of thermographic data. Some of them are characteristics of all thermographic surveys, whereas some of them are specific to building investigations. Location and orientation of examined facades concerning temperature and pressure difference between indoor and outdoor environment, the prevailing wind direction and speed, emissivity values of surface finishing materials, intensity of precipitation, relative humidity in atmosphere, cloudiness—clearness of sky, extraneous heat sources, etc. may affect the IR results and distort the thermal images (Fouad and Richter 2006). Therefore, instant data related to the indoor and outdoor environmental conditions at the time and date of field work should be measured and documented. Technical specifications and factors affecting data collection and interpretation by IR camera are summarized in Table 1.

3 Case Study

In this section, in-use thermal performance of an apartment block is studied considering the interrelations of external envelope components. A series of inspections on different scales are conducted through building enclosure, from a smaller scale to larger with a holistic approach. In this context, external wall system components which mainly comprise of RC skeleton structural frame (column, beam, shear wall and floor) with infill masonry walls (masonry unit, mortar, thermal insulation and surface finishes) and transparent sub-components (frame, glazing, window sill and joint sealants) were examined from inside and outside of the building. All obtained and recorded data, including digital and thermal images, plans and technical drawings collected and documented systematically on a site recording template to make proper assessments and interpretations.

In order to avoid misinterpretation and complication in thermographic data, IR measurements were planned on appropriate dates and time of the day in the winter period. Thermographic surveys were particularly conducted after sunset on heavily overcast days to prevent direct solar radiation. In addition, the temperature difference between indoor and outdoor environment was considered to be at least 10 °C and the interiors were heated. Before the field study, IR camera is calibrated and emissivity was set to a constant value of 0.93 for each measurement cycle. In the cases where an obstacle existed and could not be removed during the survey, (for example an inhomogeneous material, deposition of dust, highly reflective surface or other foreign matters that affect the material's surface emissivity) obstacles were taken into account evaluating the thermographic data.

The architectural features of the building were analysed from the existing drawings of application project and from the pictures taken during the construction phase. For each facade examined, the orientation of the studied surface, the approximate distance between the target point and the camera were marked on related floor plans and section drawings. Instant surface temperatures, ambient air temperature and relative humidity values were recorded and these data are arranged in a table format. IR images are combined with photographic image pairs captured from nearly the same position. In the case study part, the assembled images are presented to demonstrate thermal bridges. Further interpretation of overall assessment results are made in the conclusions.

Bursa, the fourth biggest city in Turkey, is located on the south-east coast of the Marmara Sea. According to Köppen-Geiger classification, Bursa locates in the Mediterranean climate zone. TS825 standard divides Turkey into four climatic zones depending on average temperature degree-days of heating. Bursa is in the second zone and the city's temperate climate is characterized with warm summers and mild winters. The prevailing winds are northeaster and southwester.

The investigated apartment complex is established 11 km away from downtown, in Nilufer County. As seen in Fig. 1, the total area of the gated community project is 7000 m^2 and it is surrounded by an industrial zone and residential districts. The occupants have access to site by urban transportation network.

The selected building is one of the five apartment blocks of same typology, completed in 2007 with RC structural framework on raft foundation. The project is located on the Bursa-İzmir motorway on an inclined topography. There is approximately 25 m distance between each apartment block which is adequate for receiving solar gain for each flat. The main entrance of the buildings is oriented to north, facing to the main road as seen in Fig. 2.

Examined apartment block has a basement, ground and four upper floors, including duplex penthouse apartments on the top floor. There exist two independent flats on each floor with four bedrooms and a living room. The total area of a flat is approximately 185 m^2. The complex and its facilities are naturally ventilated; each flat is separately heated by natural gas boiler and AC system optionally installed to the rooms.

Fig. 1 Aerial photograph of the case building

Fig. 2 Case building as viewed from the north façade

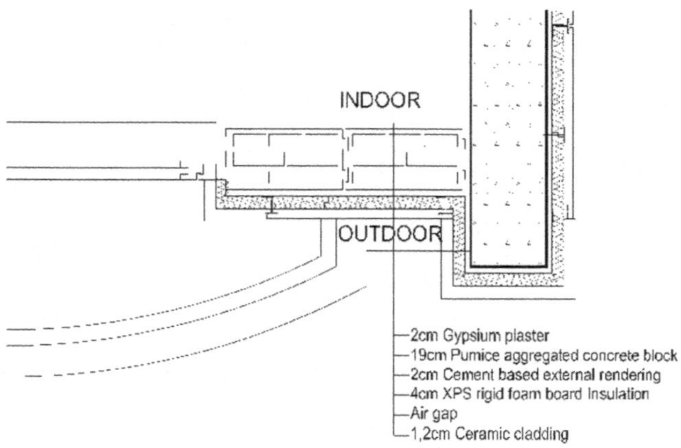

Fig. 3 Detail from external wall of the case building

The building's external wall system is illustrated in Fig. 3. External walls were built with pumice aggregated concrete blocks (PACB) in 19 cm width, with tongue and groove sides and having three-rows of hollows. Wall courses are bonded with ~10-mm cement bedding mortar, 15 cm behind the shear walls. The external envelope is insulated with 40-mm XPS rigid foam board from outside. The façade cladding is designed as curtain wall. 12-mm granite ceramic stone panels are carried by aluminium-framed stick system and attached to the building structure. Between the insulation material and the cladding, 20-mm air barrier exists that functions primarily as a vent between the exterior and adjacent wall components. Perimeter flashings are used to improve watertight performance of the curtain wall. Water that may penetrate curtain wall system is drained out of from the weep holes to the exterior. Transparent elements of façade comprises of operable PVC window and door frame, double-pane low-e insulating glazing units, marble sills and exterior blinds. 320-mm-hollow block slabs are used within the RC concrete floor.

The residential site has been visited and thermographic measurements have been performed repeatedly during the heating season. One of the apartment blocks with the highest temperature variations is selected and partly presented here. Image in Fig. 4 demonstrates the fourth floor plan of the examined building. Inspection points where thermal bridges are observed on the building envelope are marked on the plan in Fig. 4.

4 Results and Discussion

For a better understanding and more accurate interpretation of IR images exhibiting hot and cold spots, photographs, drawings and supporting measurement data are presented together in Table 2. Thermograms were taken from the outside and inside of case building, from bedroom 1 (IR Cam 1), bedroom 2 (IR Cam 2) and living room (IR Cam 3 and IR Cam 4), respectively. The difference between indoor and outdoor ambient temperature is measured as ~13 °C Table 3.

IR images in Table 2 demonstrate the examples of structural and geometric thermal bridges. Geometric thermal bridges occur at junctions, corners, inner and outer reveals around windows. Structural thermal bridges occur due to the U-value of adjoining RC building elements and discontinuity of the thermal insulation layer at roof eaves and floor slabs. Lighter colours indicate the higher external surface temperatures, which mean an increased rate of heat transfer occurs particularly in this area. While the recorded instant surface temperature of cladding panels (~10.8 °C) and outdoor temperature (~10.5 °C) are close, the outside surface temperature on these crucial points are ~4 °C higher compared to the adjacent surface temperatures. Irregularities in the thermal performance of building envelope result in increased energy loss during the heating season.

The balcony in the first IR image is built continuous with the interior RC floor slab; therefore, it creates a thermal bridge between interior and exterior space. To maintain the thermal insulation property of the envelope, additional

Fig. 4 Inspection points of the case building, fourth floor plan

measures can be taken such as insulating the floor slab from above and below, using a thermal break element or constructing a separate structure to build balconies. Heat transfer also occurs at the junctions of wall, intermediate floor and window frames as in the second and third IR image. IR image 4 represents that the wall insulation material is interrupted by eaves in the existing pitched roof detail. It is crucial to meet attic insulation with external wall insulation, especially along the roof overhang.

5 Conclusions

This research paper aims to put emphasise on building envelope details to reduce or better eliminate thermal bridges. An existing building envelope is investigated qualitatively and non-destructively using an IR camera. This tool enables to detect temperature variations emitted by surfaces and transforms them into a false-colour image. According to the colour gradient map, unintended heat flow through building enclosure can be identified and thus this qualitative IR thermography method can be practically used on site in thermal bridge diagnostics.

For the case building, the external wall system's thermal performance is a function of cladding, air gap, the aluminium frame of curtain wall, thermal insulation, PACB blocks, RC frame structure and the flashing details. The use of double-pane low-e glass coatings for glazing is positive from the energy saving and user comfort perspective. Thermal insulation was applied externally which is an important parameter to eliminate thermal bridges. Proper placement of insulation at the curtain wall perimeter also helps to reduce energy loss and potential condensation risks. On the other hand, it was observed that the continuity of the insulation was disrupted at some intersections of the envelope system. Sub-components of RC skeleton structural system have higher thermal conductance than the rest of the majority of opaque wall components due to high thermal conductivity of concrete. Since there is no thermal insulation regulating the transfer of heat by conduction, architectural

Table 2 Details, IR images and photographs of external envelope corresponding to the thermal bridges, respectively (1—balcony, 2—intermediate floor slab, 3—wall and 4—roof detail)

<table>
<tr><th>Inspection points</th><th>Thermal image</th><th>Image</th></tr>
<tr><td>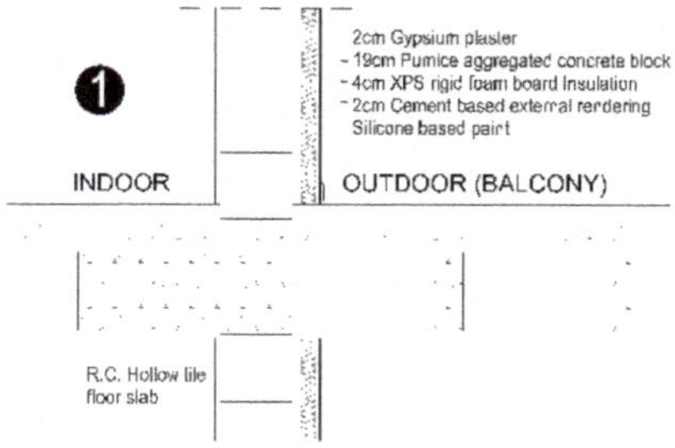</td><td>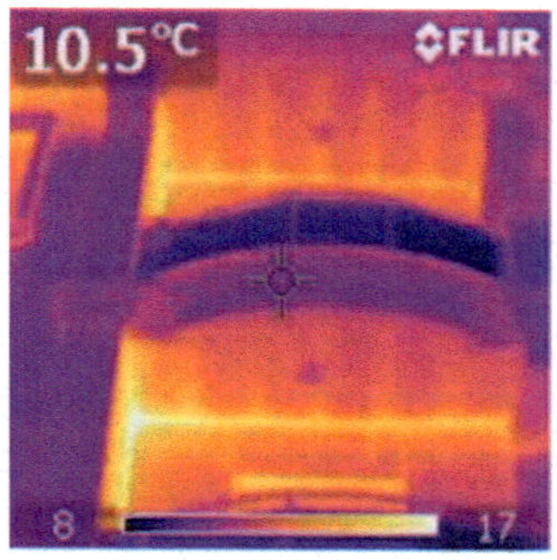</td><td></td></tr>
<tr><td>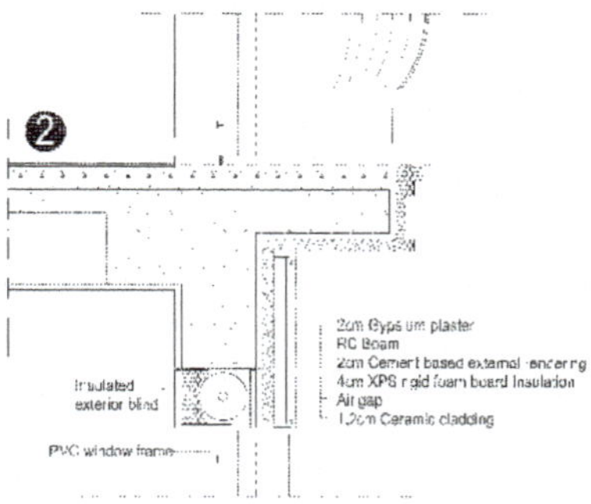</td><td>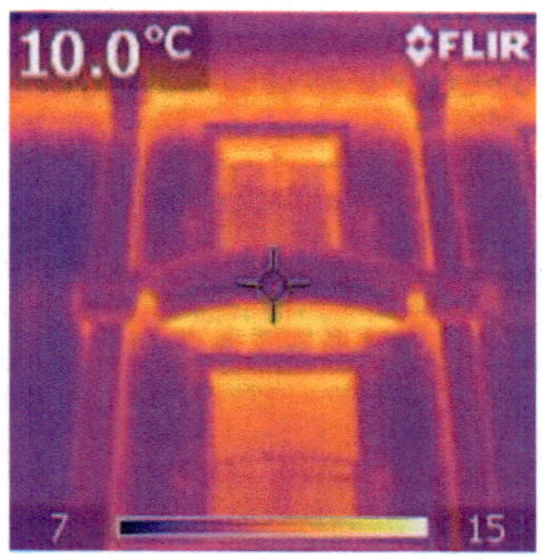</td><td></td></tr>
<tr><td>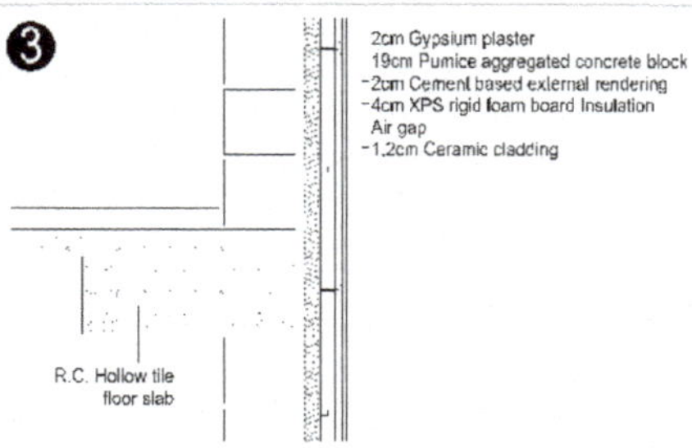</td><td>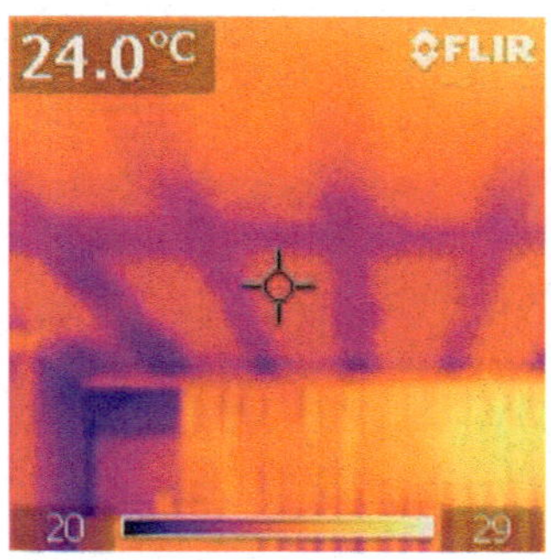</td><td></td></tr>
<tr><td>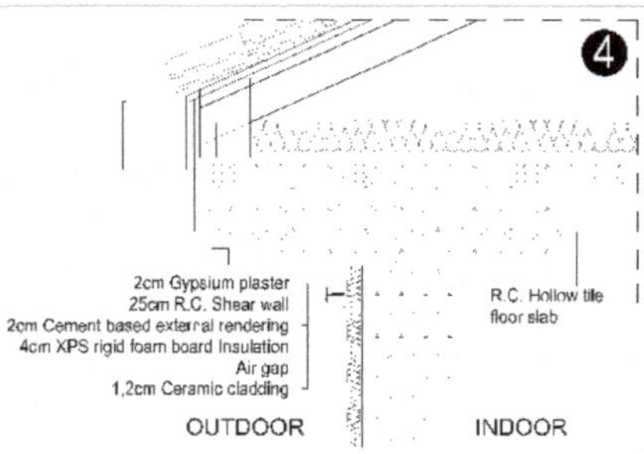</td><td>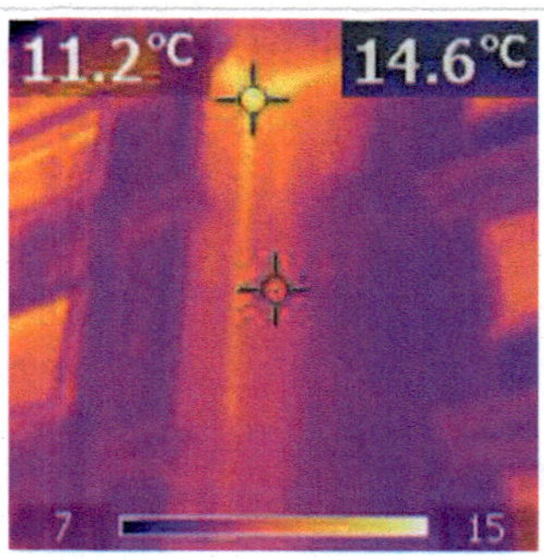</td><td></td></tr>
</table>

Table 3 Measurement data for the case building's thermographic inspection points

Measurement data	Ambient temperature		Relative humidity		Surface temperature		IR recording distance	
	Indoor (°C)	Outdoor (°C)	Indoor (%)	Outdoor (%)	Indoor (°C)	Outdoor (°C)	Indoor (m)	Outdoor (m)
IR camera 1 Bedroom 1	23.3	10.7	49	51	20.8–23.1	9.5–13.9	3.5	14
IR camera 2 Bedroom 2	24.6	10.4	65	51	22–23.6	10.2–15	4.2	13.4
IR camera 3–4 Living Room	23.8	10.5	55	51	22.5–24	10.0–14.6	4.5	12–16.6

details like eaves and floor slabs as in this case require special attention to achieve a thermal bridge free building envelope.

Both new and existing buildings need to be inspected to assess the actual thermal performance of building envelope; since, as-is thermal performance of a building may have different hygrothermal characteristics than what the architect intended during the design and construction process, and this may have been particularly changed in the long term by environmental factors. Considering existing building stock in Turkey, thermal performance of the case building can be regarded as positive. Eventhough, the indoor temperature of all rooms is above the recommended optimal comfort level (19–22 °C), the difference between the minimum surface temperatures and the interior temperature was measured less than 3 °C in accordance with TS825 standard (2013).

This study encourages the consideration of thermal bridges. Nonetheless, a building's thermal performance is not only limited to instant IR measurements of the building envelope. Further comprehensive and quantitative studies can be performed by additional simulation tools and theoretical-calculation methods to check whether the examined building meets the local regulations and standard requirements (TS EN 1745 2012; TS EN ISO 6946 2009) and to show the impact of thermal bridges on annual heating and cooling energy consumptions (BEP 2011). The research can be extended by calculating and comparing the actual and expected thermal transmittance value of external envelope configurations (U-value, W/m^2K) using the temperature data obtained by IR thermography.

References

Avdelidisa, N.P., Moropoulou A., 2002. Emissivity Considerations in Building Thermography, Materials Research Centre, University of Bath, UK, Section of Materials Science and Engineering, School of Chemical Engineering, National Technical University of Athens, Greece.

Balaras, C. A., Argiriou, A.A., 2002. Infrared Thermography for Building Diagnostics, Group Energy Conservation, Institute for Environmental Research and Sustainable Development, National Observatory of Athens, Greece.

Barreira, E., Freitas, V., 2005. Evaluation of Building Materials Using Infrared Thermography, Faculdade de Engenharia da Universidade do Porto, Portugal.

BEP, 2011. Energy Performance of Building Regulation. Turkish Official Gazette, 27911.

Cerdeira, F., Vázquez, M. E., Collazo, J., and Granada, E., 2011. Applicability of infrared thermography to the study of the behaviour of stone panels as building envelopes. Energy and Buildings, 43(8), 1845–1851.

Cho, Y. K., Ham, Y., and Golparvar-Fard, M., 2015. 3D As-is Building Energy Modeling and Diagnostics: A Review of the State-of-the-Art. Advanced Engineering Informatics, 29(2), 184–195.

Choi, G-S., Kang, J-S., Jeong, Y-S., Lee, S-E., 2006. Development of an On-site Evaluation Method for Building Thermal Performance Using an Infrared Camera, Korea Institute of Construction Technology.

Construction Products Directive (CPD), Council Directive 89/106/EEC of 21 December 1988, European Commission. Accessed on 07.06.2017. https://ec.europa.eu/growth/single-market/european-standards/harmonised-standards/construction-products_en.

Energy Performance of Buildings Directive (EPBD), Directive 2010/31/EU of the European Parliament and of the Council of 19 May 2010 on the energy performance of buildings. Accessed on 07.06.2017. http://ec.europa.eu/energy/efficiency/buildings/buildings_en.htm.

Fouad, N. A. ve Richter, T., 2006. Leitfaden Thermografie im Bauwesen; Theorie, Anwendungsgebiete, praktische Umsetzung. Fraunhofer IRB Verlag.

Fox, M., Coley, D., Goodhew, S., de Wilde, P., 2014. Thermography methodologies for detecting energy related building defects. Renewable and Sustainable Energy Reviews, 40(0), 296–310.

Kylili, A., Fokaides, P. A., Christou, P., and Kalogirou, S. A., 2014. Infrared thermography (IRT) applications for building diagnostics: A review. Applied Energy, 134(0), 531–549.

Li, Z., Yao W., Lee S., Lee C., Yang Z., 2000. Application of Infrared Thermography Technique in Building Finish Evaluation, Journal of Nondestructive Evaluation, Vol. 19, No. 1.

Martín Ocaña, S., Cañas Guerrero, I., González Requena, I., 2004. Thermographic survey of two rural buildings in Spain. Energy and Buildings, 36(6), 515–523.

Meola, C., Maio, R., Roberti, N., Carlomagno, G. M., 2004. Application of Infrared Thermography and Geophysical Methods for Defect Detection in Architectural Structures, Dipartimento di Energetica, Dipartimento di Scienze Fisiche, Dipartimento di Scienze della Terra, Universita` di Napoli Federico II, Napoli, Italy.

Molin, A., Rohdin, P., and Moshfegh, B., 2011. Investigation of energy performance of newly built low-energy buildings in Sweden. Energy and Buildings, 43(10), 2822–2831.

Ribaric, S., Marcˇetic, D., Vedrina, D. S., 2009. A Knowledge-Based System for the Non-Destructive Diagnostics of Facade Isolation Using the Information Fusion of Visual and IR Images, Faculty of Electrical Engineering and Computing & Faculty of Chemical Engineering and Technology University of Zagreb, Croatia.

Schneider, M. (2007). The Importance of IR- Thermography in Building Science and the Conservation of Energy in the Residential Sector Infrared Imaging Windsor, Ontario.

Trechsel, H., 2001. Moisture Analysis and Condensation Control in Building Envelopes, ASTM Manuel Series, USA.

TS825, 2013. Thermal Insulation Requirements in Buildings. Turkish Standards Institution. Ankara, Turkey.

TS EN 1745, 2012. Masonry and Masonry Products – Methods for Determining Desing Thermal Values. Turkish Standards Institution. Ankara, Turkey.

TS EN ISO 6946, 2009. Building Components and Building Elements – Thermal Resistance and Thermal Transmittance Calculation Method. Turkish Standards Institution. Ankara,

Wróbel, A. ve Kisilewicz, T., 2008. Detection of thermal bridges - aims, possibilities and conditions, AGH University of Science and Technology & Cracow University of Technology, Kraków, Poland.

Net Zero Energy Buildings (NZEBs) Potential in MENA Region: Critical Review on Egypt Case

Khaled Ismail, Mohamed Hamdy, and Ahmed Maher

Abstract

The aim of this paper is to assess the gaps and needs for net-zero energy buildings (NZEBS) design and implementations in MENA Region, particularly in Egypt. The paper reviews current government efforts and regulations on energy efficiency in buildings, the academic efforts in developing NZEBs concept, as well as challenges and barriers in building design phases. For illustration, the paper summarized study undertaken to analyze the potential challenges and opportunities for implement (NZEBs) in Egypt as an example of Mena region. Two case studies in Mena region E-JUST campus in Egypt and MASDAR City in UAE had been analyzed. The review and case studies show a lack of energy performance in Egyptian buildings code and optimization calculation methods, as well as limited numbers of academic work for NZEBs which studied the Egyptian case. It is concluded that the current building codes and laws need to be upgraded to include the energy performance of buildings requirements, a database for buildings materials need to be developed with studies to the cost optimal for different buildings type in Egypt, one the challenges of the NZEBs in is the vernacular environment and enhancing the implementation procedures.

Keywords

Zero-energy buildings • Zero-energy neighborhood • Energy efficiency • Gaps Scale

K. Ismail (✉)
Department of Architecture, Faculty of Fine Arts, Minia University, Minia, Egypt
e-mail: khaled.ismail@mu.edu.eg

M. Hamdy
Department of Civil and Environmental Engineering, NTNU Norwegian University of Science and Technology, Trondheim, Norway

M. Hamdy · A. Maher
Department of Mechanical Power Engineering, Helwan University, Cairo, Egypt

Nomenclature

NZEBs	Net zero energy buildings
RCREEE	The regional center for renewable energy and energy efficiency
MENA	Middle East and North Africa
GDP	Gross domestic product
AC	Air-conditioning
EE	Energy efficiency
UNDP	United Nations development program
MOEE	Ministry of electricity and energy
EEHC	The Egyptian electricity holdings companies
EPBD/EU	The recast energy performance of buildings directive
EKB	Egyptian knowledge bank
BPS	Building performance simulation
BIPV	Building integrated with photovoltaic
CO-EPL	Energy performance level
DPB	Discounted payback time
SMAA	Multicriteria acceptability analysis
RETs	Renewable energy technologies
HVAC	Heating ventilation and air conditioning
NZEN	Net zero energy neighborhood
E-just	Egyptian Japanese university
PRT	Personal rapid transit
NEAPE	National energy action plan of Egypt
EDEM	Egyptian domestic energy model

1 Introduction

According to the Regional Center for Renewable Energy and Energy Efficiency (RCREEE), the energy consumption and electricity demand in the Middle East and North Africa (MENA) countries significantly increased in the last three decades. Especially, in the last decade, the primary energy consumption and electricity demand have increased at an annual rate of 8% compared with the previous decades.

C. Alalouch et al. (eds.), *Advanced Studies in Energy Efficiency and Built Environment for Developing Countries*, Advances in Science, Technology & Innovation, https://doi.org/10.1007/978-3-030-10856-4_12

While the increase in the Gross domestic product of energy consumption was about 4%. Which mean, there is an annual gap between production and consumption about 4% (RCREEE 2015).

In Egypt, the primary energy consumption fuel mix of Egypt in 2016 was dominated by natural gas (51%), oil products (44.5%), coal (0.5%), hydroelectricity (3.5%), and renewable energy measurable impact (less than 1%), as shown in (Fig. 1), (BP Statistical Review of World Energy 2017). On the other hand, it is expected that Egypt will reform the primary energy consumption pattern by 2035, with reducing the large reliance on oil and natural gas to 62%, and increase the usage of renewable energy, coal, and nuclear energy to 14, 19, and 5%, respectively (Anhar 2017) (Fig. 2).

The consumption of electricity accounts for 21% of total primary energy consumption. The residential sector represents the largest consumption of the total use of electricity share (51.3%), followed by industry (28.7%), government and public utilities like military camps and headquarters, police stations, public educational and healthcare facilities, sewage system (11.3%), and the other shared load distributed between commercial shop, agriculture activities, public lighting, and others (MED-ENEC 2017). Thus, the residential and government buildings and urban structure in Egypt consume the largest of total electricity generated.

Moreover, energy costs are highly subsidized in Egypt. In 2013, the energy subsidies cost was approximately about 8.5% of GDP. However, In July 2014, the Egyptian government introduced a range of subsidies reforms through increases in official prices of energy in order to significantly reduce the energy subsidies (Bankaudi 2016).

The trend for high energy demand in Egypt is expected to continue over the next decade, especially in the building sector due to high population growth and significant urbanization (UN-Habitat 2014). Indeed, the annual urban population growth rates in Egypt range between 36 and 40% over the last five decades and with expected growth average of 42% in 2035. While the Egyptian population growth rate expanded at 2% per year (Tradingeconomics 2017).

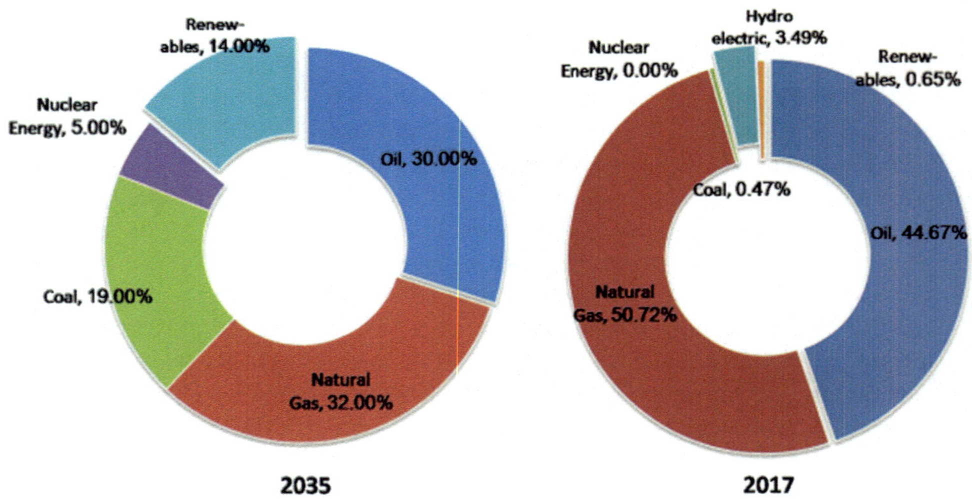

Fig. 1 Energy consumption by fuel in Egypt from 2017 till 2035

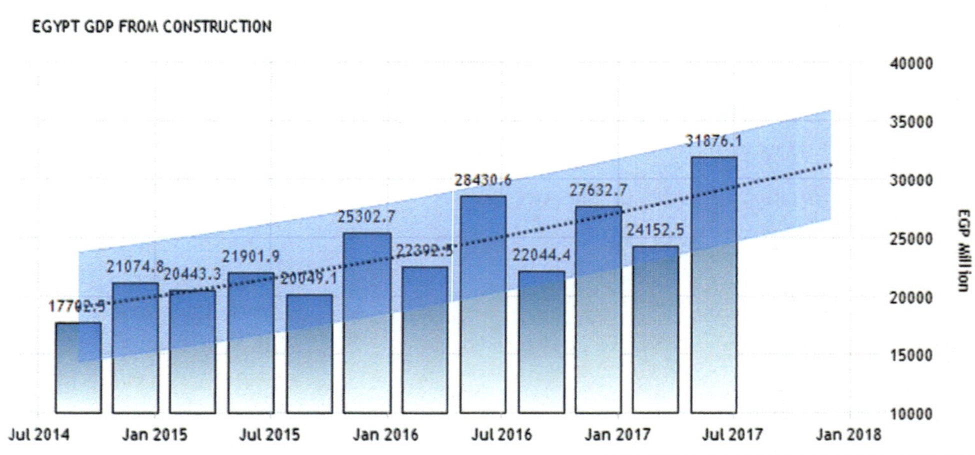

Fig. 2 Egypt GDP growth from construction from 2014 till 2018 *Source* Tradingeconomics (2017)

Furthermore, due to the urban heat island effect (UHI) and indoor high temperature, and the lack of passive buildings design solutions such as insulation of buildings envelope, the buildings sector has seen a noticeable increase in the usage of electricity-consuming cooling and refrigeration devices. For example, the use of air-conditioning (AC) units has increased severely from 1999 to 2012, the number of AC units rose from 196,000 in 1999 to three million in 2009 and then doubled to six million units by 2012 (Raslan and Mavrogianni 2013).

Moreover, Egyptian construction and buildings sector accounted for 4.8% of Egypt's GDP in 2015, with increased growth rate by 3.7–9.7% from 2011 to 2015 (Bankaudi 2016). On the other hand, Egyptian ministry of housing estimated that 500,000 new homes need to be built every year for the next five years to keep pace with a population expanding rates.

Therefore, achieving the highest energy efficiency in buildings is the most effective approach to reduce energy consumption in the building sector for a country with high construction activities as the case in Egypt. Indeed, the average life of most structural types of Egyptian buildings is fifty years (Karmany 2016). Thus, the impact of energy savings associated with energy-efficient buildings can last for numerous years, resulting in a considerable reduction in national energy consumption and greenhouse emissions.

The objective of the study is to review the ingredients that can help the NZEBs industry, the government, and decision makers in Egypt. Practically, the research aims to; (1) frame the current Egyptian energy policies, regulations, and energy codes situations in the perspective of energy efficiency and renewable energy market, (2) follow the energy efficiency and NZEBs research trends in Egypt, (3) assess the NZEBs challenges and barriers for Egypt, (4) analyze two case studies (E-JUST and MASDAR CITY) as an example for NZEBs in MENA region.

In this paper, net-zero energy buildings (NZEBs) potential is investigated for Egypt. Particular, the paper organized into three main sections: The first sections spot on the research problem with the Egyptian consumptions pattern and energy efficiency community, accordingly the literature review of the previous studies, relates the energy efficiency in buildings issues in Egypt. The second section presents a review of the current Egyptian strategic energy plan and the energy efficiency and renewable energy policy formations. In the third sections, two case studies (E-JUST and MASDAR City) had investigated and analyzed, as an example of NZEBs in Mena region. Then, discussing the findings and providing feedback, and a framework to help NZEBs experts, decision, and policymakers had been done.

2 Literature Review

In this section, the literature review aims to investigate the published academic works which illustrate the current Egyptian situation in NZEBs or energy efficiency in buildings, the challenges, and potentials for implementing and design NZEBs in Egypt for the related studies with the Egyptian situation review. It used a method comprises of three steps: (1) selection of research publications (2) selection of relevant publications (3) assessment of contributions.

Firstly, in order to identify academic researchers that published "NZEBs in Egypt"-related papers, a powerful search engine, Scopus, was used. Words like net zero energy buildings (NZEBs), energy efficient, energy performance of buildings had utilized as research keywords, by several authors research paper, in the (Task 40, IEA SHC, n.d.) and had been used in this study as the search keywords. After that, a pilot desktop search was performed under the "All field" field of Scopus, and with document type "article or review or Book or conference paper or book chapter or conference review."

Secondly, on the initial search results, a total of 75 papers were found valid for further analysis on NZEBs, only 20 papers are focused on Egypt case or have a case study in Egypt in their study, and the other research papers were relating to similar cases to Egypt such as a countries in Mena regions or developments countries like India or China, and relevant papers which was done by Egyptian researchers for analysis the NZEBs optimization in EU countries. On the other hand, the challenge for NZEBs in Egypt had been investigated through review recommendations of global experiences.

Thirdly, depending on the above selection processes, the next section showed the previous academic work relates to the Egyptian situation and NZEBs challenges.

It had been found that the academic work effort focused in four main sections; (a) Retrofitting scenarios and energy efficiency techniques for buildings in Egypt, (b) The simulation tools development, Cost optimal and multi-aid optimization for buildings, (c) Developing the sustainability concept in Egypt, (d) Net zero energy buildings (NZEBs) challenges in MENA region.

2.1 Retrofitting and Energy Efficiency Techniques for Buildings in Egypt

Early on 2003, Farghaly (2003) discussed the different strategies and techniques related to energy efficiency in buildings, as well as, a brief to highly performance buildings characteristics and its contemporary techniques and strategies, public awareness toward energy efficiency in Egypt, the best way for establishing a kind of score or rating systems for energy efficiency in buildings, had been taken over. While Attia (2010) selected a case study of a chalet in Ain-sukhna on the red sea coastline of Egypt, for study the zero-energy retrofit, to investigate the potential of achieving thermal comfort and delivering electrical demands for existing buildings on site, the study proofed that some retrofitting are cost-effectively rewarding with a payback period ranging from 2 to 7 years.

Moreover, integrated passive design solutions grounded in Egypt conditions had investigated by several authors, (El-Zafarany et al. 2013; Abdallah et al. 2014; Dabaieh and Elbably 2015; Eldabosy et al. 2015), also an evaluation of fenestration specifications through Egyptian residential energy code had been done (Mahdy and Nikolopoulou 2014) in terms of energy consumption and long-term cost-effectiveness. It recommends different combinations of window wall ratios, glass types and associated shading devices according to three climatic zones in Egypt.

Furthermore, as a reflect of photovoltaic (PV) manufacturing growth and development, building integrated with photovoltaic (BIPV) had been studied by several researchers' (Khatib et al. 2013; Kares and Singh 2016; Koo et al. 2016; Pinto et al. 2016; Tripathy et al. 2016, 2017; Youssef et al. 2016; Abdelhady et al. 2017; Gindi et al. 2017; Wang et al. 2016; Khan et al. 2017) through different methodologies and climatic zones in Egypt, Mena region, India, Brazil and Shanghai and had been reviewed in this study. An insight into the possible opportunities of applying solar energy in existing buildings had been given by (Samir and Ali 2017), based on analysis of case studies from Egypt which introduced photovoltaic in roofs, facades, skylights, and solar shades, it can produce approximately 40% of the building needs of electricity with benefit of grid connection.

The energy retrofitting with different buildings categories had investigated through several authors (Attia 2010; Dabaieh and Elbably 2015; Mauro et al. 2015; Kares and Singh 2016; Dabaieh et al. 2016; Gindi et al. 2017; Ascione et al. 2017; Dadzie et al. 2017). An integrated renewable energy supplies, refurbishment of buildings envelope, and energy efficiency considerations, the multi-level of optimization concerns costs, incentives, indoor comfort, energy demand for heating and cooling and novel approaches, had been performed for choosing the best configurations of retrofit, according to the case study conditions.

2.2 Simulation Tools Development, Cost Optimal and Multi-Aid Optimization for Buildings

In addition, a group of researchers (Attia et al. 2011) published an overview of using building performance simulation (BPS) tools among the building design professionals in Egypt, three workshops were held in Cairo, the study found a lack of interest in energy efficiency and indoor environmental quality among project developers, academic and professional educational programs, and a thermo-physical properties of typical Egyptian building materials and constructions are not available in digital databases, the study recommends that promote R & D and incentive programs in the building industry and enforce the energy standard and/or provide incentives for code compliance, indoor environmental quality and building energy efficiency need to be considered by the government. In the same time, BPS tools need to be developed.

Serval publication (Attia and De Herde 2011, Attia et al. 2012, 2013) discussed and developed design decision tools (BPS), and benchmark model for residential buildings in Egypt, as well as, ZEBO tools for design decision making, had developed and evaluated for zero-energy residential buildings in hot climates. It had been tested on several case studies in Egypt.

Respectively, a cost optimal and multi-aid optimization of buildings energy studies toward NZEBs had been investigated with different case studies and scenarios (Mauro et al. 2015; Hamdy et al. 2016, 2017; Hamdy and Sirén 2016; Hamdy and Mauro 2017), the impact of several financial parameters on the cost-optimal energy performance level (CO-EPL) had been analyzed, a large number of design options, the buildings lifetime and discounted payback time (DPB), had submitted. Some results give the CO-EPL ranges of the single-family house in Finland, with economic optimized scenarios for 4608 cases and ($\sim 3 * 109$ million) design options, as well as it enables the stakeholders to choice among optimal solutions according to different varieties.

2.3 Developing the Sustainability Concept in Egypt

The sustainability cities and green buildings trend as a strategic approaches relates to NZEBS issues for developing countries had been covered through (Dadzie et al. 2017; Darko and Chan 2016; Fahmy et al. 2017; Ghosh and Goswami 2014; Hassan and Lee 2015a, b), it shows that development countries like Egypt and China made a good efforts to promote a sustainable buildings and efficient design solutions, various sustainable development problems in cities had been discussed. Besides that, the social acceptance of renewable energy technologies (RETs) for buildings had been measures by (Jung et al. 2016), through integrated survey data and stochastic multi-criteria acceptability analysis (SMAA). It confirmed that finish residents exhibit board acceptance of multi-options RETs rather than preference for a single RET, in addition to, solar technologies and ground source heat pumps were the most preferred options.

2.4 Net Zero Energy Buildings (NZEBs) Challenges

On the other hand, net-zero energy buildings (NZEBs) trend gained a widespread acceptance as a technically feasible long-term goal in architecture implementation over the last two decade. It depends on achieving the highest energy efficiency by the exploit of renewable energies at the site. While reducing energy consumption by using passive design methods. It recruits solar energy on-site in different ways, such as enhance the daylight systems of buildings, improve the passive heating and cooling systems, and generate electricity by photovoltaic panels or other hybrid thermal systems (Voss and Musall 2013).

Furthermore, according to Athienitis and O'Brien (2015), in "Modeling, Design, and Optimization of Net-zero Energy Buildings" book, Implementations and Designing of zero-energy buildings systems had numerous challenges; an integrated and optimized design and operation of buildings need to be considered in the future NZEBs, with a satisfying comfort, and integrated design solutions of the following four buildings subsystems: (a) building fabric and envelope (b) heating, ventilation and air conditioning (HVAC) (c) solar systems/renewable, generation (d) buildings automation systems.

On the other hand, the transformation challenge from net-zero energy buildings (NZEBS) scale to net-zero energy neighborhood (NZEN) scale concept had been discussed by Marique et al. (n.d.). It is described as, "in which the annual energy consumption for buildings (or settlement) and transportation of inhabitants are balanced by the local production of renewable energy." While the development of a comprehensive, cost-effective modular system for net-zero energy neighborhood (NZEN) needs to be considered. A study by Synnefa et al. (2017) showed that the analysis conducted during this first design and optimization phase performed at building and settlement scale are able to achieve the target of having a net-regulated energy use ≤ 20 kWh/m^2 per year and renewable energy production of ≥ 50 kWh/m^2 per year and therefore be considered as zero-energy settlements.

Finally, the academic work review showed that several researchers have positive results, toward NZEBs in Egypt. Studying energy performance of buildings, energy retrofitting, and BIPV had significantly decrease the energy consumption for existing buildings. Developing BPS tools for Egypt case studies helped in expected consumption and evaluated zero-energy buildings. Nevertheless, an Egyptian building materials and constructions digital databases need to develop. The cost-optimal and optimization of buildings energy and social acceptance of renewable energy technologies depending on Egypt scenario need to be investigated in future research and the appropriate challenges to implement NZEBs in Egypt. The transformation challenge from net-zero energy buildings (NZEBS) scale to net-zero energy neighborhood (NZEN) scale concept needs to be discussed, with consideration of transportation, location, and economy.

3 Current Policies and Strategies Towards Net Zero Energy Buildings in Egypt

NZEBs received increasing attention in recent years, up to inclusion in strategic energy policy papers in several countries. In 2010, in the EU energy performance of buildings directive (Recast 2010) had been issued, it gives a platform for EU member to discuss the energy efficiency in buildings and formulate relevant definitions at the building code level more intensive.

In a developing country with limited resources and highly economic challenges like Egypt, the trend to rationalize energy and improve the energy efficiency in buildings will have a tremendous impact on national development. The following review shows the Egyptian practice, toward energy efficiency in buildings, the national energy strategic plan, the regulations framework (current policies and energy codes), the energy efficiency in buildings equipment (lighting and appliances), which formulate the bases toward NZEBs in Egypt.

3.1 Long-Term Strategic Orientation with Energy Efficiency in Egypt

In 2016, the Sustainable Development Strategy of Egypt "Egypt's Vision 2030", were adopted by the Egyptian government based on three dimensions of development; "Economic, Social, and Environmental" (SDS 2016). Moreover, the national energy action plan of Egypt 2030 (NEAPE 2030) had significantly clarified on this strategy. However, the NEAPE (2030) had not investigated an action plan to the energy performance of buildings or near zero-energy buildings (NZEBs). It had two key provisions related to energy efficiency (EE) issues:

- **"Raise up the renewable energy usage and reduce greenhouse gas (CO_2) emissions"**.

There is an action plans to expand the renewable energy capacity in Egypt to 25% of total electricity production by 2022. Moreover, it is expected to reach 30% (excluded hydro-elect and nuclear power) by 2030 and reduce the (CO_2) emissions to 5 and 10% during the next ten years (SDS 2016).

– **"Improve energy efficiency"**

The objective of the action strategy is to improve the demand side efficacy by publish a five-year energy efficiency plan with targets, adopt efficiency programs for energy-intensive industry, enforce appliance efficiency standards and building codes, introduce appropriate incentives for energy efficiency finance, phase out incandescent bulbs and rollout 10 + 50 million LED lamps, approve a plan to roll out smart meters within 5 years, conduct awareness-raising campaign (Mohamed Shaker 2015).

3.2 Egyptian Energy Efficiency Policy Framework

In line with NEAPE (2030), the new electricity law had been issued (Decree No. 87/2015) in 2015, to establish a fully competitive electricity market. Moreover, it had introduced several incentives to renewable energy private developers and commercial producers. And it had been confirmed that the newest company must have an annual action plan to manage the electricity demand in the Egyptian market, enhance the energy efficacy usage and awareness, increase the reliance on renewable energy. Furthermore, the energy efficiency standards and testing label must be set for every produced and imported electric appliance (article 49/50/51).

According to Khalil (2012), the Egyptian mandatory standards and labeling program achieve a decrease by 10–20% in refrigerator and air-conditioner energy consumption. Moreover, standards and labeling programs provide enormous energy savings potential that can direct developing countries like Egypt, towards sustainable energy use in buildings.

On the other hand, Egypt had Unified Construction Law 119 for the year 2008, which concerning urban planning, coordination, and organization of cultural construction and the preservation of real estate wealth. It gives the requirements for maximum building height in the Egyptian cities and windows-to-wall ratio (WWR), sold to void ratio, and the courts' dimensions and height. It puts the minimum passive design needs.

Egypt had several significant ingredients in energy efficiency, indeed, during the last three decades. The electricity and renewable energy laws had been restructured, as we mentioned. The energy efficiency codes of electricity, buildings, lightings had issued. As well as, the implementation of projects, to increase energy efficiency in lightings and appliances, had done.

In this subsection, the Egyptian regulations and codes related to the energy efficiency and renewable energy, as well as the Egyptian plans for energy efficiency in lightings and appliances, had clarified as follows.

3.3 Energy Efficiency and Buildings Codes

– **ECP 306-2005 The Egyptian code for energy efficiency improvement in buildings Part 1 and Part 2**

It is applicable to residential buildings and commercial buildings. Moreover, it covers the requirements related to energy efficiency in buildings through passive and active strategies. For illustration, it introduces the minimum energy performance requirements for buildings elements (such as building envelope, natural and mechanical ventilation systems, daylight requires, industrial lightings, electro-mechanics systems), the requirements of simulated energy analysis, and total energy consuming report must be issued for buildings. However, the cost-optimal energy performance and optimization requirements for the Egyptian buildings had not exposed and also the details of energy performance of buildings certificate had not clarified.

– **The Egyptian code for ventilation in buildings 2013**

This code is applicable to all residential and non-residential buildings, excluded the educational buildings, hospital, special buildings, and military buildings. It covers the requirements related to natural and mechanical ventilation in Egyptian buildings.

– **ECP 602 Egyptian code for housing design and planning**

This code includes a set of criteria for residential buildings and site design. It is considered as an essential part of Building Law (No. 119 for 2008) in Egypt. The code contains two main part, first one had investigated the minimum requirements for residential unit and buildings design (Minimum room dimensions and areas of residential unit elements), with satisfy of customs and traditions in Egypt. The second one had set the residential communities design standards, the code gives the minimum performance for exterior urban spaces, human behaviors and needs, land use density measurements.

3.4 Energy Efficiency in Buildings Equipment

Parallel with the regulations and codes formulation and issued, Egypt had an ambitious plan to improve the energy efficiency in lightings and appliances. In 2002, a ministerial decree for enforcement of the standards and labeling program for the refrigerators, freezers, washing machines and air conditioners was issued, and it had been confirmed in the newest electricity law in 2015 (Decree No. 87/2015) as a must. Furthermore, a test and calibration center for energy efficiency for appliances and renewable systems had been established under supervision of the Egyptian renewable energy organization.

On the other hand, a multi-phases projects relate to energy efficiency improvement, greenhouse gas reduction, and increase the public awareness of energy issues had been released, it executed by the Egyptian electricity holdings companies (EEHC), ministry of electricity and energy (MOEE), and supported by a grant from the global environment facility with the United Nations Development Program (UNDP) (Van den Akker 2010). Moreover, the MOEE leads a big public campaign under the title of "Enhance the energy efficiency in lightings and Home Appliances", with the support of international and national organizations and NGOs. The campaign objective is to increase the energy efficiency public awareness and encourage the public authorities to use efficient indoor lighting systems and appliances. It supported many public buildings to replace their lighting systems to an efficient one (such LED systems). Moreover, they had indicators Web site, to meter and announce their processes, and it showed more than 30% reduction in the annual electricity consumption based on building occupancy and site circumstances.

Furthermore, tracking efficient light systems in Egypt continued for twenty years as a long-term goal. Attia et al. (2017) had published a research to measure the effect of replacing new energy-efficient lighting in residences to conserve energy and reduce energy demand, during two decades of measures, and reported for 150 residential apartments, it showed that the insulation of an efficient lighting systems, like fluorescent and compact fluorescent in Egypt led to 35–40% reduction in lighting power consumption and the revealed consumption had been changed.

From the above discussion, the review generally shows that the opportunity for energy efficiency in buildings and NZEBS in Egypt significantly increased in the last decades and had been regulated with a framed policy and standards which can make good incentives to the market and future investment, towards NZEBs in Egypt. However, it had been shown an enforcement gap in the current Egypt situation. Firstly, in the Egypt vision 2035, the detailed time frame of development progress toward energy efficiency was undefined, and the energy performance of buildings or NZEBs is not considered clearly in this plan. Moreover, the current buildings energy codes give the minimum energy performance requirements for buildings elements. Besides that, the cost-optimal and optimization methods need to be considered after feed-in traffic regulation issued.

3.5 Towards Closing the Gap Lessons from EU Directive

"The recast Energy Performance of Buildings Directive (EPBD/EU) (Recast 2010) stands as an important milestone for building policies, requiring all European Member States to; (a) introduce minimum energy performance requirements for buildings, building elements, and technical building systems, (b) set these requirements based on a cost-optimal methodology taking into account the lifetime costs of the building, and (c) construct only nearly zero-energy buildings (NZEBs) from 2020 onwards." Hamdy et al. (2017) and by 2018 all new buildings occupied and owned by public authorities are NZEBs.

In order to ensure that the energy saving performance of buildings will be effective, appropriate, strengthened over time, and sustained. The EU directive (EPBD Recast) is defined as a common general framework for the calculation of energy performance of buildings, by means of three criteria, which needed to be considered in the calculation;

(a) A methodology with concern at least the following bits; actual thermal characteristics of the buildings including its internal partitions, appliances and built-in lighting installations and characteristics, natural ventilation and passive solar systems and solar protection, the design, positioning and orientation of the building, including indoor and outdoor climate, internal loads.
(b) The positive influence of the following aspects; local solar exposure conditions, active systems based on renewable sources.
(c) Classification of buildings into categories. In addition, the financial incentives and market barriers, and the necessary measures to establish a system of certification of the energy performance of buildings need to be considered in the transition to NZEBs.

From the above sections, toward the gap in current energy regulation in Egypt, the current code for energy efficiency improvement in buildings, and the housing design and planning code need to be developed. With defining to the energy performance of buildings framework, calculation methods, and considerations. In the next section, the paper aims to review the academic work related to NZEBs, to put the fundamental to assess the current gap in NZEBs implementation and regulations in Egypt.

4 Analysis of Case Studies

In this section, the research investigates and compares two case studies it investigated the Egyptian Japanese University (E-just) at Borg EL Arab New City, Alexandria, Egypt, and MASDAR city in Abu Dhabi, the United Arab Emirates, for several reasons:

- Both projects are announced to be zero-energy buildings in Mena region, designed, and developed by an international consultative alliance, with highly experienced in that type of project.
- It is expected that both projects will become pilot projects for NZEBs in Mena region.
- Both had occupations diversity and used a different passive and active design and energy efficiency solution due to the location priorities, which will be useful for the future decision making.

Moreover, comparison between E-JUST and MASDER, by examining the design conditions and parameter (location, area, climatic, partners, occupations), and the zero-energy solutions method of each project had been done in the following part (Table 1).

Table 1 Comparison between E-JUST and MASDER

		"E-JUST" campus	"MASDAR" city
Master plan		 E-JUST, master plan, "Copyrights: Arata Isozaki and Associates Co., Ltd."	 MASDAR City, master plan, "Copyrights: Foster and Partners Co., Ltd."
Location	site	Borg EL Arab New City, Alexandria, Egypt	Abu Dhabi—The United Arab Emirates
	Area	840003.8571 m^2	Area: 65000,000 m^2
Partners	Owner	Partnership between the governments of Egypt and Japan	Masdar, a subsidiary of Mubadala Development Company
	Consultings	Designer: Arata Isozaki and Associates Co., Ltd Environmental design and sustainability consultant: ARUP Group Infrastructure consultants: SHAKER consulting group	Designer: Foster and Partners Renewable energies: ETA—renewable energies, Florence, Italy Climate engineers: Transsolar Energietechnik GmbH
Occupation		E-just functional program constitutes the requirements for the academic functions and support facilities of the campus, and are divided into 12 modules I. Academic facilities (A. Faculty of engineering, B. Faculty of business and humanity) II. Centers of excellence III. Research support and media centers IV. University administration building V. Community center VI. Students service and facility center VII. Club and sports facilities VIII. Visitors center IX. Residential district X. Infrastructure departments and plants XI. Circulation network and transport systems XII. Landscaping	Masdar is a sustainable mixed-use city which planned to accommodate 45,000–50,000 people and 1500 businesses in addition to 60,000 workers are expected to commute to the city daily, and designed with eco-friendly systems, and organized into: 62% for residential zone 10% for community facilities 10% for offices 4% retail 7% research and development 4% light industrial 3% hotel and serviced apartment Institutional, Mosques, utilities, park, and open spaces

4.1 Analysis of Zero-Energy Concept and Solutions

In this section, the analysis of case studies are relay on the three stages of analysis; (1) analysis the integrated passive design solutions on buildings and urban fabric design, (2) find the integration in design between using the suitable renewable energy systems on site and the environmental solutions due to the local location, (3) analyze what happened in design development and implementation phases, as a result of reality variables in MENA region.

a. **Buildings systems and urban fabric passive design**:

E-JUST, architecture and urban design proposal, influenced by local Egyptian architecture in the form of covered paths and outdoor squares, chimneys ventilation

Furthermore, create building components follows the blueprint for the "golden ratio" esthetic in plan and façade design

However, the proposed design concept aims to achieve the zero-energy target, there is no refer to energy efficiency in buildings or insulations of buildings

An organic urban fabric and fractal buildings distribution had been developed (as shown in the figure below), and there will be neither main axis nor a forced circulation route such as a spine. Pedestrians, as well as personal mobility users, are encouraged to meander and to face with unexpected encounter

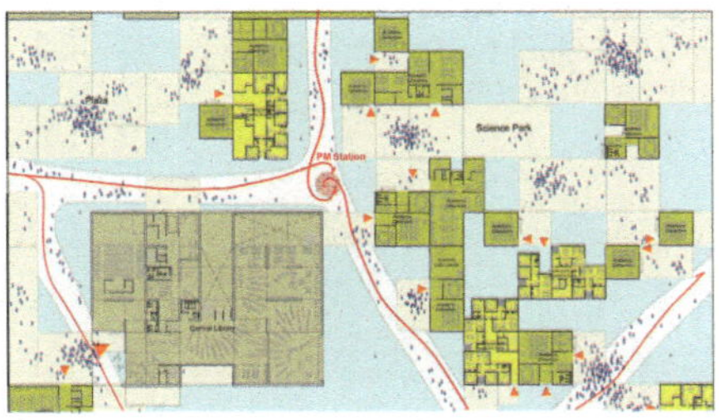

Buildings system at E-JUST Campus "Copyrights: Arata Isozaki and Associates Co., Ltd."

The urban fabric and buildings of Masdar City are specially designed to be energy efficient. In line with traditional Arab design, Masdar City's shaded paths and narrow streets are designed to create a pleasant space for walking in the region's hot climate

MASDER is oriented on southeast-northwest axis allowing for shading at street-level throughout the day

The main buildings, such as the IRENA Global Headquarters, the Siemens Building, and the Incubator Building, designed with the highly energy performance approach, with insulated building envelope, daylight systems, energy efficient, and produced hot water using solar energy. They also have angled facades to minimize the amount of glare and heat from the sun

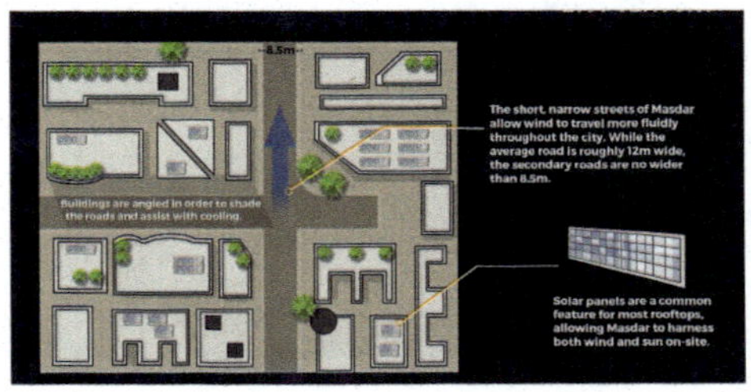

Buildings system at MASDAR City "Copyrights: ANDY THIES and Jesse Howe"

b. **Integrated environmental solutions**:

E-JUST proposed integrated environmental solutions

(1) Campus had covered with a large roof which provides peoples with comfortable shading outdoor areas and reduces the thermal heat gain on buildings envelope, the thermal comfort of peoples expected to be enhanced by the evaporation from ponds and the predominant wind blowing from the northwest

(2) Solar heating tower, provide E-just buildings and mechanical systems with the renewable resource of hot water

Integrated environmental section of E-JUST campus, Copyrights: Arata Isozaki and Associates Co., Ltd.

MASDAR City, proposed various integrated environmental systems; A 45 m-high wind tower had been developed, to cool the streets as much as 5 °C, the cool breezes from higher up are redirected down the tower to the pedestrian level, displacing and pushing hot air up and away from the courtyard and ground floors of the surrounding buildings

Furthermore, lined with buildings not exceeding five stories, and combined with the presence of street-level "solar canopies," roads receive protective shade for pedestrians throughout the city

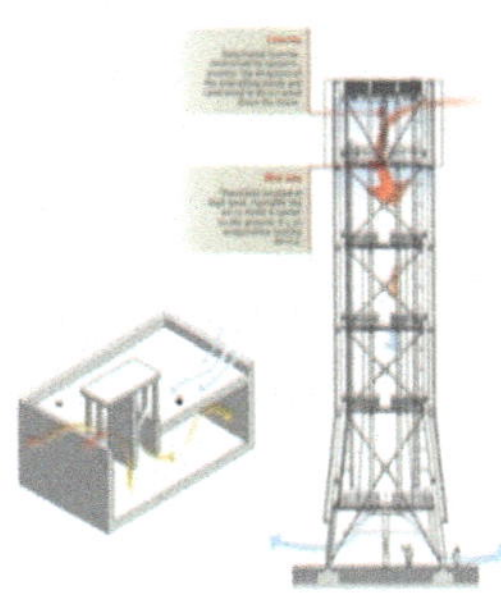

Wind tower section, Masdar city. Copyrights: "Foster and Partners Co.Ltd."

4.1.1 Integrated Passive Design Solutions Analysis

Both projects are seeking to achieve a nearly zero-energy buildings concept, based on several passive operative methods and integrated:

4.1.2 Integrates Between Renewable Energy Systems and Environmental Solutions

a. **Renewable energy systems**:

The proposed environmental control systems for E-JUST campus are comprised of the items listed below: Chimneys for natural ventilation Underground cooling/heating trench Large roof shading photovoltaic cells Reed bed system Solar hot water panel Microclimate cooling with a big pond	Masdar city had a variety of sustainable technologies, such innovative renewable energy initiatives as a 100 kW thermal power plant, solar cooling, photovoltaic's, concentrated solar power and geothermal cooling projects. The city currently receives roughly 70% of its energy through large solar fields around its perimeter, consisting of about 90,000 panels over a 50+ acre area

b. **Environmental cover**:

E-JUST had a large roof, made of smaller frames pattern each equipped with louvered solar PV panels which angled at 30°, with vary in degree of transparency, to generate clean renewable energy and also shade the outdoor urban spaces  E-JUST Environmental cover, Copyrights: Arata Isozaki and Associates Co., Ltd.	MASDAR City had an integrated canopy roof with PV solar panel, solar thermal cooler, and wind turbine, this integrated systems will provide the city with its energy need 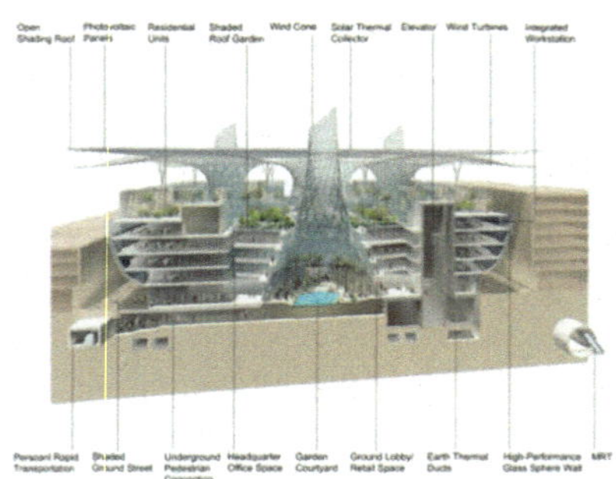 Sectional Masdar Institute building's zoning, the wind towers, and the roof canopy. Copyright: Adrian Smith + Gordon Gill

4.1.3 Design Development and Implementation Stage

Finally, the case studies analysis found that (1) the project scale needs to be discussed in NZEBs development at MENA region, and it can reinforce fully integrated and optimized solutions (2) the environmental design solutions which depend on the analysis of traditional Arabian cities integrate with innovated solar systems, such as the solar canopies (or solar roofs), wind tower, narrow streets, and organic urban fabric with outdoor areas, had a positive impact in decreasing energy consumption and enhancing the thermal comfort in buildings and outdoor areas.

We found that both projects (E-JUST and MASDER) are keen to design the building's envelope well, by use of insulation materials, double screen facade, shadings with high co-efficiency, and an ingrate's environmental roof with PV. As well as, the economic factor and the vernacular environment in E-JUST project was influential, the ingrate's environmental roof with PV had been scaled down in the design development phase, and the fractal design of building distribution in layout had changed to the geometric design.

Furthermore, a study by (Albadry et al. 2017) discuss the possible solution and optimal cost for retrofit the existing residential buildings in Egypt to net zero energy buildings, it gives a guideline of three steps:

In design development stage, the urban morphology of campus had been changed, from an organic form to semi-geometric form. The large environmental roof had fragmented to parts, to shaded the outdoor activities areas, solar chimneys had been added as part of the building, to deal with the exhaust air and reduce the heat load

Design development of buildings and urban plazas in E-JUST campus, Copyrights: Arata Isozaki and Associates Co., Ltd.

Masdar city expected to be fully implemented between 2030 into phases, the first phase included main six buildings and the Masdar Institute campus, which they are already operational right now; the Masdar headquarters, and a number of residential, commercial, leisure and retail buildings. The second phases expected to include near 15 office buildings for international companies. ("Masdar City" 2016) Building the city in phases enhanced the unique character of the city because it allowed rapidly evolving clean-tech technologies and lessons learned in earlier phases, to be incorporated into each below phases

First Phase. Masdar Institute, Masdar Headquarters, Hotel and Conference Center, The Retail District, Residential, Research and development facilities, Innovation centers and offices

5 Discussion of Results

In this section, the critical review results obtained for Egypt are discussed and the NZEBs implementation potential is summarized. The analysis of case studies gives some indicators of the stakeholders and investor's needs, as well as, possibility of zero-energy buildings implements in MENA region.

(1) Modify the current building insulation, (2) Reduce energy consumption to minimum, (3) Renewable energy generation. As well as, (Attia et al. 2012) develop a vernacular material database for building energy simulations program. Besides, Sheweka (2012) discussed the development of hybrid architectural skin and hybrid planning in Egypt, which mean designing an eco-friendly, energy-efficient

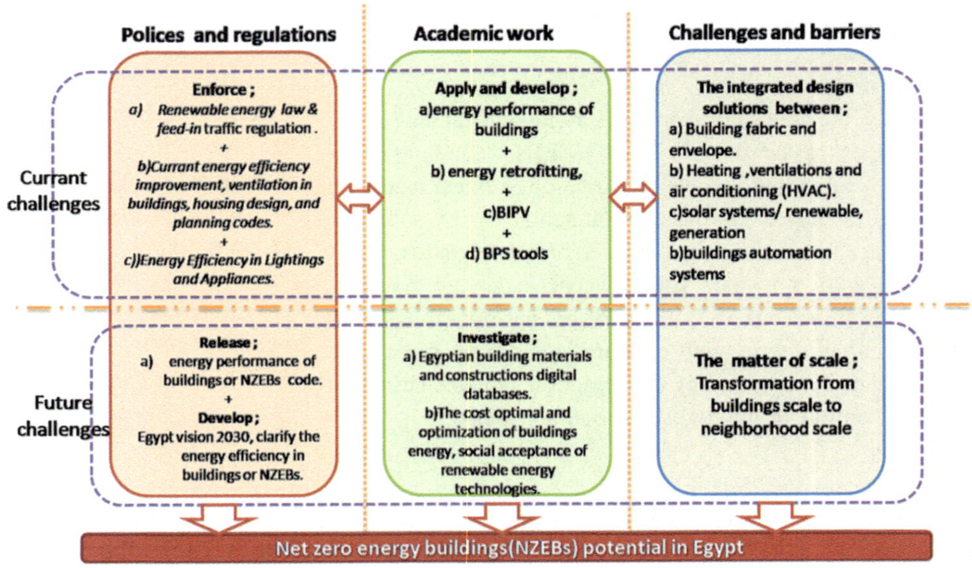

Fig. 3 NZEBs in Egypt workflow

buildings qualified for energy rating systems. It recommended several points as an action plan for new Egypt planning, such as development of the building standers and laws to include the environmental assessment as major demand for building procedures. As well as, develop recognizable low-impact durable materials from local sources as symbols of products that can be used by all in the process of design, construction, and buildings.

Nevertheless, Raslan and Mavrogianni (2013) develop a framework for the development of an Egyptian Domestic Energy Model (EDEM) supported by a representative database of building stock archetypes, the model can be used to assess the environmental and socioeconomic impacts of renewable energy technologies and fabric upgrade solutions that reduce energy demand while maintaining lifestyle, health, and comfort.

On the other hand, this study developed a framework, Fig. 3, which expected to help decision makers or stakeholder of NZEBs in Egypt, through investigating the three dimensions of this study, "policies and regulations enforcement and formulations", "academic work development", "what is the challenges and barriers of NZEBS implementations in Egypt".

- **Academic work Interest and support**.

The literature review of this study shows an increased interest from the researchers with the energy issue and NZEBs potential in Egypt. It is the main core of NZEBs application and implementations. The scientific interest with issues like energy performance of buildings, energy retrofitting, BIPV, and BPS tools needs to be developed and continued with variety in Egyptian case studies, to establish a national digital database of buildings materials and constructions.

Studying the cost-optimal and optimization of Egyptian buildings energy, and investing the social acceptance of renewable energy systems in Egypt need to be discussed in the ongoing research papers.

- **Energy efficiency efforts and policies framed**.

Egypt formulates an incentive energy efficiency policy and standard, toward an effective reduction in energy consumption in buildings, and stimulates reliance on solar energy production. The enforcement of current regulations for energy efficiency such as the energy efficiency improvement and ventilation in buildings codes, and renewable energy law (feed-in traffic) expected to support the NZEBs trends in Egypt.

Despite the fact that, energy performance of buildings or NZEBs code is needed. It is suggested to be added as an annex to the Building Law no. 119 for 2008, which defines the parties authorized to license buildings in Egypt, or as an additional chapter in ECP 306-2005. "The Egyptian code for energy efficiency improvement in buildings Part 1 and Part 2", with clarifications on national energy strategic plan of Egypt.

- **Challenges and barriers to NZEBS implementation in Egypt**.

NZEBs projects had several challenges in design and implementations phases, the integrations between systems solutions such as buildings fabric and envelope, HVAC systems, solar and renewable systems, and buildings automation systems, is the key to enhancement and improvement of energy savings in NZEBs and access the thermal comfort in indoor spaces.

On the other hand, the consideration of scale matter needs to be studied carefully in Egypt, the transformation from net zero energy buildings (NZEBs) to net zero energy neighborhood (NZEN) had great challenges which need to consider in design phases. The environmental historical analysis for traditional urban fabric and morphology transformations gives a fundamental to NZEBs in Egypt.

6 Conclusion

Through the critical review and case studies analyzed in this study, it had found that Egypt faces many challenges in NZEBs development process; firstly, the increase in annual population growth, the decreasing in energy subsidies, and the high rate of houses demand. These factors made the NZEBs concept is the solution key to energy savings and reducing the Egyptian buildings consumption pattern. Secondly, the energy efficiency buildings code and electricity regulations enforce in the new buildings or in the retrofit process, as well as, energy performance in buildings code and calculations sheets according to Egypt situations, need to be released. Thirdly, establish a scientific database and pilot projects to investigated NZEBs process in Egypt need to be studied in the future academic work with demonstrators of NZEN as indications for opportunities for intervention. Finally, in this study, a workflow diagram had been developed, to help the decision makers and stakeholders in NZEBs design in the current situations, and assess the future needs.

References

Abdallah, A. S. H., Hiroshi, Y., Goto, T., Enteria, N., Radwan, M. M., & Eid, M. A. (2014). Parametric investigation of solar chimney with new cooling tower integrated in a single room for New Assiut city, Egypt climate. International Journal of Energy and Environmental Engineering, 5(2–3), 1–9. https://doi.org/10.1007/s40095-014-0092-6

Abdelhady, S., Abd-Elhady, M. S., & Fouad, M. M. (2017). An Understanding of the Operation of Silicon Photovoltaic Panels (Vol. 113, pp. 466–475). Presented at the Energy Procedia. https://doi.org/10.1016/j.egypro.2017.04.041

Albadry, S., Tarabieh, K., & Sewilam, H. (2017). Achieving Net Zero-Energy Buildings through Retrofitting Existing Residential Buildings Using PV Panels. Energy Procedia 115:195–204

Anhar Hegazi. (2017). The Wide-Scale Deployment of Renewable Energy Technologies in Egypt Challenges and Recommendations (Renewables Readiness Assessment and Renewable Energy 2030 Roadmap Analysis for Egypt Validation Meeting)

Ascione, F., Bianco, N., De Masi, R. F., Mauro, G. M., & Vanoli, G. P. (2017). Energy retrofit of educational buildings: Transient energy simulations, model calibration and multi-objective optimization towards nearly zero-energy performance. Energy and Buildings, 144, 303–319. https://doi.org/10.1016/j.enbuild.2017.03.056

Athienitis, A. K., & O'Brien, L. (Eds.). (2015). Modeling, design, and optimization of net-zero energy buildings. Berlin: Ernst & Sohn

Attia, S. (2010). Zero energy retrofit: Case study of a chalet in Ain-Sukhna, Egypt (Vol. 7, pp. 5624–5651). Presented at the 39th ASES National Solar Conference 2010, SOLAR 2010. Retrieved from https://www.scopus.com/inward/record.uri?eid=2-s2.0-84867267547&partnerID=40&md5=a0f5fd6cc8523587ccefe4d5d0e5253f

Attia, S., & De Herde, A. (2011). Design decision tool for zero energy buildings (pp. 77–82). Presented at the PLEA 2011 - Architecture and Sustainable Development, Conference Proceedings of the 27th International Conference on Passive and Low Energy Architecture. Retrieved from https://www.scopus.com/inward/record.uri?eid=2-s2.0-84864135554&partnerID=40&md5=064ab2d5c58693e7fbb5a16981998292

Attia, S. G., Hamdy, M., Samaan, M., De Herde, A., & Hensen, J. L. M. (2011). Towards strategic use of BPS tools in Egypt. In Building Simulation 2011. Retrieved September 13, 2017, from https://pure.tue.nl/ws/files/3410583/11304156623313188.pdf

Attia, S., Gratia, E., De Herde, A., & Hensen, J. L. M. (2012). Simulation-based decision support tool for early stages of zero-energy building design. Energy and Buildings, 49, 2–15. https://doi.org/10.1016/j.enbuild.2012.01.028

Attia, S., Gratia, E., Herde, A. D., & Hensen, J. L. M. (2013). Tool for design decision making: Zero energy residential buildings in hot humid climate (pp. 3720–3727). Presented at the Proceedings of BS 2013: 13th Conference of the International Building Performance Simulation Association. Retrieved from https://www.scopus.com/inward/record.uri?eid=2-s2.0-84886705621&partnerID=40&md5=cd34ce383f9f15d32c8629e1143fed37

Attia, S., Hamdy, M., & Ezzeldin, S. (2017). Twenty-year tracking of lighting savings and power density in the residential sector. Energy and Buildings, 154, 113–126. https://doi.org/10.1016/j.enbuild.2017.08.041

bankaudi. (2016). EgyptEconomicReport. Retrieved from http://www.bankaudi.com.eg/Library/Assets/EgyptEconomicReport-2016-English-040615.pdf

BP Statistical Review of World Energy. (2017)

Dabaieh, M., & Elbably, A. (2015). Ventilated Trombe wall as a passive solar heating and cooling retrofitting approach; a low-tech design for off-grid settlements in semi-arid climates. Solar Energy, 122, 820–833. https://doi.org/10.1016/j.solener.2015.10.005

Dabaieh, M., Makhlouf, N. N., & Hosny, O. M. (2016). Roof top PV retrofitting: A rehabilitation assessment towards nearly zero energy buildings in remote off-grid vernacular settlements in Egypt. Solar Energy, 123, 160–173. https://doi.org/10.1016/j.solener.2015.11.005

Dadzie, J., Ding, G., & Runeson, G. (2017). Relationship between Sustainable Technology and Building Age: Evidence from Australia (Vol. 180, pp. 1131–1138). Presented at the Procedia Engineering. https://doi.org/10.1016/j.proeng.2017.04.273

Darko, A., & Chan, A. P. C. (2016). Critical analysis of green building research trend in construction journals. Habitat International, 57, 53–63. https://doi.org/10.1016/j.habitatint.2016.07.001

Eldabosy, M. M., Farag, O. M., & Moustafa, W. S. (2015). Mutual passive techniques impacts on human comfort for modular Unitin Cairo desert. Journal of Engineering and Applied Science, 62(2), 101–118

El-Zafarany, A., Sherif, A., & Arafa, R. (2013). Energy efficiency of external perforated window solar screens in desert environments: Effect of screen surface reflectance (pp. 817–826). Presented at the AEI 2013: Building Solutions for Architectural Engineering - Proceedings of the 2013 Architectural Engineering National Conference. https://doi.org/10.1061/9780784412909.080

Fahmy, M., El-Hady, H., Mahdy, M., & Abdelalim, M. F. (2017). On the green adaptation of urban developments in Egypt; predicting community future energy efficiency using coupled outdoor-indoor simulations. Energy and Buildings, 153, 241–261. https://doi.org/10.1016/j.enbuild.2017.08.008

Farghaly, T. A. (2003). Energy efficiency in high performance buildings related strategies and techniques. AEJ - Alexandria Engineering Journal, 42(5), 643–658

Ghosh, N., & Goswami, A. (2014). Sustainability Science for Social, Economic, and Environmental Development. https://doi.org/10.4018/978-1-4666-4995-8

Gindi, S. E., Abdin, A. R., & Hassan, A. (2017). Building integrated Photovoltaic Retrofitting in office buildings (Vol. 115, pp. 239–252). Presented at the Energy Procedia. https://doi.org/10.1016/j.egypro.2017.05.022

Hamdy, M., & Mauro, G. M. (2017). Multi-objective optimization of building energy design to reconcile collective and private perspectives: CO2-eq vs. Discounted payback time. Energies, 10(7). https://doi.org/10.3390/en10071016

Hamdy, M., Nguyen, A.-T., & Hensen, J. L. M. (2016). A performance comparison of multi-objective optimization algorithms for solving nearly-zero-energy-building design problems. Energy and Buildings, 121, 57–71. https://doi.org/10.1016/j.enbuild.2016.03.035

Hamdy, M., & Sirén, K. (2016). A multi-aid optimization scheme for large-scale investigation of cost-optimality and energy performance of buildings. Journal of Building Performance Simulation, 9(4), 411–430. https://doi.org/10.1080/19401493.2015.1069398

Hamdy, M., Sirén, K., & Attia, S. (2017). Impact of financial assumptions on the cost optimality towards nearly zero energy buildings – A case study. Energy and Buildings, 153, 421–438. https://doi.org/10.1016/j.enbuild.2017.08.018

Hassan, A. M., & Lee, H. (2015a). The paradox of the sustainable city: definitions and examples. Environment, Development and Sustainability, 17(6), 1267–1285. https://doi.org/10.1007/s10668-014-9604-z

Hassan, A. M., & Lee, H. (2015b). Toward the sustainable development of urban areas: An overview of global trends in trials and policies. Land Use Policy, 48, 199–212. https://doi.org/10.1016/j.landusepol.2015.04.029

J. Van den Akker. (2010). Energy Efficiency Improvement and Greenhouse Gas Reduction Project (FINAL EVALUATION). Retrieved from https://www.thegef.org/sites/default/files/project_documents/267_452_TE_CCM_Egypt.pdf

Jung, N., Moula, M. E., Fang, T., Hamdy, M., & Lahdelma, R. (2016). Social acceptance of renewable energy technologies for buildings in the Helsinki Metropolitan Area of Finland. Renewable Energy, 99, 813–824. https://doi.org/10.1016/j.renene.2016.07.006

Kares, M., & Singh, P. (2016). Assessment of building integrated photovolatics for the residential section in representative Urban areas in Egypt (pp. 413–420). Presented at the GHTC 2016 - IEEE Global Humanitarian Technology Conference: Technology for the Benefit of Humanity, Conference Proceedings. https://doi.org/10.1109/GHTC.2016.7857314

Karmany, H. M. (2016). Evaluation of green building rating systems for Egypt. Retrieved from http://dar.aucegypt.edu/handle/10526/4628

Khalil, E. E. (2012). Design of energy efficient commercial buildings in developing countries (Vol. 5, pp. 807–816). Presented at the Proceedings of the ASME Design Engineering Technical Conference. https://doi.org/10.1115/DETC2012-70284

Khan, M. M. A., Asif, M., & Stach, E. (2017). Rooftop PV potential in the residential sector of the kingdom of Saudi Arabia. Buildings, 7 (2). https://doi.org/10.3390/buildings7020046

Khatib, T., Mohamed, A., & Sopian, K. (2013). A review of photovoltaic systems size optimization techniques. Renewable and Sustainable Energy Reviews, 22, 454–465. https://doi.org/10.1016/j.rser.2013.02.023

Koo, C., Hong, T., Lee, M., & Kim, J. (2016). An integrated multi-objective optimization model for determining the optimal solution in implementing the rooftop photovoltaic system. Renewable and Sustainable Energy Reviews, 57, 822–837. https://doi.org/10.1016/j.rser.2015.12.205

Mahdy, M. M., & Nikolopoulou, M. (2014). Evaluation of fenestration specifications in Egypt in terms of energy consumption and long term cost-effectiveness. Energy and Buildings, 69, 329–343. https://doi.org/10.1016/j.enbuild.2013.11.028

Marique, A.-F., Penders, M., & Reiter, S. (n.d.). From Zero Energy Building to Zero Energy Neighbourhood

Masdar City: The Future of Sustainable Cities? (2016, January 27). Retrieved October 4, 2017, from http://atlaslens.com/index.php/2016/01/27/masdar-city-the-future-of-sustainable-cities/

Mauro, G. M., Hamdy, M., Vanoli, G. P., Bianco, N., & Hensen, J. L. M. (2015). A new methodology for investigating the cost-optimality of energy retrofitting a building category. Energy and Buildings, 107, 456–478. https://doi.org/10.1016/j.enbuild.2015.08.044

MED-ENEC. (2017). Realizing the Energy Efficiency Potential of Applying the Smart Grid Concept in Egypt: Energy Efficiency in the Construction Sector in the Mediterranean. Retrieved October 7, 2017, from https://www.med-enec.com/downloads/realizing-energy-efficiency-potential-applying-smart-grid-concept-egypt

Mohamed Shaker. (2015, March). "Addressing Egypt's Electricity Vision" in Egypt Economic Development Conference. Retrieved September 5, 2017, from http://www.moee.gov.eg/english_new/Presentations.aspx

Pinto, J. T. M., Amaral, K. J., & Janissek, P. R. (2016). Deployment of photovoltaics in Brazil: Scenarios, perspectives and policies for low-income housing. Solar Energy, 133, 73–84. https://doi.org/10.1016/j.solener.2016.03.048

Raslan, R., & Mavrogianni, A. (2013). Developing a national stock model to support building energy efficiency research and policy in Egypt

Recast, E. (2010). Directive 2010/31/EU of the European Parliament and of the Council of 19 May 2010 on the energy performance of buildings (recast). Official Journal of the European Union, 18(06), 2010

Regional Center for Renewable Energy and Energy Efficiency (RCREEE). Renewable Energy/Reports and Studies. http://www.rcreee.org/content/AFEXRE2015

Samir, H., & Ali, N. A. (2017). Applying Building-integrated Photovoltaics (BIPV) in Existing Buildings, Opportunities and Constrains in Egypt. Procedia Environmental Sciences, 37, 614–625. https://doi.org/10.1016/j.proenv.2017.03.048

SDS. (2016). Egypt Vision 2030 "Sustainable Development Strategy (SDS)". Retrieved September 5, 2017, from http://mcit.gov.eg/Publication/Publication_Summary/1020/

Sheweka, S. M. (2012). New Egypt with a New Hybrid Skin A New Hybrid Architecture Vision for the Egyptians 'Development Corridor'. Energy Procedia, Terragreen 2012: Clean Energy Solutions for Sustainable Environment (CESSE), 18 (Supplement C): 449–57

Synnefa, A., Laskari, M., Gupta, R., Pisello, A. L., & Santamouris, M. (2017). Development of Net Zero Energy Settlements Using Advanced Energy Technologies. Procedia Engineering, 180(Supplement C), 1388–1401. https://doi.org/10.1016/j.proeng.2017.04.302

Task 40, IEA SHC. (n.d.). Retrieved October 7, 2017, from http://task40.iea-shc.org/

The Arab Future Energy Index (AFEX). (2015). Is the First Native Arab Index Dedicated to Monitoring and Analyzing Sustainable Energy Competitiveness in the Arab Region

tradingeconomics. (2017). Egypt GDP From Construction |2007-2017| Data |Chart| Calendar. Retrieved September 2, 2017, from tradingeconomics.com/egypt/gdp-from-construction

Tripathy, M., Sadhu, P. K., & Panda, S. K. (2016). A critical review on building integrated photovoltaic products and their applications. Renewable and Sustainable Energy Reviews, 61, 451–465. https://doi.org/10.1016/j.rser.2016.04.008

Tripathy, M., Yadav, S., Panda, S. K., & Sadhu, P. K. (2017). Performance of building integrated photovoltaic thermal systems for the panels installed at optimum tilt angle. Renewable Energy, 113, 1056–1069. https://doi.org/10.1016/j.renene.2017.06.052

Voss, K., & Musall, E. (2013). Net zero energy buildings international projects of carbon neutrality in buildings. München: Institut für internationale Architektur-Dokumentation

Wang, W., Liu, Y., Wu, X., Xu, Y., Yu, W., Zhao, C., & Zhong, Y. (2016). Environmental assessments and economic performance of BAPV and BIPV systems in Shanghai. Energy and Buildings, 130, 98–106. https://doi.org/10.1016/j.enbuild.2016.07.066

United Nations. (2014). Global Report on Human Settlements 2013: Planning and Design for Sustainable Urban Mobility. United Nations Pubns

Youssef, A. M. A., Zhai, Z. J., & Reffat, R. M. (2016). Genetic algorithm based optimization for photovoltaics integrated building envelope. Energy and Buildings, 127, 627–636. https://doi.org/10.1016/j.enbuild.2016.06.018

Energy Retrofit of Existing Building Stock in Amman: State of the Art, Obstacles and Opportunities

Sameh Shamout, Paola Boarin, and Alessandro Melis

Abstract

Energy security is one of the most important challenges Jordan is currently facing, as it imports 96% of its energy resources from foreign countries. Together with other sectors, the existing building stock is responsible for high energy consumption, as average energy performance levels are much lower than the standards prescribed in the recent Jordanian Energy Efficient Building Code. Addressing issues related to energy retrofitting will assist the country to develop effective plans to reduce the energy consumption of buildings and, therefore, decrease the associated energy costs on a large scale, thus helping Jordan's sustainable development. This paper aims to provide a better understanding of the current situation of the building sector and the energy performance of existing buildings in Amman, identifying issues, obstacles and opportunities to be considered for retrofitting this stock. The paper also provides a critical reflection on interviews with six local experts in the fields of architecture and engineering, including governmental and non-governmental institutions, making recommendations and suggesting possible directions for future research.

Keywords

Existing buildings • Energy efficient buildings • Sustainability • Retrofit

1 Introduction

Jordan faces significant challenges in its sustainable development including ensuring its energy supply, as it has limited indigenous energy resources. The country relies on foreign resources, importing 96% of its gas and oil from its neighbours (Ministry of Environment and UNDP 2014; Jordan 2013). The dominance of oil and gas as primary sources of energy in the country needs to be addressed urgently, as the building sector has high energy consumption, and plans for alternative power generation have high initial costs (Shamout 2017). To address this issue, the Ministry of Environment and UNDP (2014) have defined a 20% national target for energy efficiency, and concrete measures have been proposed in many key sectors to achieve this objective by 2020. These measures deal with both demand and supply. For example, with regard to energy demand, a 10% target was set for reducing energy usage in Jordanian public buildings by 2020. Mohsen and Akash (2001) stated that the Jordanian residential building sector relies on fuel for heating, and space heating accounts for around 61% of overall energy consumption, making the achievement of energy efficiency very significant in buildings, as they "offer enormous scope for energy savings" (WCED 1987, p. 166). Energy efficiency has been given a significant amount of attention in the Jordan Green Building Guide (2013), i.e. the local green building rating system. It covers 33% of the overall scores in the assessment process. Katz (as cited in EBSCO Sustainability Watch 2010) states that the existing building stock is 80 times larger than the new construction sector. Therefore, from an economic point of view, retrofitting existing buildings is more feasible than constructing new green ones (Fan and Xia 2015). A transformation of the built environment in Jordan through a sustainable approach is an effective approach to conserving energy, water and other resources.

Perspectives on the issues related to the process of energy retrofitting of buildings, specifically the technological and non-technological challenges and feasibility, are varied.

S. Shamout (✉)
School of Architecture and Planning, Faculty of Creative Arts and Industries (CAI), University of Auckland, Auckland, New Zealand
e-mail: ssha692@aucklanduni.ac.nz; ; eng.samehshamout@live.com

P. Boarin · A. Melis
Portsmouth School of Architecture, Faculty of Creative and Cultural Industries, University of Portsmouth, Portsmouth, UK

C. Alalouch et al. (eds.), *Advanced Studies in Energy Efficiency and Built Environment for Developing Countries*, Advances in Science, Technology & Innovation, https://doi.org/10.1007/978-3-030-10856-4_14

Konstantinou and Knaack (2013) stated that there is a lack of specialised knowledge in the early strategy development for refurbishing existing residential buildings. Also, Ma et al. (2012) stated that there is still a major technical challenge regarding methods used to identify the most cost-effective retrofit measures for a particular project. Ungureanu et al. (2013) claimed that mastering the energy retrofit process requires identifying cost-optimal retrofit strategies to maximise the reduction of energy usage and carbon emissions. Their study, conducted on the building stock of Denmark, Finland, Romania, Sweden and Switzerland, resulted in recommending integrated strategies for retrofitting existing buildings to reduce primary energy use, costs and GHG emissions.

Concerning Jordanian building stock, Ouahrani (2010), in his article on the thermal behaviour simulation of a typical Jordanian flat with the highest energy consumption in Amman, argued that optimising the building elements, mainly the insulation of external walls and roof, can lead to a total of 72% energy saving on the heating and cooling loads, but optimising the window-to-floor ratio (WFR) is difficult in an existing flat as it requires altering the building's structure. When Attia and Zawaydeh (2014) compared the potential and constraints of a number of passive and active design strategies for a typical existing flat in Amman to meet the local code requirements and the zero energy retrofit target, they proved its feasibility in the case of local code compliance, but highlighted that the main barriers to retrofitting in Jordan are still economic. The study stated that some passive and active design strategies are rewarding with payback periods from three to nine years.

Most buildings in Jordan do not meet the requirements of the Jordanian Energy Efficient Building Code. Daher (2011) stated that although the Jordanian Thermal Insulation Code has always been mandatory, people in Jordan have ignored it in order to avoid additional costs accompanying it. Another study confirmed and discussed the inefficient performance of existing Jordanian buildings, through a long-term monitoring study conducted by the Housing Development & Management (HDM) and the Royal Scientific Society (RSS) (2009) on the performance of a typical building in Jordan, with the aim of defining criteria to improve thermal performance and minimise energy use. In this regard, energy retrofit can play a relevant role, considering the current challenges the country faces, including energy and water security, unplanned population growth and urban expansion, and climate change.

2 Aims of the Research

Given the scarcity of publications in the energy retrofit field in the context of Jordan, this research has involved experts from the local building sector, with the main purpose of providing the local market with useful information regarding the issues related to the construction of energy efficient buildings. The interview questions are mainly focused on the energy retrofit of existing buildings in Jordan, thus addressing one of the highest energy burdens in the country. This contribution aims at providing a better understanding of the key aspects to be considered when retrofitting existing buildings in Jordan, identifying all possible issues for reducing the building stock's energy consumption on a large scale from technological and non-technological points of view.

3 The Significance of the Study

This study is conducted to benefit the following areas. It is highly important as Jordan has been facing significant challenges for its sustainable development including ensuring its energy supply. Also, the timing of the publication is very important as the country is currently having an economic crisis and highlighting the potential reduction of the energy consumption in the building sector should be put on the table. It is also important as it fills a gap of knowledge concerning energy retrofitting of existing buildings in the context of Jordan.

4 Method of the Research

The study used a qualitative approach for data collection through interviews with a sample of local experts in the field of energy retrofit in the context of Jordan. The sample for this study was characterised by its diversity as it included architects, engineers and experts representing governmental and non-governmental organisations. Participants were recruited primarily by a third party, the Jordanian Engineers Association (JEA), which has records of local experts in many related fields in Jordan. The third party suggested the participants and invited them to take part in this study, providing them with a Participant Information Sheet. Participants who agreed to take part in the study then contacted the authors via email. Before conducting the interviews, participants were required to sign two consent forms, one to confirm their acceptance of participating in the research and one to confirm their employers' agreement. The authors conducted the interviews with six local experts working in the Jordanian building sector who were willing to contribute to the development of the local practice of energy efficient buildings. The interviews were audio-recorded, translated into English and transcripted by the authors who then returned them to the interviewees for their review and consent to use the transcripts. All participants had the right to withdraw from the study at any stage and to amend any data.

Table 1 Summary and classification of research participants

	Group 1 (Architects)		Group 2 (Engineers)		Group 3 (Governmental and non-gov. national organisations)	
Participants' codes	1a	1b	2a	2b	3a	3b
Academic background	Architect	Architect	Civil Engineer	Mechanical Engineer	Architect	Mechanical Engineer
Work field	Green buildings	Architect consultant	Engineering	Energy efficient building consultant	Green buildings	Green buildings
Years of experience	>10 years	>10 years	>5 years	>10 years	3 years	>10 years
Organisation type	Government	International organisation	–	Non-profit	Non-government	Government
Organisation size	Large	Large	Large	Large	Small	Large

Participants were classified into three groups as shown in Table 1. Two interviews were conducted for each group, and each participant was given a code to be used in the following discussion.

Research questions focused on the issues related to energy retrofit and the extent of its feasibility, any possible obstacles and potential impacts in Jordan. The focus was also on defining the state of the art of the current building stock in Jordan from technological and non-technological points of view. This paper critically discusses the answers of the participants to the interview questions and suggests recommendations to contribute to the further development of this topic in the context of Jordan.

5 Critical Discussion of the Interviews' Results

This part of the paper reviews the interview questions and critically discusses the participants' answers, and each question is studied in a separate section. Sections include: (i) the question text; (ii) a table summarising the answers; (iii) a discussion; and (iv) some conclusive reflections.

5.1 Question #1: Which of the Following Aspects Do You Consider as the Most Significant for Retrofitting the Building Stock in Jordan? Please Rank Them in Order of Importance

- Water efficiency.
- Energy saving.
- Improving the indoor environment for occupants.
- Other (please, specify) (Table 2).

It is clear that energy saving has been chosen as the most significant issue to be considered when retrofitting existing buildings in the context of Jordan. Participant 3a and participant 2a claimed that energy saving solutions do not cost as much as water saving solutions and, therefore, are more financially feasible. Participant 2b also stated that the results of energy saving solutions in buildings appear once the process is complete. Also, participant 3b emphasised that Jordan is in need of such measures as the energy problem causes more burdens on the country's economy. However, participant 1a did not prioritise energy saving solutions, in contrast to all other participants, saying that improving the indoor environment for occupants is more important due to its positive effect on occupants' productivity, especially in workplaces. The participant noted that people in Jordan are usually absent from work when there is no thermal comfort in a building and, therefore, the work environment affects their productivity.

5.2 Question #2: From Your Point of View, in Which of the Following Building Categories Would You Invest into Implement a Retrofit Process? Please Rank Them in Order of Investment Feasibility, and Provide a Reason

- Housing apartments (medium to high density).
- Villas and Dars (stand-alone dwellings).
- Commercial buildings.
- Private educational buildings.
- Public buildings.
- Other (please, specify) (Table 3).

Over 50% of the participants ranked residential buildings first, and most of them ranked commercial buildings second,

Table 2 Summary of answers to question #1

Participants		Group 1		Group 2		Group 3	
		1a	1b	2a	2b	3a	3b
1	Energy saving	2nd	1st	1st	1st	1st	1st
2	Water efficiency	3rd	2nd	2nd	2nd	2nd	2nd
3	Improving the indoor environment	1st	3rd	3rd	3rd	3rd	4th
4	Other	–	–	4th: eco-friendly materials	4th: water leakage into buildings	4th: sustainable site 5th: local materials	3rd: recycling construction materials

Table 3 Summary of answers to question #2

Participants		Group 1		Group 2		Group 3	
		1a	1b	2a	2b	3a	3b
1	Housing apartments	5th	1st	1st	3rd	5th	5th
2	Villas and Dars	3rd	4th	5th	1st	1st	4th
3	Commercial buildings	4th	2nd	2nd	2nd	2nd	1st
4	Educational buildings	2nd	5th	4th	5th	4th	2nd
5	Public buildings	1st	3rd	3rd	4th	3rd	3rd
6	Other	–	–	–	–	Industrial buildings	–

except participants 1a and 3b, who prioritised private educational buildings. Participant 1a stated that the energy retrofit process provides marketing benefits for educational buildings, as many educational institutions are keen to show their affiliation to the sustainability movement in Jordan. In contrast, participant 2b believes that retrofitting would not be suitable for educational buildings as the operational hours are usually few and in the daytime. Participant 1a thinks the energy efficiency issue is not a priority for commercial building owners, as their buildings are mostly rented and people do not prioritise this issue when renting. Also, providing their employees with a comfortable environment is not a target for this category. However, outreach programmes, more published studies in this field and publications of real cases that prove the benefits of energy saving solutions would make the current situation different and, therefore, the market value would increase when people become more aware of sustainability, and thus Jordanian buildings would have a better chance of being more sustainable.

5.3 Question #3: What Are the Obstacles and Challenges You Think the Energy Retrofit Process Might Have to Face a Diffusion in Jordan on a Professional and Cultural Level?

On the technological level, all participants claimed that there is not enough technical experience and skilled workforce available to implement the energy retrofitting process on existing building stock in Jordan. Participant 1b added that there are also not enough local experts. In contrast, other participants believe that there are enough local experts and companies, but there is a lack of experience and research on the needs of the local market. In general, lack of experience in supervision work, technical skills and retrofit-oriented design are major obstacles for the energy retrofit process in Jordan, resulting in buildings that are poorly constructed and, therefore, do not comply with local energy efficiency standards. Participant 3b claimed that most of the engineering consultancy offices in Jordan do not have the required knowledge to design green buildings. Therefore, investing in training activities to qualify local practitioners, engineers, architects, contractors and builders should be a priority to contribute to Jordan's sustainable development (Table 4).

Another important issue is that suppliers are not providing the local market with enough advanced systems, eco-friendly materials and high-performance appliances.

Auditing buildings is also a major problem in Jordan. Participant 3b stated that, unfortunately, there is a weak auditing process implemented by responsible authorities, including Greater Amman Municipality (GAM), Jordan Engineers Association (JEA) and the Jordan Green Building Council (JGBC). The main problem is the implementation and construction process does not match with project drawings. For instance, thermal insulation on site does not comply with the standards and with detailed drawings in terms of thickness and installation methods. Also, some important building elements like columns and beams are not thermally insulated in most buildings in Jordan, resulting in diffused thermal bridges and overall poor energy performances of buildings. In this regard, participants 2b and 3b also believe that there is a lack of up-to-date on-site auditing techniques, such as the use of infrared thermography to check upon thermal insulation and thermal bridges.

On the non-technological level, all participants stated that lack of awareness of the significance and benefits of retrofitting existing buildings within an energy efficient approach has been the major obstacle in Jordan. Participant 1a stated that people have lately begun to renovate existing buildings but within an architectural approach, rather than an energy efficient approach. Participants 3a and 1a noted that people in Jordan do not have enough knowledge of the advantages of the energy retrofit process.

Participant 1a said that even some building owners, who have prepared policies to save energy in their buildings, have not included solutions to improve the building envelope situation. They mostly focus on encouraging green behaviour amongst building occupants, like turning off lights and heating systems when not needed. However, participant 2b thinks that people have realised the significance of thermal insulation in buildings to save energy, but they do not have knowledge of the correct implementation pathways.

Participants 1a, 2b and 3a believe that people think more about reducing their high initial costs than investing in energy efficient systems, whether active or passive. This is another significant challenge. People will never invest in an energy retrofit to be environmentally friendly, as they pay less attention to the environment than to their budget, according to participant 1b. In fact, people in Jordan would prefer a one-off lower initial cost than a higher one with a payback period, as they are sceptical about financial solutions for funding energy efficient measures. Hence, increasing people's awareness should be a priority. According to participant 1, a possible way is to persuade them of the financial feasibility of this investment and inform them of how this could save on operational costs in their buildings.

Overall, energy retrofitting has some challenging obstacles in Jordan both from technological and non-technological points of view. Increasing people's awareness is the biggest goal, according to most participants, and a high level of cooperation between local community and the government is needed to achieve this. Regarding technological obstacles, a major achievement would be to qualify local technicians and builders by strengthening their relationships with academia and by improving research and educational activities, particularly focusing on the construction phase. Regarding funding, further work is also needed to better understand how energy retrofit activities could be funded in Jordan, taking financial feasibility into account. For instance, special loan programmes supported by local councils and the government have been shown to be a valuable support for Jordan lately and could be a pathway to explore.

Table 4 Summary of answers to question #3

Participants		Group 1		Group 2		Group 3	
		1a	1b	2a	2b	3a	3b
1	Technological barriers (professional level)	Not enough implementation techniques Shortage of technicians	Not enough experts and technologies	–	Lack of experience amongst engineers, supervisors, and technicians Lack of experts Lack of advanced auditing	Lack of technical experience for the authorised entities Resistance to implementing new technologies	Weak oversight and auditing by responsible authorities Poor implementation Lack of technological knowledge
2	Non-technological barriers (including cultural level)	Lack of funding Lack of awareness	Lack of awareness	Lack of awareness	High costs	Lack of awareness Impact of the high initial costs	Amman's topography Unplanned urban planning and expansion Jordanian traditional tribal system Traders (importers)

5.4 Question #4: How Would You Evaluate the Current Commercial Building Stock Concerning the Requirements of the Jordanian Energy Efficient Building Code of 2009?

- High performing.
- Acceptable.
- Not acceptable (Table 5).

Participant 3a said that the Jordanian Thermal Insulation Code has always been mandatory, but people in Jordan used to ignore it in order to avoid additional costs accompanying it (Konstantinou and Knaack 2013), and because government supervision has not been strict enough to ensure that the requirements are being applied. They also stated that in reality penalties for non-compliance with codes are not implemented, with an exception for some large-scale projects. However, Jordanians are becoming more aware of the significance of thermal insulation after experiencing high energy bills in buildings that are not properly thermally insulated, said participant 3a. Indeed, participant 3a claimed that most buildings built before 2010 are not insulated and new buildings that comply with local Thermal Insulation Code requirements are not being implemented using the correct methods. Participants 1a and 1b claimed that thermal bridges are rarely avoided in Jordanian buildings, due to lack of knowledge and awareness amongst designers, engineers and contractors.

Lack of awareness of correct ways to implement thermal insulation in buildings is the main reason why the performance of insulated buildings is inefficient. For example, most people focus on walls and ignore roofs when implementing thermal insulation in buildings, said participant 2b. This creates thermal bridges.

Overall, all participants stated that most Jordanian buildings, including thermally insulated buildings, do not have a complete thermally insulated envelope, and thermal bridges are a major issue. This highlights the need for implementing passive design solutions in buildings to improve their envelope performance. Most participants said that the building envelope is the most significant issue and it should be prioritised in the energy retrofit process. Further research is needed to identify current building envelope problems in the context of Jordan and the technologies available, assess their performance and suggest effective solutions to comply with building codes.

5.5 Question #5: Regarding Energy Upgrade Strategies, Please Put the Following Strategies in Order of Priority and Give Reasons for Your Ranking. If Possible, Please Provide a Reference to Precedent Research, Projects and Buildings

- Increase thermal insulation of the building envelope (mainly walls and roof).
- Improve the performance of building openings (e.g. double-glazing).
- Reduce internal loads (e.g. replace mechanical systems and electrical devices).
- Use energy generation from clean sources (solar, wind, geothermal, etc.).
- Shading devices.
- Reduce thermal bridges.
- Increase thermal mass.
- Airtightness.
- Improve natural ventilation.
- Other (please, specify) (Table 6).

Most participants consider the increase in thermal insulation in buildings as the first priority, in addition to reducing thermal bridges. Urgency should be given to strategies for saving energy in buildings, in order to reduce the energy demand for heating, cooling and hot water supply (Ouahrani 2010). Participant 3a claimed that thermal bridges are responsible for approximately 33% of heat loss in buildings and this would be avoided through insulating thermal bridges, for instance. However, participant 2b believes that reducing thermal bridges is very difficult in existing buildings and is very expensive as well. This is supported by participant 3b, who said that increasing thermal insulation could be applicable in Jordan from a technical point of view and could achieve a considerable amount of energy saving but it is costly. There is a need for identifying the amount of

Table 5 Summary of answers to question #4

Participants		Group 1		Group 2		Group 3	
		1a	1b	2a	2b	3a	3b
1	Insulated existing buildings	Acceptable	Not acceptable	Acceptable	Acceptable	Acceptable	Acceptable
2	Uninsulated existing buildings	Not acceptable	Not acceptable	Not acceptable	Not acceptable	Not acceptable	Not acceptable

Table 6 Summary of answers to question #5

Participants		Group 1		Group 2		Group 3	
		1a	1b	2a	2b	3a	3b
1	Thermal insulation	1st	2nd	1st	1st	1st	1st
2	Building openings	2nd	1st	3rd	2nd	2nd	2nd
3	Reduce internal energy loads	3rd	4th	2nd	3rd	7th	4th
4	Renewable energy generation	4th	3rd	4th	4th	8th	3rd
5	Shading devices	5th	8th	5th	–	3rd	5th
6	Reduce thermal bridges	1st	6th	–	5th	1st	1st
7	Increase thermal mass	–	7th	–	6th	6th	6th
8	Airtightness	6th	5th	7th	2nd	9th	7th
9	Natural ventilation	7th	–	6th	3rd	5th	8th
10	Other	–	–	–	–	–	–

energy saving which could be achieved through increasing thermal insulation, and further research is needed not only to illustrate the expected energy saving, but also the expected thermal comfort. Participant 1b stated that this strategy makes a difference, and participant 1a stated that it is economically feasible and would have a considerable impact not only on the internal comfort of buildings, but also on the climate of Jordan through the reduction in CO_2 emissions. This intention is supported by the literature from local authors as well. For instance, in his research conducted to identify the optimum thermal transmittance (*U*-value) for walls and roofs in a typical apartment in Amman, Ouahrani (2010) argues that it is possible to achieve an overall energy saving of up to 70% for heating and cooling, and a good thermal comfort level when the *U*-values range between 0.50 and 0.70 W/m^2 K for walls and roofs.

Improving the performance of building openings like windows and doors has mostly been ranked second. Participant 3a said that this is needed to prevent heat loss through windows. Replacing inefficient windows with energy efficient ones is becoming more feasible in Jordan, financially and technically, as the Jordanian market is developing. Participant 1a said that double-glazed windows used to be expensive (costing about 90–100 JD per m^2, equal to 140 USD), but have recently become cheaper (about 45–60 JD per m^2, equal to 60–85 USD). Participant 1b believes that this strategy should be prioritised, as windows in Jordan cause a considerable amount of heat leakage while the strategy of airtightness would not be possible because there are no quality control measures available in Jordan. However, participant 1a said that reducing air leakage from windows and doors is applicable, easy and cheap.

In Jordan, shading devices are important to take advantage of free solar gains in winter, in order to achieve thermal comfort, but overheating in summer can be avoided through the use of solar shadings, said participant 3a. Most participants agreed that external shading devices are more efficient than internal shading devices. However, participants 2b and 1a stated that external shading devices were not appropriate solutions for existing buildings because they altered the building's design, and people in Jordan pay attention to the external appearance of the building.

Regarding renewable energy generation integrated into the building, participants believed that improving the energy efficiency in buildings and reducing their energy demand should be prioritised more than generating clean energy from local sources. Participant 3a even stated that there would be no need for it if passive design solutions were applied correctly. However, participants 1a, 2a and 2b stated that there was a growing market for this strategy because the government supported it by exempting any renewable system from customs duties. However, the government allows for the installation of PV systems on empty lands (rented or owned), a method that is appropriate and efficient for commercial buildings or companies that have high energy consumption, but do not have enough space for integrating PV systems in the building itself.

Regarding natural ventilation, participant 3a said that, because of Jordan's mild climate, people use air conditioning for a short period in summer. In general, natural ventilation is better than air conditioning and it is part of the Jordanian culture, where people prefer it over air conditioning. Most participants said that Jordanians prioritise heating, not cooling, as they can deal with hot weather better than with cold weather. Participant 2b agreed that improving natural ventilation in buildings is an effective strategy, also because there is still a high smoking rate inside buildings in Jordan. This could be achieved in the case of retrofitting through operating mechanical ventilators or improving building openings, which would not be feasible as it would change the facades, said participant 1a.

Regarding increasing thermal mass, participant 1a said that increasing the thickness of roofs and walls would

Table 7 Summary of answers to question #6

Participants		Group 1		Group 2		Group 3	
		1a	1b	2a	2b	3a	3b
1	Thermal insulation material	Polystyrene	Polystyrene (5 cm thickness)	Artificial materials	Polystyrene	Polystyrene	Polystyrene
2		Polyethylene	Polyethylene	–	Rock wool	Polyethylene	Polyethylene
3		–	Foam concrete	–	–	Rock wool	Thermal blocks
4		–	–	–	–	–	AAC block

require too much space. In fact, traditional Jordanian buildings used to be built with a high thermal mass by using thick, natural stone walls, but this is not happening today because buildings are mainly built with concrete or steel, and stone is used only as an external cladding (Attia and Zawaydeh 2014). Therefore, increasing thermal mass would not be a possible solution for energy retrofitting as it is costly and occupies too much space.

In conclusion, it is evident that all participants preferred passive design solutions overactive ones. They agreed that improving building envelopes and openings have to be prioritised in the energy retrofit process to optimise energy efficiency and thermal comfort.

5.6 Question #6: Based on Your Professional Experience, What are the Most Common Insulation Materials and Installation Methodologies Used in Jordan?

Participant 2b said that the two main common thermal insulation materials in Jordan are polystyrene, both expanded and extruded, and rock wool. Expanded polystyrene is the preferred choice; it is usually blue and water resistant, and it lasts longer than extruded polystyrene, according to participant 2b. Rock wool is cheap, easy to use even in irregular-shaped buildings, and flexible. However, other participants stated that people in Jordan prefer not to use it in walls as it slips down in the cavity. Also, rock wool is not water resistant and when it is used in roofs, it must be covered by waterproofing materials, such as nylon, to prevent it from losing its insulation property. Participant 1a also stated that rock wool is mainly used to insulate ducts and pipes. Although rock wool and polystyrene have similar thermal conductivity values (*k*-value) of 0.040 and 0.035 W/m K, respectively (Ma et al. 2012), polystyrene is easier to deal with than rock wool due to its texture. However, some participants said that polystyrene is not easily used in buildings with irregular shapes or curvilinear plans because it is breakable. In many cases, polystyrene boards are broken into pieces when pumping concrete into walls, said participants 1b, 2b and 3a. Participant 3b stated that, unlike polystyrene, polyethylene is an effective material applicable to all shapes of buildings, and its foam can fill all the gaps, which results in less thermal bridges. Nevertheless, it is more expensive than polystyrene, said participants 1b and 3b (Table 7).

To conclude, polystyrene is the most commonly used insulation material in Jordan. According to the participants, natural insulation materials, like timber wool, bio-based materials, wood wool and sheep wool, are not commonly used locally.

5.7 Question #7: From Your Perspective, What Would the Priority Be in Retrofitting Between Generating Renewable Energy and Reducing Energy Demands? Can You Explain Why?

All participants preferred reducing energy demands over generating clean energy. Participant 1a claimed that the energy retrofit process might require a higher initial cost to reduce energy demand but does not require maintenance, unlike power plants and PV panels which require regular maintenance, said participants 3a, 3b and 1a. PV cells also occupy a large space and, according to participant 3b, most buildings in Jordan do not have enough space to host PV panels because their roofs are usually occupied with solar heaters and water tanks. Some participants said that this obstacle caused some building owners to rent a piece of land outside cities to install renewable generators to utilise a new system called the 'Wheeling System'. This system has been issued for people who do not have space in their buildings for the installation of PV cells, allowing them to generate electricity on other sites. Participant 1a said that this system is efficient for companies with high energy consumption rates. Another disadvantage of the renewable generators is that they have high initial costs. Also, participant 3a stated that the generators have to be replaced, unlike passive design solutions, which last as long as the buildings (Table 8).

Table 8 Summary of answers to question #7

Participants		Group 1		Group 2		Group 3	
		1a	1b	2a	2b	3a	3b
1	Generation of renewable energy	–	–	–	–	–	–
2	Reduction of energy demands	Yes	Yes	Yes	Yes	Yes	Yes

In conclusion, using passive design solutions is worthwhile, but people in Jordan do not seem to realise their significance. They do not realise that this should be prioritised over the generation of clean energy. Effective plans are needed for ensuring the safety of PV cell installations as their use is increasing in Jordan. The participants have not considered other forms of renewables other than solar energy such as wind power.

5.8 Question #8: Based on Your Professional Experience, What Are the Benefits of Upgrading the Performance of a Commercial Building in Amman, in Terms of Reducing Energy Demand and Providing Indoor Thermal Comfort for Tenants, for Both the Building Owner Specifically and the Country Generally?

On an individual level, most participants agreed on the benefit of providing thermal comfort for occupants. Some participants believe that this process improves productivity in the workplace as staff feel more comfortable and, therefore, it provides benefits to building occupants. This would reduce, for instance, employee absences in office buildings, which are often caused by building-related diseases. Therefore, improving the indoor environmental quality results in long-term benefits that are greater than reducing energy costs. Also, reducing energy costs has financial benefits on an individual level. Increasing people's awareness is important as well. For example, when people understand that thermal bridges are the reason behind mould in buildings, occupants will focus on increasing thermal insulation (Table 9).

On a national level, reducing energy consumption in buildings reduces the energy demand on a large scale, said participant 1b. This would reduce the energy problem in Jordan, said participants 1a, 3a, 2b and 3b and therefore, benefit the country. Another advantage is the reduction of the country's budget for developing new infrastructure for new areas, and therefore conservation of greenfields, due to retrofitting existing building stock.

In conclusion, there are many advantages to the energy retrofit process on both an individual and country level. The more people that understand the significance of passive design strategies, including thermal insulation, the more productive building occupants would become. These strategies have considerable impacts on individuals, including improving building occupants' health, and on the conservation of resources for the country.

5.9 Question #9: Do You Think that This Process Could Be Financially Feasible for Jordanian Building Owners, and Why?

All participants stated that an energy retrofit is a financially feasible process. Participant 2a said that it reduces costs and the time required for construction. In the case of commercial buildings, participant 1b said that it would be financially feasible only if the building owner uses the building, i.e. not renting it to other occupants. Participant 1a noted that the initial cost is usually paid once, while the energy savings would be reflected in the electricity bills and heating or cooling costs throughout the building's lifespan, and the payback period would be shorter, not exceeding five to eight years in many cases, according to participant 2b. Participant 3a also said that the process is financially feasible, especially in commercial buildings where usually high electricity bills are reduced (Table 10).

Therefore, the energy retrofit process can be financially feasible in Jordan, but this needs further work to provide the local community with successful cases and studies proving it. Also, a special focus should be given to identifying the payback period for energy retrofitting strategies.

5.10 Question #10: Concerning People's Perceptions About Green Buildings in Jordan, Including Building Certification Systems, Do You Think that This Retrofit Process Can Deliver Benefits to the Building Market? Please Give a Reason

Certified green buildings still need some time to become a common practice in Jordan, but this is already a trend in Amman amongst large corporations and national and international organisations. Participant 1a stated that some companies are developing policies to follow the sustainability

Table 9 Summary of answers to question #8

Participants		Group 1		Group 2		Group 3	
		1a	1b	2a	2b	3a	3b
1	Benefits for building owners	Increase occupants' productivity Reduce employee absence Reduce energy bills	Reduce energy bills	Reduce operational costs	Improve thermal comfort Reduce the use of HVAC systems Reduce energy bills	Higher productivity levels for occupants	Direct benefits for building users and owners
2	Benefits for the country	Reduce burdens on the country	Reduce CO_2 emissions Improve the thermal comfort of outdoor spaces Reduce energy demand on a larger scale	Move towards a greener economy Reduce the need for providing infrastructure for new areas Conserve cultivated land	Reduce burdens on the country	Reduce energy demand	Indirect benefits for the country Reduce energy demand

Table 10 Summary of answers to question #9

Participants		Group 1		Group 2		Group 3	
		1a	1b	2a	2b	3a	3b
1	Participants' answers	Yes	Yes	Yes	Yes	Yes	Yes

Table 11 Summary of answers to question #10

Participants		Group 1		Group 2		Group 3	
		1a	1b	2a	2b	3a	3b
1	Participants' answers	Yes	Yes	No	No	No	Yes

movement in the country to improve their businesses in the market. They have become more aware of the financial benefits of energy efficient buildings. However, this is not yet applicable to small-scale corporations and half of the participants said that energy retrofits do not provide marketing benefits at the moment, as they need a lot of time and work. Participant 2a gave an example, saying that if someone needs to rent an office, the initial cost would probably be prioritised in the buying decision regardless of whether the building is sustainable or not. The government in Jordan has issued new incentive systems to encourage people to invest in energy efficient buildings (GAM 2015). Participant 1b said that people's buying decisions in the building market might be influenced by a trusted green building certificate issued by a local independent party. The best practical marketing tool would be using real cases of energy retrofitted buildings as models, showing the amount of energy and operating costs that could be saved (Table 11).

In conclusion, much more time and work are needed to increase people's awareness of the significance of energy efficiency in buildings, and this will definitely provide marketing benefits to the building market.

5.11 Question #11: Do You Think This Upgrade Could Result in Energy Retrofitted Buildings that Meet Similar Standards for New Buildings that Already Adhere to Jordan's Energy Efficient Building Code?

- Yes.
- Maybe.
- No, it is not possible (Table 12).

On the one hand, half the participants believed that the energy retrofit process could ensure existing buildings have a similar performance to new energy efficient buildings. Participant 1a said that it could not achieve more than 90% of the Jordanian Energy Efficient Building Code

Table 12 Summary of answers to question #11

		Participants		Group 1		Group 2	
				Group 3			
	1a	1b	2a	2b	3a	3b	
1	Yes	Yes	Yes	–	–	–	Yes
2	–	–	–	Maybe	–	–	–
3	–	–	–	–	No, it is not possible	No, it is not possible	–

requirements as there are some recommendations that cannot be achieved in existing buildings, such as the building orientation and opening sizes. However, it can meet the mandatory requirements when implementing the passive design solutions using the correct methodology and ensure thermal bridges are avoided. Participant 2a claimed that an energy retrofitted building can achieve the new local Jordanian Green Building Guide rating system with a level (A) depending on technical aspects (GAM 2015). On the other hand, other participants think that it is challenging for energy retrofitted existing buildings to have energy efficient performances similar to new green buildings. The reason is that a new building is designed from the ground up, sometimes starting from site selection, so it is easier to deal with these projects from the early stages.

In conclusion, the success of the energy retrofitting process depends on the quality of buildings, their age, materials and many other factors, so some could achieve a similar performance to new energy efficient buildings while others would not. In general, retrofitting buildings gives a chance to control the growth of cities, improve the lifespan of buildings and preserve the embodied energy in buildings.

6 Conclusions

Amongst the issues discussed in this paper, a number of points can be highlighted.

First, it is clear that the performance of most Jordanian buildings in regard to energy efficiency and thermal comfort is poor. They do not comply with the standards of the local Energy Efficient Building Code for new buildings, so it is significant to define a target for making existing building stock meet those standards by encouraging all sectors towards retrofitting. This will save energy on a large scale, starting from the building level, through to the city, the district and the whole country. Energy efficiency is a fundamental issue and it "*should be the cutting edge of national energy policies for sustainable development*" (WCED 1987, p. 163). Since all participants stated that buildings consume the greatest amount of energy in Jordan, it is recommended to develop strategies encouraging people to invest in the improvement of their buildings through energy retrofitting.

Second, retrofitting existing buildings provides many advantages to improve the quality of life and well-being inside buildings, as well as in the surrounding environment. Uncontrolled urbanisation in the main northern cities in Jordan threatens vegetation and water resources (Saleh and Al Rawashdeh 2007) and, therefore, retrofitting the main existing urban areas through a sustainable approach could be an effective strategy to reduce the demand for new urbanisation and to protect arable land. Hence, retrofitting is an opportunity to control the unplanned growth of urban areas in Jordan, to improve building lifespans and to preserve their embodied energy. Wilkinson et al. (2014) said that the longer the lifecycles of buildings, the less carbon will be released. In other words, buildings would be responsible for fewer GHG emissions throughout their lifecycle.

Current practices in the Jordanian construction market give the lowest priority to environmental aspects and first priority to affordability. Therefore, the financial factor seems to be the greatest obstacle to the development of the energy retrofit concept in Jordan. Overcoming this obstacle requires effective cooperation between the local community and the government to achieve the aim of reducing energy usage in buildings. Since the government cannot provide enough financial support to people to encourage them to undertake retrofitting, it is recommended that strategies are developed to encourage the private sector to invest in such processes, especially the commercial sector. For example, since the main barriers to the spread of energy efficient solutions in Jordon appear to be mainly economic, Attia and Zawaydeh (2014) conducted research to prove the economic success of compliance with the Energy Efficient Building Code using different design strategies for an existing apartment in Jordan.

Moreover, there should be more focus on developing strategies to strengthen the supervision and auditing of works by corporate parties, including the Jordanian Engineers Association, Municipalities and the Ministry of Public Works and Housing. Royal Scientific Society (RSS) and Friedrich-Ebert-Stiftung Amman (FES Amman) (2013) stated that the lack of supervision and laxity in adopting green building codes for construction plans in Jordan is a major challenge.

Regarding the participants' answers, it could be observed that their academic backgrounds and work fields have an influence on their answers. For instance, architects have prioritised occupants' needs and local practices when answering the questions, while participants with engineering backgrounds have focused on the technological aspects. The third group have particularly focused on building regulations.

7 Recommendations and Further Research Proposals

Energy upgrade strategies have been discussed briefly in this study, so there is a recommendation for further studies that further examine the efficiency of each energy retrofit strategy when used in the context of Jordan. Furthermore, the issue of thermal bridges in Jordan's buildings was discussed in the article, but further studies are needed to analyse and develop strategies to reduce thermal bridges in existing insulated Jordanian buildings.

It is important to deliver the right information to the public regarding the technological and non-technological feasibility of the energy retrofit process. This requires effective outreach programmes to illustrate the financial feasibility and the expected payback period for retrofitting projects. Developing a documented experiment of a building that has been retrofitted in Amman is recommended not only for identifying the actual payback period but also to provide a full analysis of comparisons between the building situation before and after the retrofit, such as monitoring interior temperatures over a year.

As buildings in Jordan consume the greatest amount of energy, developing strategies to encourage citizens to invest in improving the condition of their buildings through energy retrofitting is highly recommended.

On the policy side, the Jordan Green Building Guide needs further work to be applicable to existing building retrofits. Regarding the ongoing challenges that Jordan is facing, further research is needed to develop resilient strategies for transferring the existing built environment to become resilient against ongoing stresses and potential risks including shortages of energy resources.

Acknowledgements This research was funded by the New Zealand government's Overseas Aid and Development Programme, year 2015. This research was approved by the University of Auckland Human Participants Ethics Committee (UAHPEC), on 1 April 2016, for three years. Reference Number 016827. Authors would like to thank all the participants who agreed to take part in this research.

References

Ministry of Environment, UNDP. Jordan's Third National Communication on Climate Change. Amman, Jordan; 2014.

Jordan GBC, Situational analysis for home owners and housing companies: "Building a smart program for saving water & energy through design and construction". Amman: Jordan GBC; 2013. Retrieved from https://jordankmportal.com/resources/introduction-on-building-a-smart-program-for-saving-water-and-energy-through-design-and-construction-arabic.

Shamout, S. Energy Retrofit of the Existing Building Envelope in Jordan: A Study on Mixed-Use Buildings (Retail and Office) in Amman, The University of Auckland, Auckland, New Zealand; 2017.

Mohsen, M., Akash, B. Some prospects of energy savings in buildings. Energy conversion & management, 42(11); 2001. p. 1307–1315.

World Commission on Environment and Development (WCED). Our Common Future. Oxford University Press; 1987. p. 163–166.

Jordan National Building Council. Green Building Guideline of Jordan. Amman, Jordan: Ministry of Public Works and Housing; 2013.

EBSCO Sustainability Watch. LEED Strategies for Existing Buildings. 2010. Retrieved from ebscosustainability.com: https://ebscosustainability.files.wordpress.com/2010/07/leeds-strategies-for-existing-buildings.pdf.

Fan, Y., Xia, X. A Multi-objective Optimization Model for Building Envelope Retrofit Planning. Energy Procedia, 75; 2015. p. 1299–1304.

Konstantinou, T., Knaack, U. An approach to integrate energy efficiency upgrade into refurbishment design process, applied in two case-study buildings in Northern European climate. Energy and Buildings, 59; 2013. p. 301–309.

Ma, Z., Cooper, P., Daly, D., Ledo, L. Existing building retrofits: Methodology and state-of-the-art. Energy and Buildings, 55; 2012. p. 889–902.

Ungureanu, V., Bollinger, R., Fulop, L., Jakob, M., Kiss, B., Maneschi, D. Integrated Strategies for Retrofitting Buildings to Reduce Primary Energy Use, GHG Emissions and Costs. Central Europe towards Sustainable Buildings CEBS 13; 2013.

Ouahrani, D. Towards energy efficient buildings in Amman, Jordan: Defining thermal requirements by mean of thermal simulations. 44th Annual Conference of the Architectural Science Association, ANZAScA 2010, Unitec Institute of Technology, Auckland, New Zealand; 2010.

Attia S, Zawaydeh S. Strategic Decision Making For Zero Energy Buildings in Jordan. Proceedings of the ASHRAE Conference on Energy and Indoor Environment for Hot Climates. Doha, Qatar; 2014.

Daher, M. I. Comparing Green Structures with Different Thermal Efficiency Designs. Jordan International Energy Conference, Jordan Engineers Association (JEA), Amman, Jordan; 2011.

Housing Development & Management (HDM) and the Royal Scientific Society (RSS). Climate Conscious Architecture and Urban Design in Jordan; 2009.

GAM. Foundations of the application of green building standards and incentives offered by the Greater Amman Municipality. Amman; 2015. Retrieved from Greater Amman Municipality official website: http://www.ammancity.gov.jo/site_doc/green.pdf.

Saleh, B., Al Rawashdeh, S. Study of Urban Expansion in Jordanian Cities Using GIS and Remoth Sensing. International Journal of Applied Science and Engineering, 5(1); 2007. p. 41–52.

Wilkinson, S. J., Remoy, H., & Langston, C. Sustainable Building Adaptation: Innovations in Decision-making. Somerset, NJ, USA: John Wiley & Sons, Incorporated. 2014.

Royal Scientific Society (RSS), Friedrich-Ebert-Stiftung Amman (FES Amman). Green Building Development in Jordan. Amman: the Royal Scientific Society of Jordan and the Friedrich-Ebert-Stiftung; 2013. Retrieved from http://library.fes.de/pdf-files/bueros/amman/10678.pdf.

Part III
Alternative and Renewable Energy Resources Quest in Architecture and Urbanism

In the last part of this book, the authors provide highly technical solutions to issues concerning sustainable development in urban environments. The chapter titled "Biofuel in Developing Countries—Ethical Concerns" discusses possible ethical, social, economic, environmental, and technical concerns associated with extracting biofuel. The chapter focuses, however, on the ethical issues that arise as a consequence of the methods of current production of biofuels such as human rights, solidarity, sustainability, stewardship, and justice.

In "Solar Photovoltaic (PV) Power Systems in Jordan: The Past, the Present and the Future?", the authors review the history of PV power systems in Jordan since the early eighties of the past century. Then, the chapter critically discusses the progress in the total installed PV capacity after the Renewable Energy and Efficiency Law (REEL) became in place in Jordan in year 2014. It concludes with speculations for the future of PV market in Jordan and the expected technical limitations. Similarly, in the chapter titled "The Application of a Solar Parabolic Trough Collector for the Melting of Plastic Waste", the author assesses the ability of developing countries to control plastic waste using thermal energy. This affects their ability to recycle and thus achieve higher standards of sustainability. The authors, therefore, test the feasibility of applying a parabolic trough collector for melting plastics using solar thermal collectors.

Aiming to reaching the sustainable development goals in Jordan, the authors of the chapter titled "The Application of Ground-Source Heat Pumps for a Residential Building in Jordan" discuss heating and cooling loads as the main drivers for energy consumption and one of the major sources of pollution in Jordan. They, therefore, examine geothermal energy as one of the solutions that can be used to minimize pollution particularly in Jordan, where there are abundant geothermal resources that can be utilized effectively. In this chapter, a typical Jordanian residential building is analyzed to find the optimum size of heating and cooling systems that work by using a ground heat exchanger. Also taking a closer look at Jordan, the authors of the chapter titled "Wind Resource Assessment for a Proposed Wind Farm" discuss the fact that 96% of the country's energy need comes from oil and natural gas importation. This has motivated the government to develop a large investment plan for the utilization of renewable and nuclear power for energy generation. This chapter assesses the wind data available for a proposed site to examine the possibility of building 100 MW wind farm located in Ajloun, north of Jordan.

Biofuel in Developing Countries—Ethical Concerns

Aiman Albatayneh, Yaqoub Al-Khasawneh, Firas Alawneh, Atif Alkhazali, and Sulaiman Mohaidat

Abstract

The uses of energy are vast, and the demand for it increases every year. With this, energy is becoming a scarce commodity, and therefore, scientists are coming up with different sources to derive energy from. Energy is obtained from many different types of sources commonly oil, coal and natural gas. Renewable sources are also existent and include but not limited to wind, solar and hydropower. But without efficient use of these resources, energy sources are slowly being depleted causing a demand to find more energy sources. Biofuel is energy obtained from biomass or biowaste, basically plant and animal material. It has been a great solution for the energy problems that the world is facing, but there have been concerns with its production. The concerns are ethical, social, economic, environmental and technical issues that arise out of the methods of the current production of biofuels. Social and ethical issues that arise are such as human rights, solidarity, sustainability, stewardship and justice. In this paper, the principle outlines that the environment should be able to sustain biofuels which should also help in the reduction of greenhouse gas emissions. Trade principles relating to biofuels should be fair and benefits be distributed equitably. Its usage should not come as an expense of the communities' basic rights of food, water and work rights. Since biofuel is obtained from plants and animals, too much production of it may lead to reduced food distribution to the people, which denies them the basic right of food.

A. Albatayneh (✉) · Y. Al-Khasawneh · F. Alawneh
Energy Engineering Department, School of Natural Resources Engineering and Management, German Jordanian University, Amman, Jordan
e-mail: Aiman.Albatayneh@gju.edu.jo

A. Alkhazali
Industrial Engineering, Faculty of Engineering, The Hashemite University, Zarqa, Jordan

S. Mohaidat
Engineering Department, Jordan University of Science and Technology, Irbid, Jordan

Keywords

Ethics • Biofuels • Food and energy

1 Introduction

Biofuel production has increased over the recent years in a bid to meet the increasing energy needs that the other fossil and nuclear sources cannot meet. Petroleum is the lifeblood of our civilization, but for the last century, petroleum-derived fuels have been the major source of the world's needs of energy. However, predictions that oil will be used up in the near future and environmental concerns have trigged the examination of alternative energy sources, as shown in Fig. 1. Future projections indicate that economics and energy needs will increase the focus on the production of synthetic fuels derived from non-petroleum sources including biomass and waste products among others (Tashtoush et al. 2007).

One way to solve the problems above is to look for alternative and renewable energy sources. There are many forms of renewable energy such as ethanol which is renewable since it can be produced from biomass and can be used in gasoline engines. Photovoltaic cells are solar plants that generate electricity from sunlight. Wind turbines are machines that generate energy from wind. Hydropower is produced from flowing water. Producer gas is a gas made from waste wood that can be used as a renewable fuel. Each of these renewable energy sources serves a specific purpose. One of the most promising alternative energy sources is biomass. It is renewable, widely available, and containing much less sulphur and nitrogen compared to fossil fuels thus making it more environment-friendly.

Biomass extract from living organisms such as animal fat, oil crops, sugar plants, starchy plants and cellulosic biomass to produce liquid and gaseous biofuels. In some power plants, biomass can either be combusted by itself or with other fossil fuels to generate electricity; also, it can be

C. Alalouch et al. (eds.), *Advanced Studies in Energy Efficiency and Built Environment for Developing Countries*, Advances in Science, Technology & Innovation, https://doi.org/10.1007/978-3-030-10856-4_13

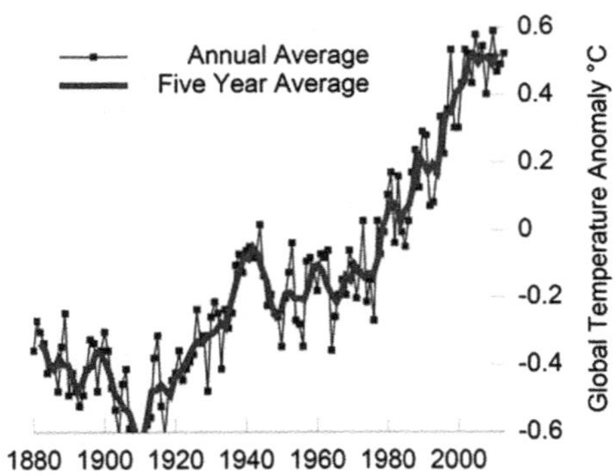

Fig. 1 Instrumental temperature record (NASA)

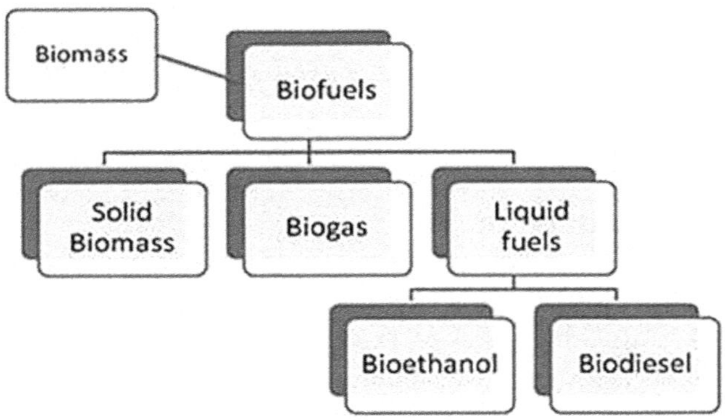

Fig. 2 Different types of biofuels (IGEM 2017)

transformed to liquid for easier transportation or gas as biogas for gas power plants as shown in Fig. 2.

1.1 Biofuels

"A biofuel is a type of fuel whose energy is derived from biological carbon fixation. Biofuels include fuels derived from biomass conversion, as well as solid biomass, liquid fuels and various biogases" (Demirbas 2008).

Biofuel has a great deal of advantages. They produce less greenhouse gases than fossil fuels when burned allowing countries with no oil reserves to have fuel security. For example, biodiesel is a liquid transport that is easy to transport making it efficient as it also does not need a large storage space. Biofuels are also much more easily renewable as crops can be grown. An economic advantage would be that they are less expensive compared to the fossil fuels.

1.2 Types of Biofuels

There are two types of biofuels: first-generation biofuels and second-generation biofuels. First generation is made from sugar, starch and vegetable oil. They are made from unsustainable feedstock and constitute the greater part of biofuels which are presently in use. The second-generation biofuels are obtained from the sustainable feedstock, but it is still underdeveloped and under research. Examples of second generation are methanol, biohydrogen and cellulose ethanol. Others include the algae-based biofuels.

Among biomass sources, waste vegetable oils (WVOs) and waste animal fat (WAF) have attracted much attention as potential renewable resources for the production of an alternative for petroleum-based diesel fuel. Food consumption around the world produces large amounts of WAF and WVO that, in many world regions, are disposed of in harmful ways.

Alternative engine fuels are considered to be one way of reducing the growing pressure from traffic on environmental and energy demands. They could create an alternative for the classic fossil fuels such as petrol, resources of which will be depleted in a few decades.

For instance, biodiesel is a fuel made from VO or WVO or animal fat or WAF, we use WAF and WVO to make biodiesel, and it is a stable fuel performs reliably in all diesel engine, safe to handle and work with all diesel fuel storage and pumping systems. No engine modifications are needed to use biodiesel in a diesel engine. Biodiesel can be poured straight into the fuel tank of any diesel vehicle.

This method involves a diesel engine that can run on straight vegetable oil as long as the engine is started on diesel fuel and the vegetable oil is heated. Used cooking waste vegetable oil (WVO), new oil, or WAF can be used as a fuel, but it required installing a vegetable oil tank and modified the heater hoses of the diesel engine.

Once the tank is installed and the heater hoses have been modified, the diesel engine started on diesel or biodiesel fuel and then switched to hot straight vegetable oil. When the engine is running, the engine coolant is used to heat the vegetable oil so that it has a similar viscosity to diesel fuel when it is time to turn the engine off; it is cooled down on diesel or biodiesel fuel.

1.3 Food Versus Fuel

There is a risk in redirecting farmland crops for biofuel production. Some products such as corn, sugarcane and vegetable oil can be either used as feeds or for biofuel productions. This increases the prices of such commodities and their related counterpart as demand arises for them. It is, however, a complicated situation that brings about contradictory views on which are requires greater priorities.

On the other hand, the criticism is established on an naive interpretation of the ethical concerns of the food security consequences of biofuel. Three ethical accounts can be raised at the food vs. fuel argument. First, market drivers of biofuels undeniably compete with the consumption of agricultural grains were more farmers encouraged towards biofuels for more profit gains. Second, the criticism does not reflect an satisfactory sympathetic of the way that increasing food prices and modifications in agricultural technology affect the ability of poor people to secure their daily food. Third, increasing food prices could be beneficial to poor farmers in the developing countries (Thompson 2012).

2 Ethics Discussion

As mentioned above, the production of biofuels raises various ethical issues which shall be discussed below.

2.1 Rights

"Biofuels production and development should not be at the expense of people's essential rights". (Nuffield Council on Bioethics 2011).

One such right is the right for sufficient food. There has been a rise in the demand for food and water for the processed of biofuel productions, and this has threatened the essential right of food. The increased demand results in increased food prices globally making the commodities less affordable to people with lower standards of living.

The attractive prices of biofuel crops will lead farmers to focus more on growing them than their usual crops, creating scarcity of such types of food and ultimately resulting in hiked prices not only for the biofuel crops but other food crops as well. In this chain reaction, the prices of meat would also become higher because the livestock feed is used for biofuel production. This high food prices can result in drastic consequences such as malnutrition in children and famine.

Before 2008 financial crisis, almost every region of the world has experienced drastic food price inflation this year. Retail prices are up 17% in Sri Lanka, 18% in China and 10% or more throughout Latin America and Russia. Zimbabwe tops the chart with a more than a 25% increase. That inflation has been driven by double-digit price hikes for almost every basic foodstuff over the past 12 months. Dairy products are as much as 200% more expensive since last year in some countries. Maize prices hit a 10-year high in February. Wheat is up 50%, rice up 16% and poultry nearly 10% (Kingsbury 2007).

The right of good health is also essential for humans, and it is being abused greatly by the production of biofuels. The forests are being cleared, and the carbon emitted in this process results in air pollution which is dangerous for people living in the surrounding areas. Growing biofuel crops also requires the use of fertilizers and pesticides which results in water pollution of the surrounding water bodies making people ill.

If such crops are consumed by humans, the dangerous chemicals used to grow them will have adverse effects to their health and well-being in general. With increase in prices, food will not be reasonable and hence many people will not afford a balanced diet and resulting in malnutrition.

Humans have a right to employment, and some have argued that the production of biofuels will result in employment opportunities, and thus, people will have more jobs. There will be a need for more farmers, bioscientists, engineers, environmental entrepreneurs and construction and industrial workers.

2.2 Duties

There is an ethical duty to develop biofuels whose production will not harm the environment, local people, and economy or cause a change in climate of the area. The cost of the development should not outweigh the rewards obtained from their uses. The competition of food as a biofuel source versus it as a source of nutrition to the people must be kept at a minimum so that food prices are kept in check.

There is a need to search other options of energy production that are also better at reducing gas emissions as well as maintaining the ethical principles required. Ways to produce significant energy amount with minimum resources should be researched so as to use the crops for alternative uses. All uncertainties that arise from the technological methods of biofuel production should be addressed, and efforts are made to reduce them. Regulations should be put up to ensure that the ethical duties are maintained.

2.3 Virtues

"Costs and benefits of biofuels should be distributed in an equitable way" (Nuffield Council on Bioethics 2011). Policies must encourage equity in distribution and consumption both in local and international companies. Developing countries should not be disadvantaged in any way, politically, environmentally and economically. They do not use biofuels as much as the developed county, yet they pay the highest prices for its use. The ethical virtues are to protect the people whose rights may invoke by the biofuels development in the area.

2.4 Utilitarianism

The simple definition of utilitarianism is that the summation of all pleasure that results from an action, minus the misery of anyone involved in the action (Bentham 1972). Where the finest action is the one that make the most of the well-being of the individual perceptive in another words it considers the intrests of all beings equally.

Biofuel production should aim to provide maximum happiness to all the parties socially, economically, politically and environmentally. In the developed world, investment in biofuels which reduce gas emissions results in a better climate change but in less developed countries where small-scale production is done, and the communities tend to be disadvantaged.

A limited number of people are likely to enjoy the benefits of its production, whereas the greater number suffers from increased food prices, especially in the developing countries.

2.5 Environmentally Sustainable

The environment should be able to sustain biofuel production. The diversity of biological organisms is threatened when forests and peatlands are cleared to create land for the growth of biofuel crops. Biofuel crops generally produce low amounts of energy and so more and are needed to expand production resulting in reduced agricultural land forcing farmers to seek other areas.

Although the use of biofuels reduces gas emissions, the use of pesticides are chemicals in the process of intensification reduces the quality of water and causes soil erosion and degradation in the area. Second-generation biofuels that use non-food crops should be developed in order to produce fuels without violating the environmental principles.

Standards should be set to ensure that environmental pollution is reduced by facilitating and promoting less emittance fuel and more environmentally sustainably biofuels to be developed considering all the issues disscussed above.

2.6 Stakeholders and Their Responsibilities

The stakes are high in the production of biofuel, and each stakeholder has their own responsibility. The farmers should focus on growing food crops and not shift to biofuel crops due to the attractiveness of their prices. The biofuel producer on the other hand should carry out advanced research on new methods of production that reduces pollution and other negative effects. They should work together with the environmental representatives.

The government should pass laws that protect the environment and the humans regarding biofuel productions. These laws are to be observed by all the players involved in the industry, including the consumers. The consumers should use biofuels reasonably and effectively.

3 Recommendations

There is ethical duty to develop a new biofuels (second-generation biofuel) which can be produced in commercial scales without harming the environment, economy, local people and humanity because it is:

- Highly beneficial for local people as they do not need to use the premium land to grow first-generation biofuel crops; they can use any location to grow second and third-generation biofuel crops because it needs less resources (water, land and labour).
- Minimize the competition with food production which reduces the food hiking inflation prices.
- Significant amount which can be produced with minimum resources will be highly economically beneficial for investors.
- Maintain the current jobs; even more jobs will be created especially if the farming extends to new non-farming lands (lower-quality lands).
- There are ethical virtues to protect vulnerable people foods supply and availability and their health from production and using biofuels.
- Develop a sustainable future and protecting the environment and ensuring sustainable use and production of biofuels.
- Use lower-quality land to produce biofuel crops and keep the premium land for human food production. For biofuel producer duty to use latest technology (second-generation biofuel) and support research to produce crops with less usage of land and water.
- Ensure the production, and usage of biofuel does not harm the environment (environmental life cycle risk assessment for biofuel).
- Use the biofuel responsibly and in efficient ways.
- The biggest loser between all the stock holder, and they have the right to oppose using food as a biofuel.
- Legislations to protect premium land are being used by biofuel producer.
- Legislation to protect environment.
- Support research and development for second-generation biofuel and more efficient way to use energy.
- Legislation to protect humanity rights to access food even if they are in other countries.

3.1 Future Research Proposals and Suggestions

Instead of using fresh food products, we can use wast food such as waste vegetable oil (WVO) and waste animal fat (WAF) to produce biofuel which I worked on and published the report "Factorial analysis of diesel engine performance using different types of biofuels" in Journal of Environmental Management 84 (2007) 401–411. By using this method, we can:

- Minimize the competition with human food resources.
- Significant amount of WVO and WAF can be collected from local market such as expired oil product and from restaurant especially if we establish an efficient collection method to collect WVO and WAF which at least will reduce the demand on human food to produce biodiesel.
- Provide effective disposal way (recycle) for the WVO and WAF instead of dump it.
- Reduce food inflation prices by reducing the demand on the products which used to produce biofuel.
- Using non-food crops for biofuel such as second-generation biofuel which can be produced in commercial scales without harming the environment, economy, local people and humanity.
- Biofuels during all productions and usage stages should contribute to greenhouse gasses reductions by using efficient way to extract biofuel fuel from crops and use the biofuel responsibly and efficiently to reduce GHG emissions.

Finally, there are no clear and agreed solutions for biofuels issues, but the main framework should include:

- Biofuels usage and production should consider human rights for food, health and work.
- Biofuels should be environmentally sustainable and reduce greenhouse gasses emissions.
- Costs and benefits of biofuels should be distributed in an equitable way to deal with biofuels.

But instead of using fresh food product which consumed by human, we can use waste vegetable oil (WVO) and waste animal fat (WAF) to produce biofuel, non-food crops for biofuel and second-generation biofuel instead of first-generation biofuel.

4 Conclusions

Biofuel is energy attained from plant and animal waste. It has been an inordinate solution for the energy issues that the world is facing, but there have been worries with its production. The concerns are ethical, social, economic, environmental and technical issues that arise from the ways of the current production of biofuels. Social and ethical issues that arise are such as human rights, solidarity, sustainability, stewardship and justice.

Biofuels and its advancement in developing countries have its advantages and disadvantages, and the stakeholders are coming up with various solutions regarding its production and uses. Although discussions on the issue are taking place, there are no clear and agreed solutions yet.

Nonetheless, the main framework should include five key issues. First, biofuels and its production should be environmentally sustainable and legislative laws should be passed to ensure so. Secondly, the costs and benefits of biofuels should be equitable. The costs should not outweigh the benefits. Thirdly, biofuels reduce gas emissions making them preferable to conventional fuels such as fossil fuels. Fourthly, more research should be conducted to facilitate the use of second-generation biofuels for production of energy, and lastly, ethical duties and virtues are to be observed to ensure that human and environmental rights are not invoked.

There is risk in transmitting farmland crops for biofuel production. Some products such as corn, sugarcane and vegetable oil can be either used as feeds or for biofuel productions. This increases the prices of such commodities and their related counterpart as demand arises for them. It is, however, a complicated situation that brings about contradictory views on which are requires greater priorities.

Waste vegetable oils (WVOs) and waste animal fat (WAF) have a great potential as renewable resources for the production of an alternative for petroleum-based diesel fuel. Food consumption around the world produces large amounts of WAF and WVO that, in many world regions, they disposed of in harmful ways. They could create an alternative for the classic fossil fuels such as petrol, resources of which will be depleted in a few decades without competing with the human main food sources.

Acknowledgements The authors acknowledge the financial support of the Deanship of Scientific Research at the German Jordanian University for the presented work in this paper.

References

Tashtoush GM, Al-Widyan MI, Albatayneh AM. Factorial analysis of diesel engine performance using different types of biofuels. Journal of environmental management. 2007 Sep 30;84(4):401–11.

IGEM, Institution of Gas Engineers & Managers, Biofuel Production from Biomass. (n.d.). Retrieved June 20, 2017, from http://www.igem.org.uk/technical-standards/research/biofuels/production.aspxDemirbas A. Political, economic and environmental impacts of biofuels: a review. Applied energy. 2009 Nov 30;86:S108–17.

Kingsbury K. After the oil crisis, a food crisis?. Time Magazine. 2007 Nov;16.

Thompson PB. The agricultural ethics of biofuels: the food vs. fuel debate. Agriculture. 2012 Nov 6;2(4):339–58.

Nuffield council on bioethics, oN BioEThiCs, NUffiEld CoUNCil. "Biofuels: ethical issues." (2011).

"Read "Renewable Fuel Standard: Potential Economic and Environmental Effects of U.S. Biofuel Policy" at NAP.edu." National Academies Press: OpenBook. N.p., n.d. Web. 31 Mar. 2017. <https://www.nap.edu/read/13105/chapter/6#164>.

"Liquid Biofuels for Transport Prospects, risks and opportunities." Biofuels: 1. What are biofuels? N.p., n.d. Web. 31 Mar. 2017. <http://www.greenfacts.org/en/biofuels/l-2/1-definition.htm>.

House of Commons Environmental Audit Committee. "Are biofuels sustainable." The Government Response (2008).

Bentham J. An Introduction to the Principles of Morals and Legislation (Chapters I–V). Blackwell Publishing Ltd; 1972.

Demirbas A. Biofuels sources, biofuel policy, biofuel economy and global biofuel projections. Energy conversion and management. 2008 Aug 1;49(8):2106–16.

Solar Photovoltaic (PV) Power Systems in Jordan: The Past, the Present and the Future?

Firas Alawneh, Aiman Albatayneh, Mohammad Al-Addous, Yaqoub Al-Khasawneh, and Zakariya Dalalah

Abstract

Photovoltaics (PVs) is the direct conversion of sunlight into electricity. It has many advantages compared with other energy sources as PV has no moving parts and the fuel source (sunlight) is free, abundant and widely distributed, available to every country and person in the world. For the last two decades, PV has been the fastest growing industry of its size. Continuing at the present, growth rate of 40% for the next two decades will allow PV to be the world's largest energy source. For this reason, Jordan started recently installing PV power systems in a wide range. This paper will discuss the history of PV power systems in Jordan since the early eighties of the past century, in addition to the progress achieved so far in the total installed PV capacity after the Renewable Energy and Efficiency Law (REEL) became in place in Jordan in year 2014 besides discussing the future PV market in Jordan. In the early eighties of the past century, PV applications, namely solar home systems, telecommunication, water pumping and brackish water desalination, were limited in remote areas as off-grid or standalone PV installations. The total installed capacity of these installations did not exceed 1.5 MW, according to the records of the Ministry of Energy and Mineral Resources (MEMR) in Jordan, before the REEL became in place in Jordan in year 2014. This paper will present an overview about these off-grid PV installations and the reasons behind the limited market. In addition, the paper will show the present status of PV market in Jordan after the REEL became in place. According to recent records obtained from the Energy and Minerals Regulatory Commission (EMRC) in Jordan, the total PV installed capacity in Jordan exceeded 300 MW distributed in large-, medium- and small-scale projects according to the kind of installation, whether it is energy net metering, power wheeling or power purchase agreement. These kinds of installations will be discussed briefly in this paper along with the amount of installed capacities. The future PV market in Jordan will be discussed at the end of the paper to show the opportunities in this market and the expected technical limitations especially those related to the capacity of the national grid to absorb the power that will be produced from PV projects under planning.

Keywords

Photovoltaics (PVs) • Off-grid PV system • On-grid PV system • Renewable energy and efficiency law • National grid • Cumulative solar PV capacity • Specific solar PV electricity yield • Jordan's Renewable Energy and Efficiency Law (REEL) • Jordan's energy strategy (2007–2020) • 10% renewable energy target

1 Introduction

Jordan is facing a real challenge in securing energy supply represented by almost no indigenous energy resources, highly dependent on imported energy (97% import in 2012), high cost (the energy imports accounted for 18% of GDP in 2012) and high growth of primary energy demand (2008–2020: electricity demand growth 7.4% and primary energy demand growth 5.5%). The volume of electricity generated in 2015 reached 18,911 GWh, and the peak load of the electricity system has recorded 3300 MW (Annual Energy Report 2015).

Jordan has a huge potential of renewable energy utilization (solar and wind) and energy resources (oil shale and uranium). In regard to renewable energy projects, a target of 10% renewable energy input into the energy mix by 2020 is set in the national energy strategy (2008–2020), mainly aiming for about 1000 MW of wind and 600 MW solar as Jordan has great potential sources of renewable energy,

F. Alawneh (✉) · A. Albatayneh · M. Al-Addous · Y. Al-Khasawneh · Z. Dalalah
Energy Engineering Department, German Jordanian University, Amman, Jordan
e-mail: firas.alawneh@gju.edu.jo

C. Alalouch et al. (eds.), *Advanced Studies in Energy Efficiency and Built Environment for Developing Countries*, Advances in Science, Technology & Innovation, https://doi.org/10.1007/978-3-030-10856-4_15

particularly solar and wind energy. Jordan lies in the earth–sun belt area and has vast solar energy potential with an average global solar radiation, ranging between 5 and 7 kWh/m^2 per day, one of the highest figures in the world. The country, with an estimated 330 days of sunshine per year, is also blessed with relatively moderate temperatures and low dust and humidity levels: ideal conditions for the use of solar PV energy (Unlocking thr Sunbelt 2010).

To achieve this 10% renewable energy target by 2020, the Renewable Energy and Efficiency Law (REEL) was passed in April 2012 that has been later amended in 2014. This law was the first in the region that all owed local and foreign investors to identify and develop grid-connected electricity production projects through five different approaches, namely (Updated Master Strategy of Energy Sector in Jordan 2007);

- Direct proposals.
- Competitive bidding.
- Engineering, procurement and construction (EPC) turn-key.
- Energy net metering.
- Electrical power wheeling.

The REEL was a turning point in the widespread of market penetration of PV systems in Jordan. This paper will discuss a survey regarding cumulative operational, committed (contracted) and planned solar PV capacity in Jordan during 1980–2020 based on the records available at the Ministry of Energy and Mineral Resources (MEMR 2014), Energy and Minerals Regulatory Commission (EMRC), National Electric Power Company (NEPCO), in addition to local and international papers and press releases in this regard. The percentage of contribution of PV electricity in the total electricity demand in Jordan is calculated based on Jordan's typical annual specific PV electricity yield and annual growth in total electricity demand (Annual Report for the National Electric Power Company 2015).

The barriers facing the penetration of PV systems in Jordan represented technically by the incapability of the national electric grid to withstand the demanding PV power installations in the next years. The Green Corridor Project managed by the National Electric Power Company (NEPCO) will be discussed in this paper based on local and the international papers and reports published about this issue (Annual Report for the National Electric Power Company 2015).

2 PV Systems Survey (1980–2012)

Since early 1980s until 2012, many PV systems for different off-grid applications (remote telecommunications, solar home systems, water pumping, water irrigation, brackish water desalination, etc.) were installed over Jordan, especially in remote areas suffering from the lack of water and electricity networks. Many of these PV systems were installed to demonstrate the technical and economic feasibility of PV electricity generation in comparison with diesel generators under the Jordanian weather conditions that were mainly funded by the local and international funding agencies due to the high cost of off-grid PV systems with battery storage (>US$7000/kW installed) at that time. In 2012, the total installed capacity of PV systems reached only about 1.5 MW including some grid-connected PV systems without any battery storage which were installed only for the purpose of self-consumption and demonstration without injecting any kWh into the utility grid before the REEL became in place (Enabling PV in the MENA Region 2014).

3 PV Systems Survey (2013–2016) and Projection (2017–2020)

After 2012, after the REEL became in place, the total installed PV capacity so far has reached about 310 MW until the preparation of this paper, and the government is planning to build a new high-voltage transmission line connecting Jordan's south with the large electricity consumption in the north. This would allow for another 300 MW of solar PV to be connected. Operational, committed (contracted) and planned PV systems in Jordan will be discussed in this chapter. The prices for PV projects went down below US $1500/kW which stimulated the market a lot (Developing renewable energy projects 2016).

As mentioned in Introduction of this paper, Jordan is engaged with a 5-track approach to develop RE projects. These approaches or schemes will be discussed below with some details.

3.1 Direct Proposals

Direct proposals were submitted through three rounds, namely Round 1, Round 2 and Round 3.

In Round 1, twelve (12) solar PV power plants with a total nominal capacity of 200 MW were erected and commissioned in years 2015 and 2016; ten (10) projects with a total capacity of 170 MW located near Ma'an city in the southern part of Jordan, one (1) PV power plant with a capacity of 20 MW located west to Mafraq city in the northern part of Jordan and another one (1) PV power plant with a capacity of 10 MW located in Aqaba city in the far south of Jordan. Power purchase agreements (PPAs) for the twelve (12) projects were signed in March 2014, and financial close was finished by the end of May 2015. The twelve (12) projects costed around US$500 million and offered more than 150 job opportunities which are considered the largest commercial projects of its kind in the region. The feed-in tariff Round 1 was very attractive for the investors, which was set by the Government of Jordan through the Energy and Minerals Regulatory Commission (EMRC) at 100 JD fils/kWh or approximately 14 US$ cents/kWh (EMRC 2017).

In Round 2, which was launched in August 2013, 83 applications were received on 14 November 2013 for PV projects (50 MW capacity each), out of them 45 MOUs had been signed. A total of 34 solar PV proposals were received in February 2015, 24 of them technically were qualified in May 2015. PPAs had been signed with the top four ranked bidders (50 MW for each project) in terms of feed-in tariff with a total nominal capacity of 200 MW. The average of the top four tariffs in this round is over 50% lower than the Round 1 tariff, reflecting a dramatic improvement in the cost of solar in Jordan. The winning tariffs range approximately between 6 and 8 US$ cents which are considered as the new benchmark for utility-scale solar projects in the stable markets within the Middle East.

Regarding Round 3, it was first launched in 2014 and cancelled in same year, because the studies showed later that the national grid will not be able to withstand the transmission of the generated power from the solar and wind projects in the southern regions of the country to the electricity consumption centres in central and northern Jordan.

So, the idea of the Green Corridor Project arose as a vital for the development of renewable energy in Jordan because it will allow solar and wind power generated to be transmitted as mentioned before. The National Electric Power Company (NEPCO) is responsible for the Green Corridor project by adding two 400/132 transformers with a capacity of 800 MVA, thirteen 400 kV bays and eleven 132 kV bays to Ma'an substation 400/132/33 kV and is expected to complete and operate the project in the year 2018 to bring 650 MW of transmission capacity. The project will include another phase by installing connections between the Ma'an substation and Qatraneh substation to add 1500 MW of transmission capacity by 2020. The total cost of the Green Corridor project is expected to reach approximately US$160 million.

Accordingly, Round 3 was relaunched in December 2016 for 200 MW solar PV projects to be installed within Ma'an Development Area (MDA) called (MDA 2) next to the 170 MW project of Round 1 called (MDA 1).

3.2 Competitive Bidding

Several projects are under investigation by the Ministry of Energy and Mineral Resources (MEMR) to be tendered on public lands (state-owned) on due time, pending on grid capacity availability (Green Corridor Project) in year 2018.

3.3 Engineering, Procurement and Construction (EPC) Turn-Key

In 2015, two solar PV projects with total capacity of 5.2 MW were constructed at Azraq located 90 km eastern Amman, in cooperation between with the Government of Jordan and the Spanish Government, and are currently operational.

In 2016, a 100 MW solar PV project near Quweira town within Aqaba Governorate, funded through a grant from Abu Dhabi Fund (US$150 million) is already contracted with an Abu Dhabi-based EPC contractor Environment Power Systems and Spain's TSK Group to build the project. Also, Masdar, a clean energy developer based in Abu Dhabi, UAE, in October 2016 signed a power purchase agreement (PPA) with Jordan's National Electric Power Company (NEPCO) for its 200 MW solar PV plant Global Market Outlook for Photovoltaics, 2012.

3.4 Energy Net Metering

Net metering is a billing mechanism that credits costumers for excess self-generated electricity they feed back into the grid. Each month the consumers pay the utility the net difference, on kWh basis, between what they consumed and what they fed back into the grid. If what they produced exceeds what they consumed, the excess is rolled over to the next month on kWh basis. Characteristics of Jordanian scheme: projects larger than 5 MW are unable to make use of the scheme. The size of the solar system making use of the scheme cannot exceed the consumer's aver age monthly consumption. In the case of electrical energy surplus production, the financial settlement is made at the end of the year at a price of 120 JD Fils/kWh or approximately 17 US$ cents/kWh.

According to the records of the Energy and Minerals Regulatory Commission (EMRC), the accumulated net metering PV-operated systems at the end of November 2015 reached 1285 systems with a total nominal capacity of 33 MW. At the mid of June 2017, 5317 systems are operational with a total nominal capacity of 105.74 MWp. The relative increase in the total nominal installed capacity is 220% between both above-mentioned records (EMRC 2017).

3.5 Electrical Power Wheeling

The transformation of electricity through transmission and distribution lines from off-site generation to area of consumption. The Jordanian wheeling regulation is applicable for projects regardless of size. A wheeling charge is paid to the relevant utility according to EMRC instructions. The total installed capacity of PV systems under wheeling scheme, so far, exceeded 50 MW. There are many PV wheeling projects belonging to medium- and large-scale industries (telecom, companies, banks, private hospitals and schools, etc.) that are under construction and development over Jordan during the next years.

4 Results and Analysis

The figures listed in Table 1 summarize the cumulated capacity of operational, contracted and developed grid-connected PV systems in Jordan during 2011–2020, in addition to some indicators regarding the contribution of these systems in meeting the total electricity demand of Jordan.

The cumulated capacity of PV systems will reach 1443 MW by 2020 as illustrated in Fig. 1. The total amount of PV electricity that will be generated in year 2020 will amount around 2250 GWh which will cover around 10% of the total electricity demand of Jordan at the said year. The specific energy yield for PV systems is assumed 1700 kWh/kWp/year on average during the first five years of operation.

5 Conclusion

In this paper, the PV market in Jordan was surveyed based on the actual installations on the ground besides the contracted and developed PV projects that will be installed and operated at the end of year 2020.

Net metering and power wheeling schemes have been very helpful in expanding small-distributed PV systems in Jordan. Given the high solar yield and steep electricity tariffs for large customers, further projects in net metering and power wheeling are expected to be implemented in the near future, especially in the commercial and industrial electricity consumer market segments.

The total cumulated capacity of PV systems that will be installed over Jordan in year 2020 will reach 1400 MW which will exceed the predetermined target of the government of 600 MW by more than twice which means that PV systems are very attractive for the investors in this field.

After the Green Corridor Project is implemented, it will permit more large-scale PV systems to be implemented in Jordan besides other renewable energies, especially, wind.

The Government of Jordan was be successful in leading and developing the renewable energy market in Jordan during in all stages starting from legislations and regulations

Table 1 Survey and projection results about PV systems in Jordan (2011–2020)

	Years*									
Project type	2011	2012	2013	2014	2015	2016	2017	2018	2019	2020
1. Direct proposal (on-grid)	0.0	0.0	0.0	0.0	30.0	210.0	210.0	410.0	410.0	610.0
2. Competitive bidding (on-grid)	0.0	0.0	0.0	0.0	0.0	0.0	0.0	0.0	0.0	0.0
3. EPC turn-key (on-grid)	0.0	0.0	0.0	0.0	5.2	5.2	5.2	105.2	305.2	305.2
4. Net metering (on-grid)	0.0	0.0	2.6	12.0	33.0	75.0	105.0	147.0	205.8	288.1
5. Wheeling (on-grid)	0.0	0.0	0.0	0.0	5.0	30.0	50.0	83.3	138.9	231.5
6. Others (self-consumption off-grid/on-grid)	1.5	2.0	2.4	2.9	3.5	4.1	5.0	6.0	7.2	8.6
Total accumulated nominal capacity of PV systems (MW)	1.5	2.0	5.0	14.9	77	324	375	752	1067	1443
Total electricity demand (GWh)	14,273	14,274	14,588	15,418	18,891	19,390	19,975	20,570	21,200	23,063
Estimated total PV electricity production (GWh)**	2.6	3.4	7.7	23.2	119.6	506.0	585.3	1172.3	1664.6	2251.7
Percentage of estimated PV electricity production from total expected electricity demand (%)	0.00	0.02	0.05	0.15	0.63	2.61	2.93	5.70	7.85	9.76

* Cumulative installed PV capacity in year 2011 represents the period 1980–2011

** Assumed specific PV electricity yield = 1700 kWh/kWp

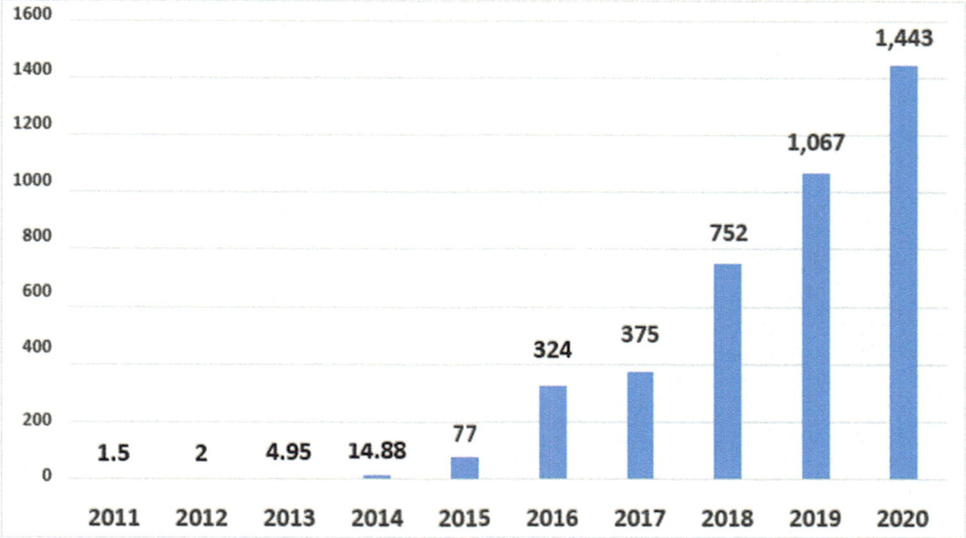

Fig. 1 Cumulative capacity of PV systems in megawatts (MW) over Jordan until year 2020

and ending with stimulating investors to implement projects. This will be a good learning lesson to other countries in the MENA region.

Acknowledgements The authors acknowledge the efforts of the Ministry of Energy and Mineral Resources (MEMR), Electricity Regulatory Commission (EMRC) and the National Electric Power Company (NEPCO) in disseminating online most of the information presented in this paper. We would like to thank public institutions and private companies for publishing press releases about their PV projects which were very useful and helpful in preparing this study.The authors acknowledge the financial support of the Deanship of Scientific Research at the German Jordanian University for the presented work in this paper.

References

Annual Energy Brochure, published by the Ministry of Energy and Mineral Resources (MEMR) in Jordan, 2017.

Annual Energy Report, published by the Ministry of Energy and Mineral Resources (MEMR) in Jordan, 2015.

Annual Report for the Energy and Minerals Regulatory Commission (EMRC) in Jordan, 2014.

Annual Report for the National Electric Power Company (NEPCO) in Jordan, 2015.

Developing renewable energy projects - A guide to achieving success in the Middle East, published by Eversheds and pwc, Third Edition, January 2016.

Enabling PV in the MENA Region - The Emerging PV Market in Jordan, published by the German International Cooperation Agency (GIZ), 09/12/2014.

Global Market Outlook for Photovoltaics 2013–2017, published by the European Photovoltaic Industry Association (EPIA), Second Edition, 2012.

Unlocking thr Sunbelt - Potential of Photovoltaics, published by the European Photovoltaic Industry Association (EPIA), Second Edition, October 2010.

Updated Master Strategy of Energy Sector in Jordan for the period (2007–2020), published by the Ministry of Energy and Mineral Resources (MEMR) in Jordan, December 2007.

The Application of Ground-Source Heat Pumps for a Residential Building in Jordan

Yaqoub Al-Khasawneh, Aiman Albatayneh, and Sa'ed Althawabiah

Abstract

Heating and cooling loads are the main drivers for energy consumption and one of the major sources of pollution in Jordan. Geothermal energy is one of the solutions that can be used to cover these loads effectively; however, it is not exploited properly in Jordan even though it is available abundantly. In this paper, we questioned the potential of utilizing geothermal energy to cover the heating and cooling loads for a typical Jordanian residential building. The heat extracted from the ground will be transferred to the surface using a special heat pump designed for this purpose. This system is called a ground-source heat pump (GSHP). In summer, heat is transferred from the surface to the ground and the opposite will happen in winter. The design and layout of the ground-coupled heat exchanger were investigated and calculated using Earth Energy Designer software. Then, an economic comparison with conventional heating and cooling systems was performed using RETScreen Clean Energy Management Software (RETScreen). The results showed that a GSHP is an energy-efficient and environmentally clean option. A reduction of more than 60% in annual operating and maintenance costs was achieved, which will result in a huge amount of savings over the lifetime of the project. Furthermore, the GSHP will save around 157 tons of CO_2 emissions which is 70% less than the emissions produced by the conventional systems.

Keywords

Ground-source heat pump • Geothermal • Ground loop • District heating and cooling • Heat pump • Jordan

Nomenclature

ASHP	Air-source heat pump
CDD	Cooling degree days
COP	Coefficient of performance
EER	Energy efficiency ratio
GSHP	Ground-source heat pump
HDD	Heating degree days
HDPE	High-density polyethylene
HVAC	Heating, ventilation, and air-conditioning
kCal/h	Kilo calories per hour
RT	Refrigeration tons
tCO_2	tons of carbon dioxide

1 Introduction

Jordan is a poor country in terms of fossil fuel resources. Consequently, it has to dedicate a sizable chunk of its gross domestic product to import crude oil and natural gas. Furthermore, the costs of generating electricity are increasing each year, and they are already above the average citizen's financial capabilities. This situation drives us to find solutions that will help in improving the economic situation in Jordan.

The majority of the imported energy sources go to produce electricity for industrial and municipal uses. For instance, in 2009, the residential sector in Jordan accounted for 30% of energy consumption (Al-Azhari and Al-Najjar 2012). A large portion of this electricity is used to run district heating and cooling systems.

The energy consumption of heating and cooling systems depends mainly on the energy efficiency of these systems. In Jordan, air-source heat pumps (ASHPs) are commonly used for air-conditioning. They are less efficient in heating, so a diesel boiler is used for district heating in winter in most situations (Akash and Mohsen 1999).

Y. Al-Khasawneh · A. Albatayneh (✉) · S. Althawabiah
Energy Engineering Department, German Jordanian University, Amman, Jordan
e-mail: Aiman.Albatayneh@gju.edu.jo

C. Alalouch et al. (eds.), *Advanced Studies in Energy Efficiency and Built Environment for Developing Countries*, Advances in Science, Technology & Innovation, https://doi.org/10.1007/978-3-030-10856-4_16

Geothermal energy is the heat energy generated by the earth due to its formation and the radioactive decay of materials inside it (Dye 2012). There is a recent interest in geothermal energy due to its convenience in acting as a stable source of energy around the year. In 2015, the estimated electrical operating capacity worldwide was 12.8 GW (Geothermal Energy Association 2015).

As we go deeper into the ground, the temperature becomes less sensitive to weather fluctuations. Actually, for depths below 8 m, the temperature remains almost constant throughout the year (Mean Annual Air Temperature 2017). Depending on the location and geological formation, the temperature could be maintained between 15 and 20 °C, which is an ideal temperature for air-conditioning application (Hamdhan and Clarke 2010).

This constant ground temperature can be exploited—in district heating and cooling—by making a loop of underground pipes either vertically or horizontally or as a combination of both (Geothermal Heat Pump Loop Fields 2017). A working fluid is pumped through these pipes to absorb heat from the ground in winter. The direction of heat transfer will be reversed in the summertime during which the ground will act as a heat sink. A special type of heat pumps is installed above the ground to transfer the heat between the ground loop and the building. This system is called a ground-source heat pump (GSHP) (Rafferty 1997). According to the US Department of Energy, the total thermal capacity of GSHPs installed worldwide was 15.4 GW in 2009 (Goetzler et al. 2009).

1.1 Aims and Objectives

This research deals with energy issues in Jordan through the use of geothermal energy for district heating and cooling for residential buildings.

A GSHP system will be sized for a typical Jordanian residential building. The size and specifications of the ground loop and the heat pump will be found using Earth Energy Design software. Next, an economic comparison between GSHP and conventional heating and cooling systems will be made. The conventional systems are represented by two systems. One system uses an ASHP for both heating and cooling. The other one will be a system that uses ASHP for cooling and a diesel boiler for heating.

This research is intended to find the economic and technical benefits for utilizing geothermal energy as a substitute for the conventional heating and cooling systems in Jordan.

2 Literature Review

Gamage and Fahrioglu (2014) studied the feasibility of using GSHP instead of the conventional HVAC system in the dormitory building II in the Middle East Technical University, Northern Cyprus Campus (METU-NCC). They have found that the use of GSHP can result in 65% reduction in the annual energy cost and 62% reduction in CO_2 emissions.

Abu-Hamatteh el al. (2010) studied the possibility of electrical generation using the thermal wells in Jordan. Sawarieh (2008) concluded that the current technology is not mature enough to harness the available power from thermal wells in Jordan.

The energy consumption of a typical Jordanian residential building was discussed in many papers. Ghandoor et al. (2008) and Akash and Mohsen (2003) gave reference guidelines to most of the assumptions made in this paper about the typical household energy consumption in Jordan.

Another work by Akash and Mohsen (1999) contains energy analysis of Jordan's residential sector. The paper also provided a guideline for formulating the proper assumptions of household specifications such as the number of residents, the area of the house, and the consumption of different types of fossil fuels.

Wood et al. (2010) used Earth Energy Designer (EED) software to study the performance of GSHP that utilizes the concrete foundation piles as a ground-coupled heat exchanger. The same software will be used in this study. Hellström et al. (1997) proved that EED is a useful tool to design ground-coupled heat exchangers.

Several studies (e.g., Mirzahosseini and Taheri (2012) and Recalde (2010)) used RETScreen Clean Energy Management Software (RETScreen) for economic evaluation of renewable energy projects. The tool is also used here to make an economic comparison between GSHP and other systems.

3 Sizing Outline

In Jordan, more than 80% of the population are living in houses with an area ranging from 50 to 200 m^2 (Akash and Mohsen 1999). For this study, a GSHP system for a single-floor residential building which is located in Amman with an area of 150 m^2 will be sized. The annual distribution of surface temperature of the study area was obtained from the national energy research center.

3.1 Heating and Cooling Loads

The proposed house is considered as a worst-case scenario in terms of heat conservation where minimal shading, low insulation, and single-glazed windows are assumed. Furthermore, it is assumed that there are five persons living in the house. This is the average number of persons per house based on regional demographics (Jordan Average Household Size 2017).

The building specifications of a typical Jordanian residential building were taken from Jaber and Ajib (2011). The thermal heat specifications of building materials were taken from the Jordanian Thermal Insulation Code (2009).

Table 1 HDD and CDD for Jordan

Month	Cooling degree days (for a base temperature of 23 °C)	Heating degree days (for a base temperature of 15 °C)
January	0	255
February	0	215
March	6	123
April	48	50
May	104	13
June	155	3
July	260	0
August	202	0
September	127	2
October	86	15
November	17	80
December	0	218
Summation	1005	974

Heating and cooling loads were found using Carrier's Hourly Analysis Program (HAP). The building requires 9.962 kW of cooling as a base value assumed for an empty house. The following sources of energy were also added:

- A house of this type is usually located in the suburbs, and there are no buildings to cast shade on it. We need 10% 0.996 kW which is 10% of the base load.
- The heat from a human body can be approximated as 120 W/person (ASHRAE 2001). For five persons in the building, the total power produced will be 600 W.
- Losses from leaks and overventilation in Jordanian houses are estimated to be 7.500 kW.

The total load adds up to be around 19.34 kW, which is equivalent to 5.5 refrigeration tons. It will be rounded to 6 tons taking into account safety factors. The heating load was found to be 15 kW.

The total energy consumption calculation is based on heating degree days (HDD) and cooling degree days (CDD), which reflects the amount of energy needed to heat or cool a building when the outside temperature differs from a base value. In this study, the base temperature is taken as 23 °C for cooling and 15 °C for heating. HDD and CDD data are taken from degreedays.net Web site. They are presented in Table 1. The heating and cooling load distribution is illustrated in Fig. 1.

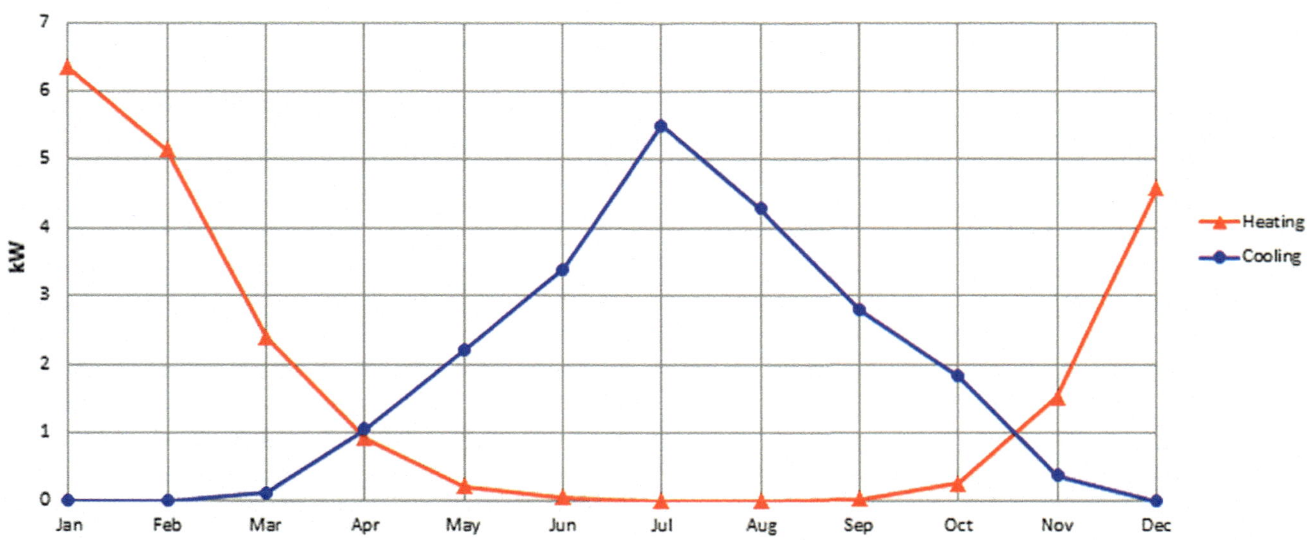

Fig. 1 Heating and cooling load distribution

3.2 Heat Pump Sizing

Now that the number of refrigeration tons needed is calculated, the size of the heat pump can be determined. After searching in the American market, as it is the dominant market for geothermal heat pump, a heat pump that fits the need was identified. It is manufactured by Carrier's company with the product number: GT072VTLCDET1XX1. It costs USD 6810, and according to the product datasheet, it provides 21 kW for cooling and 16 kW for heating. The coefficient of performance, which is the ratio between the heating capacity and the electrical energy input, is 4.3. The energy efficiency ratio, which is the ratio between the cooling capacity and the electrical energy input, is 25.4 at part load (Carrier 2014).

The heat pump should be located in an area that is built specifically for it to provide a protection against weather conditions like rain and dust to increase its life span.

3.3 Ground-Loop Sizing

There are two major types of ground loops, vertical and horizontal. Both of the configurations are closed-loop systems; i.e., water will be circulated through the system with no exchange with the environment. The vertical type is preferred as it requires less land area and it is more efficient due to increased depth which stabilizes the temperature around the year (RETScreen 2005).

In a study made by Alnawafleh et al. (2013), we can find that the ground type in Amman is mostly a Mesozoic sedimentary rock. Furthermore, an important factor for the sizing of a ground loop is the ground temperature. Al-Sarkhi el al. (2008) studied the thermal gradient in different locations in Jordan. For Amman, the average earth temperature value is 17 °C at a depth of 20 m.

Earth Energy Designer software will be used for the ground-loop sizing. The main factors to be determined in the system are the length of the loop and the number of boreholes. They depend on many factors related to the ground properties. The main parameter will be the ground thermal conductivity k (W/m K) which will help to determine specific heat extraction rate (W/m) (Rybach 2001). The average ground thermal conductivity for limestone, which is the major rock type in the Mesozoic sedimentary rocks, is 2.4 W/(m K) (ASHRAE 1995).

The pipes used are high-density polyethylene (HDPE). They are durable, and they have high-pressure rating and high chemical resistance which means they will not degrade or corrode easily. On the contrary, the only drawback of utilizing them is the high initial cost.

The diameter of the tube is 32 mm which is installed in a specially designed grout to prevent any leakage of the circulating fluid to the ground. The grout's thermal conductivity is 2.45 W/(m K).

4 Economic Consideration

After sizing the ground heat exchanger, an economic comparison between GSHP and other systems is performed using RETScreen software. This tool is used widely for evaluating the economics of renewable energy projects. All of the calculations were made in the currency of US dollar (USD).

4.1 Capital Cost

The capital cost of the ground loop is high. First, you will need special permissions to drill into the ground that will cost USD 2450. The drilling itself will cost USD 20 per meter for a hole diameter of 0.15 m in a medium rock. The power rating of the circulating pump is usually calculated as 17 W for each 1000 W of the heat pump capacity used by the system. For the circulating fluid, we need a mixture of antifreeze and water as the temperature in Jordan sometimes reaches freezing point. Propylene glycol will be used due to its low toxicity with a cost of USD 2600/m^3. The cost of HPDE is USD 3 per meter. And finally, for fittings and valves, all connections must be thermally fused to ensure a tight connection that will last for a very long time. The cost can be approximated as USD 12 per kW of system capacity (RETScreen 2005).

4.2 Operating and Maintenance Cost

The attractiveness of GSHP comes from its low operating and maintenance cost. This is due to the high efficiency when compared to other solutions. A considerable amount of energy can be saved each year that will make it profitable to implement GSHP. The electricity cost in Jordan is illustrated in Table 2 (National Electric Power Company 2017). A considerable reduction in electricity consumption is expected that will also reduce the unit price of electricity by moving to a lower tariff block. For operating the diesel boiler, the price of diesel is $0.68 in 2017 in Jordan.

The maintenance cost for a geothermal heat pump is the lowest among all types of heat pumps (Martin et al. 1999). Bloomquist (2001) found that the maintenance cost for a GSHP is USD 1.4/m^2. For an ASHP, it equals USD 3/m^2.

Table 2 Electricity prices in Jordan

Tariff block (kWh/Month)	Price (USD/kWh)
1–160	0.05
161–300	0.1
301–500	0.12
501–600	0.16
601–750	0.22
751–1000	0.27
More than 1000	0.37

Table 3 Summary of economic considerations

Component	Value
Drilling permission (USD)	2450
Drilling cost (USD/m)	20
Antifreeze (USD/m^3)	2600
HPDE (USD/m)	3
Fittings and valves (USD/kW)	12
Maintenance cost of GSHP (USD/m^2)	1.4
Maintenance cost of ASHP (USD/m^2)	3

Table 4 Ground-loop properties

BOREHOLE	Value
Borehole depth (m)	80
Borehole spacing (m)	7
Borehole diameter (mm)	110
U-pipe diameter (mm)	32
U-pipe thermal conductivity (W/(m K))	0.42
U-pipe shank spacing (mm)	70
Filling thermal conductivity (W/(m K))	2.45

The average maintenance cost of a diesel boiler is USD 366 which is based on actual projects reported by the members of homeadvisor.com Web site. A summary of economic considerations is shown in Table 3.

5 Results and Discussion

The technical results of the ground-loop system were made by Earth Energy Designer software, and they are shown in Table 4. The number of boreholes needed is 2 with a depth of 80 m. The boreholes are set 7 m apart from each other which is the recommended distance to prevent thermal linkage and efficiency loss. The flow rate is 2 l/s.

Table 5 Initial cost of ground loop

Item	Cost (USD)
Heat pump	6810
Drilling and grouting	3223
Circulating fluid	74
Pipes	967
Circulating pump	704
Fittings and valves	237
Total	12,015

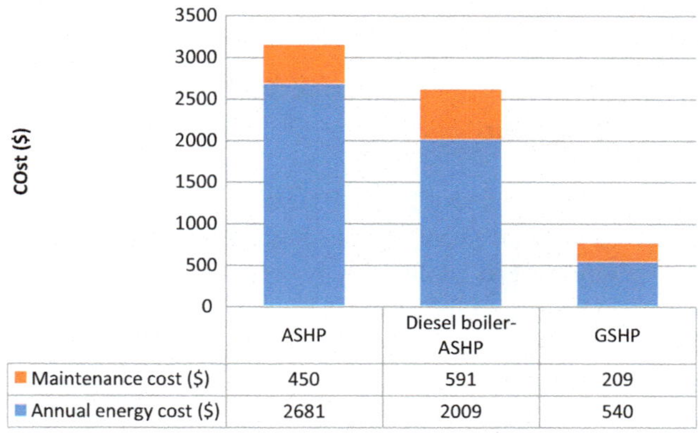

Fig. 2 A comparison between the operating and maintenance costs of a GSHP and other systems

The initial cost of the system is estimated to be USD 12,015 as shown in detail in Table 5. The high cost is justified by the low running cost compared to other systems. This will make GSHP the most economical option in the long run. The reduction in energy consumption will also decrease electrical tariff. The average tariff in the GSHP case is USD 0.11. In the case of a system that uses ASHP for heating and cooling, the average tariff will be USD 0.18. Figure 2 shows a comparison between the operating and maintenance costs of a GSHP and other systems.

The economic evaluation made by RETScreen software shows that the GSHP is a feasible option. A reduction in annual energy consumption by approximately 60% is achieved by using GSHP systems. By comparing this system with a system that uses a diesel boiler for heating and ASHP for cooling—which is the nearest option in terms of running cost—we get the accumulative cash flow diagram shown in Fig. 3. The internal rate of return is 17.4% and the equity payback period is 5.9 years, which indicates a good investment.

RETScreen also found that installing a GSHP will result in a reduction of greenhouse gas emissions by an amount of 6.3 tCO_2 each year which is equivalent to saving 1.4 acres of

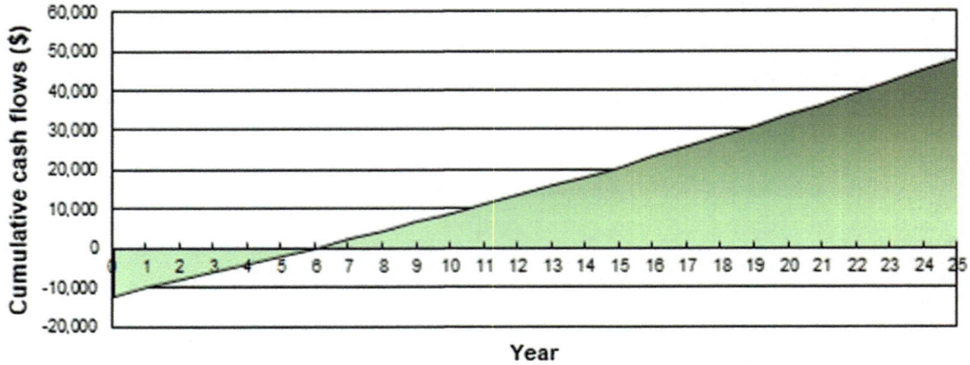

Fig. 3 Accumulative cash flow of a GSHP project against diesel boiler–ASHP system

forests. Over the lifetime of the system, the amount of greenhouse gas saved is 157 tCO_2.

6 Conclusion

The research yielded a positive result regarding the use of GSHP as a replacement of conventional systems. A reduction in heating and cooling energy costs by more than 60% was found with accumulative profits of USD 48,000 over the lifetime of the project. The system is also very beneficial for the environment as it reduces greenhouse gas emissions resulted from high electricity consumption.

On the other hand, GSHP requires high initial cost and elaborate efforts to install. Workers should be experienced especially in making thermal fusion connection to prevent any leakage while running the system. Another hindrance is the existing laws and protocols that need to be more supportive especially for the drilling process. Furthermore, more awareness is needed among the Jordanian population about geothermal technologies in general. Further studies are required for the direct use of geothermal energy such as the generation of electricity.

References

Abu-Hamatteh, Z., Al-Zughoul, K., & Al-Jufout, S. (2010). Potential Geothermal Energy Utilization in Jordan: Possible Electrical Power Generation. International Journal Of Thermal And Environmental Engineering, 3(1), 9–14. http://dx.doi.org/10.5383/ijtee.03.01.002.

Akash, B. A., & Mohsen, M. S. (1999). Energy analysis of Jordan's urban residential sector. Energy, 24(9), 823–831.

Akash, B., & Mohsen, M. (2003). Current situation of energy consumption in the Jordanian industry. Energy Conversion And Management, 44(9), 1501–1510. http://dx.doi.org/10.1016/s0196-8904(02)00146-2.

Al-Azhari, W. & Al-Najjar, S. (2012). Challenges and Opportunities Presented by Amman's Land Topography on Sustainable Buildings. in Proc. ICCIDC-III Conf., 2012.

Alnawafleh, H., Tarawneh, K., & Alrawashdeh, R. (2013). Geologic and economic potentials of minerals and industrial rocks in Jordan. Natural Science, 05(06), 756–769. http://dx.doi.org/10.4236/ns.2013.56092.

Al-Sarkhi, A., Akash, B., Abu-Nada, E., Nijmeh, S., & Al-Hinti, I. (2008). Prospects of geothermal energy utilization in Jordan. Energy Sources, Part A, 30(17), 1619–1627.

ASHRAE. (1995). Commercial/Institutional Ground-Source Heat Pump Engineering Manual, American Society of Heating, Refrigerating and Air-Conditioning Engineers, Inc., 1791 Tullie Circle, N. E., Atlanta, GA, USA.

ASHRAE. (2001). Fundamentals. American Society of Heating, Refrigerating and Air Conditioning Engineers, Atlanta, 111.

Bloomquist, R.G. (2001). The Economics of Geothermal Heat Pump Systems for Commercial and Institutional Buildings, Proceedings of the International Course on Geothermal Heat Pumps, Bad Urach, Germany, September 2001.

Carrier. (2014). GT Performance Series Geothermal Heat Pump Sizes 024, 036, 048, 060, 072 [product datasheet]. Retrieved from: http://s7d2.scene7.com/is/content/Watscocom/Gemaire/carrier_gt072vtlcdet1xx1_article_1422491346054_en_datasheet.pdf?fmt=pdf.

Dye, S. (2012). Geoneutrinos and the radioactive power of the Earth. Reviews Of Geophysics, 50(3). http://dx.doi.org/10.1029/2012rg000400.

Gamage, K. J., & Fahrioglu, M. (2014). The current technological status of ground source heat pump systems and their potential use in Northern Cyprus. Int J Wind Renew Energy, 3(3), 39–48.

Geothermal Energy Association. (2015). The International Geothermal Market At a Glance – May 2015.

Geothermal Heat Pump Loop Fields. (2017). Geothermalgenius.org. Retrieved 29 July 2017, from http://www.geothermalgenius.org/how-it-works/geothermal-ground-loop-fields/.

Ghandoor, A., Hinti, I., Akash, B., & Nada, E. (2008). Analysis of energy and exergy use in the Jordanian urban residential sector. International Journal Of Exergy, 5(4), 413. http://dx.doi.org/10.1504/ijex.2008.019113.

Goetzler, W., Zogg, R., Lisle, H., & Burgos, J. (2009). Ground-Source Heat Pumps. Overview of Market Status, Barriers to Adoption, and Options for Overcoming Barriers. Navigant Consulting, Inc., Chicago, IL (United States).

Hamdhan, I. N., & Clarke, B. G. (2010, April). Determination of thermal conductivity of coarse and fine sand soils. In Proceedings of World Geothermal Congress.

Hellström, G., Sanner, B., Klugescheid, M., Gonka, T., & Mårtensson, S. (1997). Experiences with the borehole heat exchanger software EED. Proc. Megastock, 97, 247–252.

Jaber, S., & Ajib, S. (2011). Optimum, technical and energy efficiency design of residential building in Mediterranean region. Energy And Buildings, 43(8), 1829–1834. http://dx.doi.org/10.1016/j.enbuild.2011.03.024.

Jordan Average Household Size, Department of Statistics. (2017). Jordan Statistical Yearbook 2017. [online] Available at: http://dosweb.dos.gov.jo/products/statistical_yearbook2017/ [Accessed 7 Feb. 2019].

Martin, M. A., Durfee, D. J., & Hughes, P. J. (1999). Comparing Maintenance Costs of Geothermal Heat Pump Systems with Other HVAC Systems in Lincoln, NE Public Schools: Repair, Service, and Corrective Actions. ASHRAE Transactions, Vol. 105, No. 2.

Mean Annual Air Temperature. (2017). Icax.co.uk. Retrieved 1 August 2017, from http://www.icax.co.uk/Mean_Annual_Air_Temperature.html.

Mirzahosseini, A. H., & Taheri, T. (2012). Environmental, technical and financial feasibility study of solar power plants by RETScreen, according to the targeting of energy subsidies in Iran. Renewable and Sustainable Energy Reviews, 16(5), 2806–2811.

National Electric Power Company. (2017). Nepco.com.jo. Retrieved 29 July 2017, from http://www.nepco.com.jo/en/electricity_tariff_en.aspx.

Rafferty, K. (1997). An information survival kit for the prospective residential geothermal heat pump owner. Geo-Heat Center Quarterly Bulletin, 18(2).

Recalde, M. (2010). Wind power in Argentina: Policy instruments and economic feasibility. International journal of hydrogen energy, 35 (11), 5908–5913.

RETScreen. (2005). Clean Energy Project Analysis: RETScreen® Engineering & Cases Textbook: Ground-source Heat Pump Project Analysis Chapter.

Rybach, L. (2001). Design and performance of borehole heat exchanger/heat pump systems. Proc. European Summer School of Geothermal Energy Applications, Oradea/Romania (CD-ROM).

Sawarieh, A. (2008, May). Geothermal Water in Jordan. In Workshop for Decision Makers on Direct Heating Use of Geothermal Resources in Asia, UNU-GTP, TBLRREM and TBGMED, Tianjin, China.

Wood, C. J., Liu, H., & Riffat, S. B. (2010). An investigation of the heat pump performance and ground temperature of a piled foundation heat exchanger system for a residential building. Energy, 35(12), 4932–4940.

The Application of a Solar Parabolic Trough Collector for the Melting of Plastic Waste

Daniela Mewes and Libbis Sujessy

Abstract

Municipal waste, comprised partially of plastic, has negative ecological impacts on a global scale. Developing countries face unique challenges in collecting and recycling their waste, where even in developed countries, recycling rates of plastics rarely approach 100%. Reshaping plastics into usable products requires a considerable amount of heat energy and infrastructure, which particularly impacts the accessibility of recycling technologies to developing countries. In regions with sufficient sunlight, solar cookers designed for the 150–200 °C range most useful for food preparation may be able to reach temperatures hot enough to also melt plastics, where the melting point of most common plastics is below 200 °C. The aim of this study was to test the feasibility of applying a parabolic trough collector for melting plastic waste. The lack of published data on this particular application of solar thermal collectors highlights a significant research gap. Two different prototypes were tested after improvements in the sheet metal material and parabolic shape were realized. Results demonstrated that a non-evacuated tube solar cooker constructed with simple methods and locally available materials could reach temperatures as high as 139 °C within one hour of testing in Stockholm, Sweden. ABS plastic waste from a 3D printer was successfully melted in this process. The final prototype was tested in Ioannina, Greece in collaboration with the refugee-aiding FabLab, Habibi.Works. Temperatures inside the collector were measured at a maximum value of 211 °C in Greece. The results were examined in the context of developing areas and refugee camps and suggest that this may be feasible and affordable for developing countries. Further research should be conducted on the potential of solar cooker designs to be applied for waste reduction in these areas of the world.

Keywords

Sustainable energy • Renewable • Solar cooker • Concentrated solar power • Recycling • Plastic waste • Solar • Municipal solid waste • Thermal • Melting • Plastics • Developing • Appropriate technology • Parabolic trough • 3D printing • Sustainable

1 Introduction

Plastic waste is a predominant component of the waste produced particularly in developing countries and transient locations such as refugee camps. Without regulated means to recycle this waste, it becomes a major contribution to pollution. This results in waste aggregating within municipalities or being informally burned by community members, releasing pollutants into the atmosphere, as largely practised in developing areas (Pariatamby and Tanaka 2014; Guerrero et al. 2013). Similarly, comparing data on the composition of waste from both developing and developed countries, plastics contribute 10% of the total yearly household waste as per 2009 (Rajaeifar et al. 2017). Although the amount of generated waste has increased, developing countries still produced half as much waste per capita ($\sim$0.65–0.95 kg person^{-1} day^{-1}) in comparison with developed countries (Karak et al. 2012; Rajaeifar et al. 2017). Thus, roughly up to 35 kg/person/year can be attributed to plastics in developing countries (including municipal and industrial sources of plastics).

Recycling one ton of plastic is equivalent to conserving 16.3 barrels of oil, 23 cubic metres of landfill space and 5.77 MWh of energy (Paśnikowska-Łukaszuk and Wójcicka-Migasiuk 2017). As such, a community of 1000 people in a developing country saves up to 202 MWh by recycling their plastics. However, collection logistics and

D. Mewes (✉) · L. Sujessy
School of Industrial Engineering and Management, KTH Royal Institute of Technology, Stockholm, Sweden
e-mail: mewes@kth.se

C. Alalouch et al. (eds.), *Advanced Studies in Energy Efficiency and Built Environment for Developing Countries*, Advances in Science, Technology & Innovation, https://doi.org/10.1007/978-3-030-10856-4_17

infrastructure requirements are a barrier for recycling in rural areas and developing cities (Guerrero et al. 2013). The idea examined in this study would be intended to replace the majority of the energy used for recycling plastics with solar thermal energy via solar collectors at the community level. In developing countries, clustered around the equatorial region, sunlight hours are often abundant and solar-based solutions are a practical idea for sustainable energy applications (Eggers-Lura 1979; Quarshi and Hussain 2005).

One often under-utilized technology is that of solar thermal energy on a non-industrial scale, for uses other than steam generation or hot water production (Yettou et al. 2014). When considering the applications of solar thermal energy, solar cookers offer a low-cost alternative (Oelher and Scheffler 1994).

The contribution of plastic to municipal solid waste (MSW), although significant, is less than that of paper-based products. However, there are more underlying complexities with recycling plastic in comparison with paper or metal waste (Karak et al. 2012). For instance, plastic is a polymeric material consisting of multiple plastic types with different physical properties. The extensive variety of plastics can be problematic due to thermodynamic differences in structure (Duval 2014). The most common plastics found in MSW in developing countries are primarily (Areeprasert et al. 2017): polyethylene terephthalate (PET), high-density polyethylene (HDPE), polyvinyl chlorine (PVC), low-density polyethylene (LDPE), polypropylene (PP), polystyrene (PS), along with other plastics such as acrylic, acrylonitrile butadiene styrene (ABS), polycarbonate (PC), and polylactic acid (PLA). The typical melting temperature of commercial plastics is within the range of ~130–170 °C (Table 1). However, melting temperature depends on the specific polymer composition and reported values vary somewhat between sources. Plastics pass through several transition temperatures before decomposition that may be less distinct in some polymers, suggesting that the glass transition temperature may be an additional useful value to consider (Piringer and Baner 2008; McKeen 2007; Brydson 1999). This is particularly applicable to PVC and ABS, which are categorized as amorphous plastics.

Despite advancements in waste management, plastic waste recycling faces challenges due to processing requirements of mixed-plastic waste. Normally, once sorted, plastic waste is recycled through a mechanical process. Due to the individual material characteristics and low purity, some plastic waste cannot be mechanically recycled. Feedstock recycling is a viable alternative for these scenarios, since it converts low purity plastic waste into its fuel-based constituents through thermal cracking (Hopewell et al. 2009). Another alternative is chemical recycling which has been specifically applied to PET plastics, where PET is depolymerized to a monomer, which can be re-polymerized to produce virgin polymer (Shen and Worrell 2014; Hopewell et al. 2009). These types of recycling are, however, sophisticated and less-affordable technologies and thus not appropriate for the small-scale recycling facilities that often exist in developing countries. Additionally, these processes consume energy and water in significant quantities and thus may not be reasonable for these environments.

An additional solution may lie in a renewable energy source. Solar cookers were originally invented for more sustainable cooking, reducing both energy consumption and pollutants. The technology was later developed for various purposes including hot water production and industrial steam generation (Saxena et al. 2011). There are many notable advantages of using solar cookers aside from their low cost (Nandwani 1996; Saxena et al. 2011): improvement of health conditions and household environments, reduction of deforestation, replacement of fossil fuel-based cooking, construction possibilities with local materials and the potential to reach temperatures comparable to oven and stove appliances. Open fires are used throughout a large portion of the developing world both for cooking and to reduce MSW on an individual or community level (Martins et al. 2009). Women and children are often in the closest proximity to open fires used for either cooking or waste reduction and have the greatest risk of related respiratory disease (Fullerton et al. 2008). Additionally, according to the World Health Organization, there are more than 3 million deaths per year globally caused by conditions associated with exposure to open fires (WHO 2016). Replacing open fire cooking and fire-based waste reduction instead with solar thermal-based processes would help improve living conditions in developing communities that have adequate sunlight for this technique.

Table 1 General range of melting temperatures of common commercial plastics (Piringer and Baner 2008; McKeen 2007; Brydson 1999)

Plastic	PET	HDPE	PVC	LDPE	PP	PS	Acrylic	ABS	PC	PLA
Melting temp. (°C)	255	135	80[a]	108	140–170	230–270	162–193	100[b]	143–154[c]	175–262

Glass transition temperatures are listed in the cases where melting temperature is not defined by literature

[a]Refers to glass transition temperature as stated by (Brydson 1999)

[b]For unfilled ABS (McKeen 2007). Based on glass transition temperature

[c]Based on Vicat softening temperature from (McKeen 2007)

Commercial plastics such as HDPE, LDPE and polypropylene that melt below 200 °C could be potential targets for recycling efforts with solar cookers, where properly constructed solar cookers can easily reach temperatures over that necessary to boil water. Industrial solar cookers, often utilizing focal points such as those of parabolic dishes and parabolic troughs, can reach temperatures in excess of 400 °C (Yettou et al. 2014). Other plastics with higher melting points could thus also be melted with a solar cooker since even PET, commonly found in plastic bottles, begins to transition below its melting temperature of ~255 °C (Table 1). Not only would working with solar applications for waste reduction contribute towards transitioning to 100% renewable energy following the UN Sustainable Development Goals, it also tackles a major environmental concern in a sustainable manner (Harmim et al. 2014; Nandwani 1996; UN 2017).

Parabolic-shaped solar concentrators focus incoming solar radiation, utilizing curved mirrors to reflect the rays at an angle to converge at a specific focal point. This allows for a higher temperature and greater efficiency than other types of solar cooker designs (Cuce and Cuce February 2013). Typically, the medium being cooked is located at the focal point, and the concentrator must be re-aligned throughout the day to track the sun. The collector located at the focal point consists of a conductive metal pipe within an outer transparent pipe, typically made of glass. High-performance collectors are evacuated, containing no air between the conductive pipe and the glass, which can offer an increase in efficiency over non-evacuated tubes due to lower conduction losses (Kumar and Kumar 2015). However, low-cost applications can use non-evacuated tubes.

This paper examines the construction of a prototype solar cooker based on a parabolic trough collector design intended to melt plastic waste. The prototype was first tested in Sweden, and later tests were then repeated at a community workshop space termed a “FabLab”, or digital fabrication laboratory, where tools and knowledge are shared based on a model that originated at MIT (2017). This FabLab in particular, Habibi.Works, cooperates with refugees living in the vicinity of Ioannina, Greece. This allowed the feasibility of melting plastic waste to be studied within the context of the resources available to and plastic waste produced by those in a transient environment.

2 Methods

The prototype was constructed in Stockholm, Sweden, where initial tests and adjustments were performed on the prototype, through the period of June to August 2017. It was then tested in Ioannina, Greece, during the first week of September 2017.

Stockholm, the capitol city of Sweden (59° 19′ 46″ N 18° 4′ 7″E), has a humid continental climate with an average global horizontal radiation of 4.39 kWh/m^2 in August (Boxwell 2010). The mean temperatures for August were 17.4 and 17.7 °C in 2016 and 2017, respectively (NOAA 2017). These temperatures ranged from 9.4 to 26.7 °C in 2016 and 11.1 to 25 °C in 2017.

Ioannina, in north-western Greece (39° 40′N 20° 51′E), has a humid subtropical and Mediterranean climate with an average of 227.9 h of sunlight in September and 4.80 kWh/m^2 global horizontal radiation (Boxwell 2010; Matzarakis and Katsoulis 2006). The average temperature in the region for the month of September was 18.1 °C in 2016 and 20.7 °C in 2017 (NOAA 2017). Temperatures ranged from 5 to 31 °C in 2016 and 5 to 34.4 °C in 2017. The prototype was tested at an NGO located adjacent to the previous site of the Katsikas refugee camp.

Temperature averages were calculated using datasets provided by the National Centers for Environmental Information, through NOAA. Since data was recorded at 3–8 h intervals, all temperatures were included over the 24 h period. Averages presented here are not indicative of the temperature during only the hours of daylight. Thus, minimum and maximum operating temperatures are important to note for the design of equipment that is not stored in climatized environments, where expected temperatures would be 20 °C.

2.1 Technical Design

Two different prototypes were constructed and subsequently tested. Although both prototypes were built with the same design specifications, there were slight differences in the materials used and degree of precision (Fig. 1). The second prototype was built as an improvement on the first with increased precision in the parabolic profile and a manufactured glass sleeve over the collector (Fig. 1b).

The two prototypes used similar and commonly available materials (Table 2). There were some differences in specifications between prototypes due to available materials. The materials were selected based on several parameters including availability, price, thermal characteristics and previous research indicating appropriate thickness and reflectivity. A 0.1 mm stainless steel sheet with a bright finish was used for the first prototype over a 0.4 mm thick aluminium roofing material to hold the shape of the thinner steel. The second iteration used a bright annealed 0.7 mm thick stainless steel sheet. The mirror finish was noticeably much clearer on the second version as the bright finish of the first version only offered a dull, blurred reflection. A perfect mirror was not employed for this study in order to maintain the previously mentioned specifications of accessible materials.

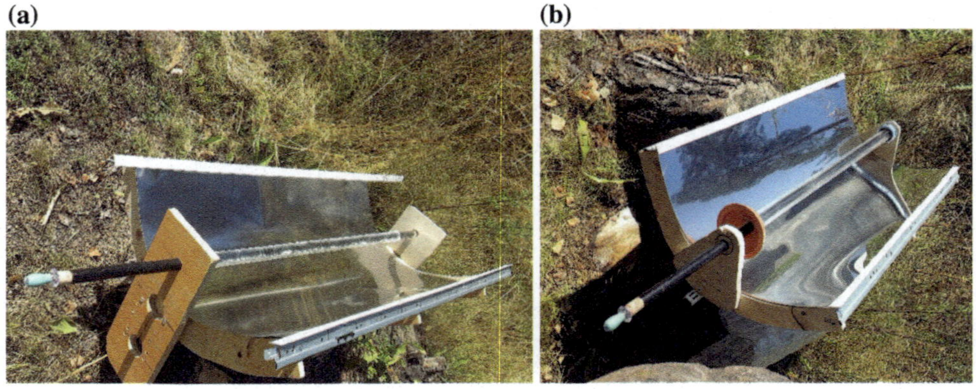

Fig. 1 **a** First prototype during testing. **b** The second prototype during testing

Table 2 Summary of technical properties of the prototypes

	Prototype I	Prototype II
Sheet metal	Stainless steel sheet Bright finish, reflective but not mirror Dimensions: 1000 × 600 × 0.1 mm 0.4 mm aluminium backing	Stainless steel sheet Bright annealed finish, near-mirror Dimensions: 1000 × 600 × 0.7 mm No backing
Rib support	Material: Plywood Manually cut with a jigsaw Thickness: 0.13 mm	Material: Plywood Laser cut for precision Thickness: 0.19 mm
Collector	Non-evacuated rectangular glass tube, glued Width: 26 mm Thickness: 2 mm Inner pipe: copper, 22 mm, painted black	Non-evacuated cylindrical quartz tube Outer diameter: 38 mm Thickness: 2 mm Inner pipe: copper, 22 mm, painted black

The ribs and supporting structures were cut with a jigsaw out of 0.13–0.19 mm thick plywood (Table 2). To increase precision in comparison to the first prototype that simply utilized a paper template, the ribs were laser cut to more carefully follow the parabolic profile. The rib design was based on designs provided in the Solar Parabolic Trough Concentrator Home Experimenter's Manual (Plhak et al. 2017), which included parabolic rib templates sized for 60 cm wide sheets, one of the most commonly available metal sheet widths. A length of one meter was chosen in order to match some commonly available material lengths while also maximizing the area of the parabola.

The chosen collector type was a non-evacuated tube that has the advantage of being more easily constructed in developing regions since evacuated tubes require manufacturing precision, leading to increased material costs (Kumar and Kumar 2015). Although previous research indicates a potential loss of efficiency in non-evacuated tubes due to convection and related conduction losses, this depends on the size of the air gap which can be minimized to reduce convection (Kumar and Kumar 2015; Rabl et al. 1980).

The first glass tube was constructed of rectangular pieces of clear 2 mm thick and 26 mm wide borosilicate glass glued together in a rectangular prism and bonded with silicone glue. Normal "glass glue" provided at hardware stores to seal cracks in windows is unstable at temperatures close to 100 °C, whereas silicone glue can withstand temperatures up to 300 °C. Initial tests with the glass glue resulted in the glue melting at the seams, which was not a problem with the silicone-based glue. The rectangular prism shape was tested as an alternative to a cylindrical glass tube that may be difficult to acquire in developing regions. Later tests used a quartz rod with an outer diameter of 38 mm and thickness of 2 mm that was closed on one end. The quartz rod, sold as a UV light quartz sleeve, was not available locally although strips of borosilicate glass were. Quartz offers advantages over other glass types in that it has a high degree of transparency to UV, particularly those sold as UV lamp sleeves, allowing nearly all UV spectrum light through.

The diameter of the quartz rod was based on the thickness of the inner collector, a 22 mm copper pipe. A 6 mm air gap was assumed based on calculations performed by Kumar and Kumar (2015) which demonstrated the size of the air gap appropriate for non-evacuated tube collectors. The copper rod diameter was based on the largest diameter available locally in Swedish construction stores. It was coated with a high-temperature resistant automotive black spray paint which is stable to temperatures of 650 °C. Copper was selected as the material based on its high thermal conductivity in comparison with steel and aluminium (Coccia, Di Nicola, & Hidalgo, 2016). Steel and aluminium are often used in evacuated tube collectors in order to reduce costs.

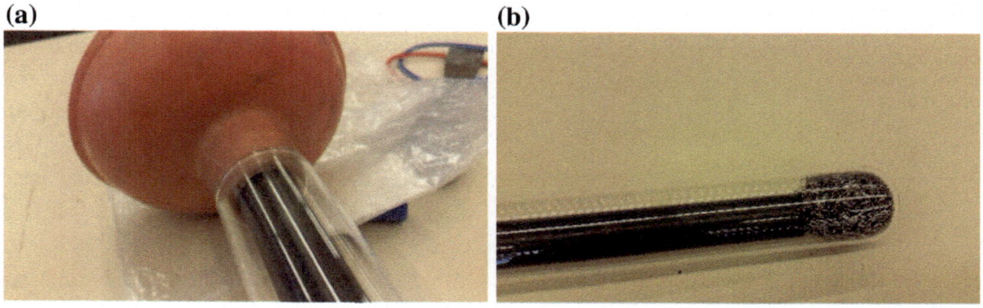

Fig. 2 **a** The open end of the quartz pipe was closed with a silicone stopper **b** Steel wool was positioned inside the quartz tube to hold the copper collector pipe in place and maintain the calculated air gap

The glass tube was sealed with silicone fittings, whereas glue was used to seal the ends of the glass prism. Ends of the copper tube were plugged with wood. The copper tube within the quartz sleeve was filled with steel wool on the closed end of the quartz sleeve, which also surrounded the copper tube, keeping it in place while simultaneously protecting the glass (Fig. 2).

An inner tray of 0.2 mm thick hollow stainless steel rod was cut in half lengthwise in order to support the plastic for melting within the collector. This emulated similar designs of solar cookers which typically use stainless steel trays within the collectors to cook food. Angles could be maintained by rotating the collector about its horizontal axis with a locking clamp mechanism and pre-marked angles on the support pieces. The focal point was pre-determined on the templates for the side support structures and was manually adjusted with the turning mechanism installed on the parabolic cooker.

2.2 Testing

A laser infrared thermometer, Lasergrip 1080, with an accuracy of ±2 °C, allowed for temperature measurements without the use of a thermocouple inside of the rod. The measurement area was proportional to the distance the thermometer was held from the collector tube, with 20 cm leading to a measurement area of 2.3 cm. For the purpose of these tests, a 10 cm measurement height was maintained to ensure that only the temperature of the collector and not of the surrounding air or metal sheet was being measured. Points were measured at the end of the pipe within the glass tube plugged with steel wool, the center of the pipe, the portion of the pipe exiting the tube, and the end of the pipe extending beyond the structure. Tests in Stockholm were conducted in the afternoon from 11:00 to 16:00, where the ambient temperature was usually the warmest, and the collector was situated at an angle to maintain the focal point. Data was collected every 10–15 min except when weather conditions were rapidly changing (such as cloud cover) in which case tests were conducted every 2–5 min until conditions stabilized. The second prototype was examined for a period of 1 h in Stockholm, and the copper pipe within was measured separately after this period as it was extracted from its quartz enclosure. Additionally, waste ABS 3D-printer filament was wrapped in aluminium foil and tested in the second prototype during this trial.

In order to test the accuracy of the parabola and adjust its focus accordingly, a laser pointer was aimed at various points along the parabolic trough in a dark room to observe if its rays were reflected to the focal point and the collector tube. If the parabola was found to be inaccurate, a rasp was used to further refine the profile of the side support structures suspending the collector tube above the trough.

A K-type thermocouple paired with an Arduino Nano and thermocouple amplifier were employed during the testing of the second prototype in Greece, with steel wool at the end of thermocouple wire to plug the end of the copper pipe. Data was recorded at 10 s intervals. Additional measurements using the laser infrared thermometer were taken at 10–15 min intervals when possible at the same formerly mentioned points along the collector.

3 Results

The results from this study are based on quantitative tests that were conducted in two different countries, Sweden and Greece. The maximum observed collector temperature for each trial and prototype version is detailed in Fig. 3. Temperatures within the copper pipe were not measured as part of this study except in a final deployment in Greece.

The initial test in Stockholm, Sweden on the 14th of August (red plus signs in Fig. 3), with solar noon occurring at 12:54, served as a control test of the copper rod without a parabolic collector and without plugs on either the glass cover or copper pipe. Temperature varied along the length of the rod and was warmest at the rod edges with an initial

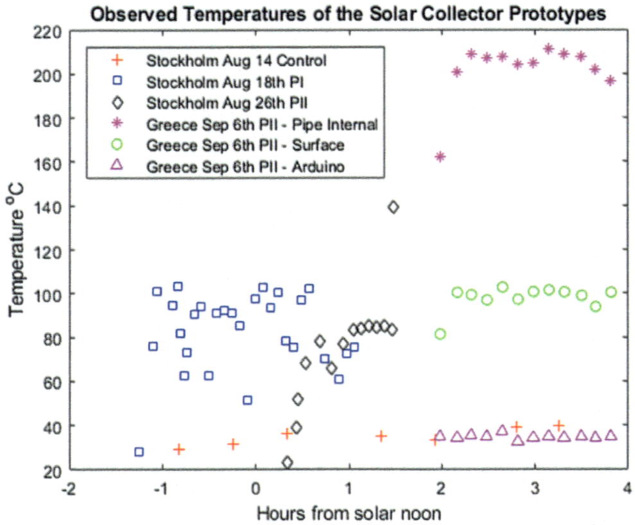

Fig. 3 A depiction of the different sets of collected temperature data including a control with just the copper pipe and tests with Prototype I (PI) and Prototype II (PII) plotted against hours from solar noon at the location/date corresponding to the data set. Only maximum values are depicted for data sets that included temperatures for multiple points along a profile

temperature of 29.1 °C at 12.05, 49 min before solar noon, when the center of the rod was 24.0 °C. At 12.40 the center had reached 31.2 °C, and the edges, 30.7 °C. The following maximum measurements of each data set occurred at the ends of the collector, outside the glass enclosure, until 16:10. The highest end point temperature, 39.1 °C, was recorded at 15:42, 3 h before solar noon. At this time, the center was measured at 29.5 °C. This was followed by a maximum recorded collector temperature of 39.8 °C at 16:10, measured at the middle of the rod, when the maximum end temperature was 37.0 °C. Note that Fig. 3 only depicts the maximum of the temperatures observed at a particular time along the collector and does not demonstrate temperatures along the collector for each time. The standard deviation of the maximum temperatures was 3.93 °C with a mean of 34.83 °C.

In Sweden, temperatures for the first prototype on the 18th of August (blue squares in Fig. 3), with solar noon occurring at 12:53, demonstrated quick response times to changing weather conditions. The partly cloudy day resulted in interchangeable periods of sunlight and cloud cover. The initial temperature of 27.9 °C at 11:38 and ambient temperature of 22 °C increased rapidly to the highest observed temperature of 103 °C at 12:03, 51 min before solar noon. The impact of clouds was noted from 12:07 to 12:23, just before solar noon, when the temperature dropped to 62 °C. With increased sunlight, the temperature rose at 12:09 to 73 °C and at 12:14 to 90 °C before dropping again to 62 °C at 12:23 with passing clouds. Similarly, at 13:08, 100 °C was noted but with clouds and a breeze at 13:13, the temperature decreased. When the sun re-emerged between 13:23 and 13:28, readings of 96.7 and 102 °C were recorded, respectively. Continuously cloudy conditions following 13:38 were coupled with decreasing temperatures. The mean was 82 °C with a standard deviation of 18 °C.

Hot glue was initially used to seal the rectangular glass prism and not under the glass itself. It was notably melting at 12:00. The EVA-based hot glue stick had a melting point of <88 °C according to manufacturer data sheets, which supports the observed temperature range.

The temperatures recorded for the second prototype on the 26th of August (black diamonds in Fig. 3), in Sweden, depended on the portion of the collector that was measured, although weather conditions were stable at 20 °C with clear skies and a mild breeze. Temperatures depicted in Fig. 3 demonstrate the maximum recorded temperature at that time along the collector. An initial temperature of 23 °C was recorded at 13:12, 21 min after solar noon, after which they continued to rise with the exception of clouds at 13:40 when the temperature decreased to 66 °C before rising again and reaching a relatively stable temperature range from 13:54 to 14:19 with a mean during this period of 84.0 °C and a standard deviation of 0.78 °C. On the section of the collector within the glass tube, the highest temperature recorded was 84.8 °C at 14:14. The mean including the initial temperatures was 69.8 °C with a standard deviation of 20.0 °C. When the trial concluded at 14:19, the point where the copper pipe entered the quartz sleeve, under the silicone plug, was measured at a maximum of 100.5 °C, which is not depicted in Fig. 3.

Aluminium foil containing small pieces of waste 3D printer filament had been placed inside the copper pipe of the second prototype. This was examined at the end of the one hour trial and it was found that the ABS plastic within had melted into a clump of plastic with each plastic filament joined to neighboring filaments (Fig. 4). While extracting the copper tube from the quartz sleeve, temperatures were observed ranging from 105 to 139 °C. This appears as a maximum point for the dataset and an outlier in Fig. 3 that is not used in calculations, but was a direct measurement of the surface of the copper rod without the sleeve. These measurements of the surface of the copper rod agreed with sources stating that the initial melting of ABS would occur around 100 °C (Table 1) (McKeen 2007).

On the 6th of September, the second prototype was deployed in Greece with the same design configuration as in the previous trial in Sweden. The temperatures were observed to stay within a higher range consistently than those of the tests in Sweden, ranging from 81 to 102.5 °C on the glass surface (green circles, Fig. 3). However, the first prototype with a glued rectangular glass outer pipe demonstrated a higher maximum glass surface temperature of 103 °C.

Fig. 4 Resulting melted waste ABS

The control data from the Arduino which was collecting temperature data is also depicted in Fig. 3 (denoted by purple triangles) with a mean of 34.5 °C and a standard deviation of 1.09 °C.

Additional data from the inside of the copper rod was collected using a K-type thermocouple on September 6th 2017. The recorded temperature inside at approximately the center to the last half of the pipe ranged from 161.8 to 211 °C (magenta asterisks, Fig. 3). Although the initial observed temperature during the trial was the minimum temperature both inside the pipe and on the surface of the glass, there was no observed correlation between the maximum temperature on the glass surface and the maximum temperature inside the pipe. From 14:46 to the end of the observation period at 16:25 with data collected every 10 s, the mean observed internal temperature was 206 °C with a standard deviation of 3.61 °C. The mean temperature on the surface of the glass during this time interval was 99.0 °C with a standard deviation of 2.52 °C.

4 Discussion

The prototypes tested in this study were intended to be built in developing countries with limited materials and tools. Two main differences between the constructed prototypes were the reflectiveness of the sheet metal and the degree to which the ribs followed the parabolic profile. Although the second criteria could be improved without the use of a laser cutter, it would take a very careful approach to ensure accurate results. Reflective sheet metal is more difficult to obtain. It requires specific finishes to reach a near-mirror surface or access to polishing equipment to achieve this. However, this type of metal manufacturing would be common in most countries in urban industrial zones, and a perfect mirror is not necessary. The degree to which the quartz sleeve in the second prototype contributed to the observed maximum temperature in comparison to the glued glass prism of the first prototype is not clear. The rectangular prism of glued glass strips demonstrated similar surface temperatures and may offer a viable alternative to the quartz sleeve in regions without access to manufactured cylindrical glass sleeves. Further studies on the internal temperature could confirm whether it would be a suitable replacement for the collector sleeve despite fragility.

The differences in maximum recorded temperature observed on the surface of the quartz sleeve and the surface of the copper rod when extracted from the quartz sleeve suggest that the infrared thermometer did not accurately record the temperature of the copper collector pipe through the quartz enclosure. A thermocouple would need to be employed in future tests to more accurately measure this variable. This may indicate that the temperatures of the first prototype were also higher than measured if the difference was not partially due to the curved nature of the quartz tube.

Additionally, price comparisons of selected materials were obtained in order to gauge the expense of building a similar setup in different parts of the world. Table 3 demonstrates reported prices of similar materials and the corresponding lower cost for developing countries. Example prices were collected from distributor websites and through contacts in the locations in question reporting the costs at the distributor outlet stores, which are outlined in the table footnotes. The prices suggest that a parabolic trough would be affordable for developing countries to construct using locally-sourced materials, but would necessitate further economic comparisons including data from a wider variety of material sources, particularly in cases where items such as glass sleeves are made to order. It is also possible that a parabolic trough could be built utilizing waste or recycled materials, which requires further studies on what is available at each location.

In Greece, the second prototype was more extensively tested using a thermocouple to further examine the maximum achievable temperature. The maximum glass surface temperature was not observed in Greece but rather with the first prototype in Stockholm using thinner glass sealed with silicone glue. Since the first prototype with the rectangular glass prism pipe did melt hot glue but did not melt ABS plastic, suggesting lower internal temperatures, it is possible that the thicker quartz glass sleeve of the second prototype was more insulating and allowed for the higher internal temperature observed on August 26th and September 6th along with less heat loss observed at the glass surface expressed through lower glass temperatures. Although this cannot be confirmed by thermocouple temperatures on August 26th, it is possible to compare the observed 131 °C

Table 3 Price comparisons of materials at common suppliers in 4 different countries, in USD, as of 2017

	Sweden	United States	India	Indonesia[1]
Steel sheet metal (1000 × 600 mm)	$47[2]	$38[3]	$19[4]	$30
Copper rod, 20–22 mm, 1000 mm	$14[5]	$5.50[6]	$3[7]	$7.50
20–22 mm steel rod, 1000 mm	$18[1]	$9[3]	$5[3]	$14
Plywood, 0.13–0.19 mm thick	$56[4]	$27[3]	$5[8]	$16
Glass tube 30–40 mm, 1000 mm	$110[9]	$38[10]	$20[1]	$15[11]
Black high temperature paint	$12[12]	$4[7]	–	$4

[1]Depo Bangunan, Bali, Indonesia
[2]Aston Carlsson
[3]McMaster Carr
[4]IndiaMart
[5]Bauhaus, Stockholm
[6]Home Depot
[7]BuildKar
[8]eConstructionMart
[9]Rann-UV, Poland (imported)
[10]Amazon.com
[11]Quote from local glass worker
[12]Biltema, Stockholm

on August 26th at the pipe surface with the internal recorded temperatures on September 6th. The temperatures reached at least 131 °C internally. However, to establish whether the glass surface temperature can be correlated with the internal temperature would require additional thermocouple measurements inside the copper pipe. For a nearly 30 min period on August 26th, a mean surface temperature of 84.0 °C was recorded with a standard deviation of 0.78 °C, demonstrating the potential temperature stability of the second prototype. This is particularly evident in comparison to the first prototype, where although a mean of 82 °C was observed on August 18th, the standard deviation was 18 °C. This fluctuation in glass surface temperatures observed in the less insulated model suggests that heat loss may be a greater factor in the first prototype.

Notably, the maximum internal temperature of 211 °C observed in Greece did not correlate with the maximum glass surface temperature. It did, however, signify the potential of the second prototype to melt most common plastics with the exception of PET. Plastics that should be examined for ability to be melted within the collector include LDPE, HDPE and Polypropylene, which all melt below 200 °C (Table 1). The stable temperatures observed with a mean of 206 °C and standard deviation of 3.61 °C, maintaining the temperature required to melt common plastics for a period of over an hour, suggest that it would be possible to use similar designs for the purpose of plastic re-forming. Additionally, the role ambient temperature and transient cloud cover play in altering the internal pipe temperature requires further analysis in order to be able to moderate the temperatures for plastic melting applications.

5 Conclusions

This study verifies the possibility of building a low-cost solar collector following the concept of parabolic trough solar cookers for applications such as re-forming low melting point plastic waste. The design was specifically altered to meet conditions for developing countries and continued testing in conjunction with an NGO for refugees in Greece, for instance the use of a non-evacuated tube collector instead of the more expensive evacuated tube models. Accuracy in the parabolic shape sufficient for melting plastic waste was obtained with only minimal power tool usage, and waste ABS plastic from a 3D printer softened and demonstrated that it had reached the glass transition temperature. Temperatures sufficient for melting LDPE, HDPE and Polypropylene were maintained for a period of 1.5 h. These plastics are commonly found in plastic bags, containers and hard plastic bottles. Two separate prototypes tested variations of glass sleeves for the non-evacuated collector pipe and demonstrated the potential to use borosilicate glass pieces in a rectangular prism configuration in the absence of cylindrical quartz sleeves for low-temperature solar cooking or water heating applications. In both cases, a non-evacuated tube was employed, which allowed for a design more suitable for developing countries, where evacuated collector tubes may be the most cost-prohibitive factor in building a solar cooking device. Thus, this study has demonstrated the potential to melt and re-form low melting point plastic waste in developing regions using solar cooker-type devices. Additional studies are required to examine the parameters influencing heat loss observed at the surface of the glass pipe and related differences with the internal pipe temperatures in order to further verify results.

This application could provide a sustainable solution to plastic waste recycling in developing countries following the UN Sustainable Development Goals. In cities with few recycling services, the majority of waste accumulates in landfills or is burned. A solar cooker designed to operate in the temperature range intended for the melting of plastic waste could thus reduce CO_2 emissions and improve local living conditions such as air quality. The option to eventually use this method to re-form plastics into other usable shapes and products could also potentially benefit the economies in these regions by allowing local merchants and artisans to more sustainably craft waste into a profitable item. However, the feasibility of reusing the resultant melted plastic and the ability to re-form it cannot be confirmed by this study and would also require further investigation. Additional current data is needed on plastic consumption and waste composition of developing countries. Alternative uses for solar cookers outside of either cooking or industrial temperature ranges should be examined further, particularly for developing countries where they are most applicable.

Acknowledgements This project would not have been possible without funding provided by the KTH Royal Institute of Technology Alumni Association's Opportunity Fund and KTH Sustainability's MHU på Tvärs "Environment and Sustainability Without Borders" grant. Teammates that assisted with the project development included Rohit Agarwal, Ehimenmen Michael Inegbedion, Aiswarya Kumar, Anna Adrianarski, Kasra Khatami and Harald Sundberg. Their work was instrumental to the prototyping process. The KTH Kista Mentorspace along with KTH Elab provided equipment for construction and allowed for storage of materials throughout the study. Additionally, Habibi.Works, a project of Soup and Socks e.V., hosted the project during its final testing phase in Ioannina, Greece.

References

A. Pariatamby and M. Tanaka, Municipal Solid Waste Management in Asia and the Pacific Islands, Singapore: Springer, 2014.

L. A. Guerrero, G. Maas and W. Hogland, "Solid waste management challenges for cities in developing countries," Waste Management, vol. 33, pp. 220–232, 2013.

M. A. Rajaeifar, H. Ghanavati, B. B. Dashti, R. Heijungs, M. Aghbashlo and M. Tabatabaei, "Electricity generation and GHG emission reduction potentials through different municipal solid waste management technologies: A comparative review," Renewable and Sustainable Energy Reviews, vol. 79, pp. 414–439, November 2017.

T. Karak, R. Bhagat and P. Bhattacharyya, "Municipal Solid Waste Generation, Composition, and Management: The World Scenario," Critical Reviews in Environmental Science and Technology, vol. 42, no. 15, 2012.

M. Paśnikowska-Łukaszuk and D. Wójcicka-Migasiuk, "Selected Aspects of Development Towards Energy Efficient Buildings," Journal of Ecological Engineering, vol. 18, no. 5, pp. 137–143, 2017.

A. Eggers-Lura, Solar Energy in Developing countries: an overview and buyers' guide for solar scientists and engineers, vol. 1, Pergamon Press, 1979.

M. Quarshi and T. Hussain, "Renewable Energy Technologies for Developing Countries: Now and to 2023," Islamic Educational, Scientific, and Cultural Organization, 2005.

F. Yettou, B. Azoui, A. Malek, A. Gama and N. Panwar, "Solar cooker realizations in actual use: An overview," Renewable and Sustainable Energy Reviews, vol. 37, pp. 288–306, September 2014.

U. Oelher and W. Scheffler, "The use of indigenous materials for solar conversion," Solar Energy Materials and Solar Cells, vol. 33, pp. 379–387, April 1994.

O. G. Piringer and A. L. Baner, Plastic Packaging: Interactions with Food and Pharmaceuticals, Second Edition, Wiley Online Library, 2008.

L. McKeen, Effect of Temperature and other Factors on Plastics and Elastomers, William Andrew, 2007.

J. Brydson, Plastics Materials (Seventh Edition), Butterworth-Heinemann, 1999.

C. Duval, "Plastic Waste and the Environment," in Environmental Impact of Polymers, Wiley Online, 2014.

C. Areeprasert, J. Asingsamanunt, S. Srisawat, J. Kaharn, B. Inseemeesak, P. Phasee, C. Khaobang, W. Siwakosit and C. Chiemchaisri, "Municipal Plastic Waste Composition Study at Transfer Station of Bangkok and Possibility of its Energy Recovery by Pyrolysis," Energy Procedia, vol. 107, pp. 222–226, February 2017.

J. Hopewell, R. Dvorak and E. Koisor, "Plastics recycling: challenges and," Philosophical Transactions of the Royal Society, vol. 364, pp. 2115–2126, 2009.

L. Shen and E. Worrell, "Chapter 13 - Plastic Recycling," in Handbook of Recycling, Boston, Elsevier, 2014, pp. 179–190.

A. Saxena, Varun, S. Pandey and G. Srivastav, "A thermodynamic review on solar box type cookers," Renewable and Sustainable Energy Reviews, vol. 15, no. 6, pp. 3301–3318, August 2011.

S. S. Nandwani, "Solar cookers—cheap technology with high ecological benefits," Ecological Economics, vol. 17, no. 2, pp. 73–81, May 1996.

P. Martins, J. Rosado-Pinto, M. do Céu Teixeira, N. Neuparth, O. Silva, H. Tavares, J. Spencer, D. Mascarenas, A. Papoila, N. Khaltaev and I. Annesi-Maesano, "Under-report and underdiagnosis of chronic respiratory diseases in an African country," Allergy, vol. 64, no. 7, pp. 1061–1067, July 2009.

D. G. Fullerton, N. Bruce and S. B. Gordon, "Indoor air pollution from biomass fuel smoke is a major health concern in the developing world," Transactions of the Royal Society of Tropical Medicine and Hygiene, vol. 102, no. 9, p. 843–851, September 2008.

WHO, "Household air pollution and health," February 2016. [Online]. Available: http://www.who.int/mediacentre/factsheets/fs292/en/.

A. Harmim, M. Merzouk, M. Boukar and M. Amar, "Solar cooking development in Algerian Sahara: Towards a socially suitable solar cooker," Renewable and Sustainable Energy Reviews, vol. 37, pp. 207–214, September 2014.

UN, "Sustainable Development Goal 7," 2017. [Online]. Available: https://sustainabledevelopment.un.org/sdg7.

E. Cuce and P. M. Cuce, "A comprehensive review on solar cookers," Applied Energy, vol. 102, pp. 1399–1421, February 2013.

D. Kumar and S. Kumar, "Analytical performance investigation of parabolic trough solar collector with computed optimum air gap," International Journal of Energy and Environment, vol. 6, no. 1, pp. 87–96, 2015.

MIT, "Fab Central," 2017. [Online]. Available: http://fab.cba.mit.edu/.

M. Boxwell, Solar Electricity Handbook: A Simple, Practical Guide to Solar Energy-Designing and Installing Photovoltaic Solar Electric Systems, Greenstream Publishing, 2010.

NOAA, "NNDC Climate Data Online," November 2017. [Online]. Available: https://www7.ncdc.noaa.gov/CDO/cdoselect.cmd.

A. Matzarakis and V. Katsoulis, "Sunshine duration hours over the Greek region," Theoretical and Applied Climatology, vol. 83, no. 1–4, pp. 107–120, January 2006.

G. Plhak, "Solar Parabolic Through Concentrator Home Experiment's Manual," FastForward Media, March 2017. [Online]. Available: http://www.ffwdm.com/solar/gen2-book.htm.

A. Rabl, J. O'Gallagher and R. Winston, "Design and test of non-evacuated solar collectors with compound parabolic concentrators," Solar Energy, vol. 25, no. 4, pp. 335–351, 1980.

G. Coccia, G. Di Nicola and A. Hidalgo, Parabolic Trough Collector Prototypes for Low-Temperature Process Heat, Springer, 2016.

Wind Resource Assessment for a Proposed Wind Farm

Mohammad Al-Addous, Hussein Al-Taani, Zakariya Dalalah, Firas Alawneh, and Aiman Albatayneh

Abstract

Jordan is a country that is dependent on foreign energy sources, with 96% of the country's energy need come from oil and natural gas importation. This has made the government to make an investment plan of $15 billion for renewable and nuclear energy. Also in an attempt to curb these problems, a National Energy Strategy for 2007–2020 was formulated and targeted at boosting reliance on homemade energy sources, from 4 to 40% at the end of the decade using shale oil and some sources of renewable energy, which include the wind and solar energy. Wind resource assessment is the most important step in planning a wind project, because it is the basis for determining initial feasibility, and cash flow projections, and is ultimately vital for acquiring financing. This paper assess the wind data available for a proposed site to examine the possibility of building 100 MW wind farm located in Ajloun north of Jordan. The wind data were analysed using Weibull Distributions for VESTAS V90 2.0 MW wind turbine which showed that the number of hours per year the turbine does not generate any electricity is around 1825 hours per year, which is about 21% of the whole year. While the wind turbines were generate electricity for most of the time during the studied year (around 79%). These results can be used for acquiring project finance.

Keywords

Wind farm • Wind resources • Weibull distributions • Wind assessments

M. Al-Addous · Z. Dalalah · F. Alawneh · A. Albatayneh (✉)
Energy Engineering Department, German Jordanian University, Amman, Jordan
e-mail: Aiman.Albatayneh@gju.edu.jo

H. Al-Taani
School of Basic Sciences and Humanities, German Jordanian University, Amman, Jordan

Nomenclature

C	Scale factor
IRR	Internal rate of return
K	Shape factor, higher k is better for wind farm than lower value of k
$P(U)$	The probability density function
U	Long-term average wind speed
$u(z)$	Wind speed at 80 m height
$u(z_r)$	Wind speed at 3 m height
Z_o	Surface roughness [some trees hedges $z = 1.5$ m (Vestas 2013)]
σ_u	Standard deviation of the individual wind speed averages

1 Introduction

In Jordan, the overdependence of the country on foreign import of oil consumes a large part of the country gross domestic product (GDP). This has made the government of the country to make an investment plan of $15 billion for renewable and nuclear energy. Also in an attempt to curb these problems, a National Energy Strategy for 2007–2020 was formulated and targeted at boosting reliance on homemade energy sources from 4 to 40% at the end of the decade using shale oil and some sources of renewable energy which include the wind and solar energy (Union of Concerned Scientists 2017).

Back in 2015, only three operational wind plants are available in the country, one at Ibrahimyah, at Haifa and Tafila. The Ibrahimyah plant is located at around 80 km north of Amman, and it has four wind turbines which can produce only 0.08 MW for each. The Haifa plant is however constructed within 92 km north of Amman, and it consists 0.225 MW capacity each of five wind turbines. The Tafila Wind Farm is located in Tafilah Governorate in southwest Jordan which generates 117 MW (IEA 2016).

C. Alalouch et al. (eds.), *Advanced Studies in Energy Efficiency and Built Environment for Developing Countries*, Advances in Science, Technology & Innovation, https://doi.org/10.1007/978-3-030-10856-4_18

A significant number of regions in Jordan are associated with considerable wind speed values for the generation of electricity. The greatest potential areas lie in the northern, central and southern parts of the country. Commissioned in 1988, Jordan's first wind farm had a gross installed capacity of 320 kW and was located in Ibrahimiya, close to Hofa. The project was implemented as a pilot study, and it showed successful technical and financial outcomes producing an average annual electrical generation of about 650 MWh. The experience gained from this project and the mutual cooperation with the German Government resulted in a new wind-electric power installation of 1.125 MW nominal capacity, in Hofa. This project was connected to the national grid in 1996 and became fully operational in 1997. According to published wind maps, there is good potential in Khamsheh, Fujeij, Tafilah, Wadi Araba and Ibrahimiya. Throughout those sites, some of the current wind generation power plants already exist and some pilot projects have been set up. In Ibrahimiya, for example, data is available and has been collected for the past 20 years, since the establishment of some small wind turbines in that area. The wind potential in Jordan is promising and cost-effective in many sites according to preliminary measurements and studies. However, there is an obvious need to develop a national wind atlas that covers all promising sites in the country.

The most commonly used distribution for the description of wind regimes is Weibull. The Weibull distribution is considered appropriate for the evaluation of temperate zone areas and, for an altitude of up to 100 m, and uses two main parameters, namely the scale c and the shape k factors, so as to determine the probability density $p(U)$ of a specific wind speed. More precisely, higher values of k correspond to lower dispersion of wind speeds and thus greater concentration of the latter around the mean wind speed. Moreover, it is found that regarding wind speed distributions, k normally takes values above 1 ($k > 1$) (Kaldellis 2008).

Wind energy is a clean and renewable source of energy. All fossil fuel by-products which include coal and oil produce carbon dioxide which depletes the atmospheric components. However, wind power has been of great help in reducing the emitted carbon dioxide, resulting in minimizing the effect of climate change (CEGC 2015).

Wind energy is and effective and consistent energy source. In addition, the maintenance and operating cost is relatively low when compared with other power sources and there is high efficiency and reliability in the use of wind turbines to generate electricity. And likewise, wind energy can be generated at low cost. There is a relatively low price for electricity generated by wind farms, and the price is competitive compared with gas-fired power plants, etc. Also, there is an expectation that it becomes lower as technology advances.

2 Methodology

The construction of the wind farm project is intended to go through various stages of assessment which starts with the initial assessment of the project, followed by the detailed site characterization, and then the validation of data collected and the final wind data assessment results can be used for acquiring finance.

2.1 Site Assessment

The proposed location for this project is Ajloun. It is a hilly town in the northern part of Jordan located 50 km from the capital of Jordan Amman. It is shown in Fig. 1.

In this location, there is an availability of a large expanse of land which can be used for the project; a road infrastructure extends toward the site and a nearby grid transmission line. The elevation of the proposed site is from 1000 to 1100 m above sea level including an excellent wind potential plus; i.e., there are good wind resources (average = 8.157 m/s). According to recorded data, it was observed that the resources have high wind energy potential.

There is already the presence of established infrastructure with grid transmission line very close to the place as shown in Fig. 2. The road construction that leads to the site will be useful for the construction and further maintenance of the site.

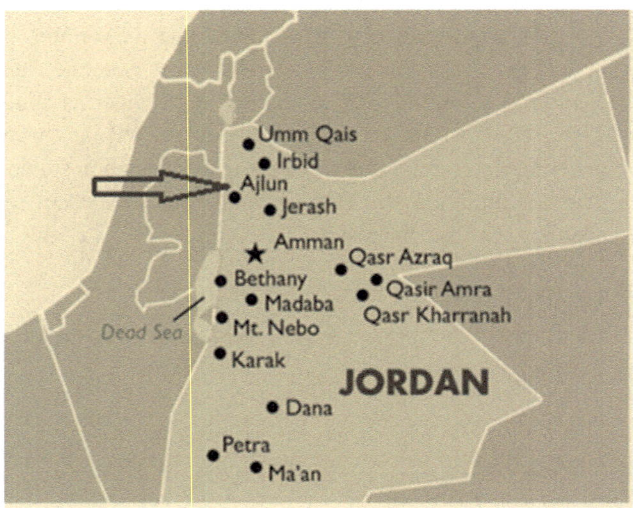

Fig. 1 Proposed location for the wind farm in Ajloun north of Jordan (Jordan Vacation 2017)

Fig. 2 Transmission lines and generators in Jordan (NEPCO 2017)

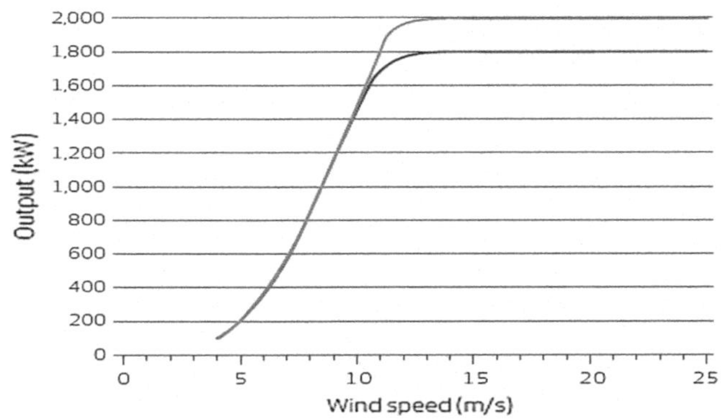

Fig. 3 Power curve for VESTAS V90 wind turbine, the blue curve for 2.0 MW turbine, and the red curve for 1.8 MW turbine

The terrain of the land has a few obstacles, and a few highlands that are present in the location will be useful for building the wind energy farm, as the plain surface will cause low turbulence and higher wind speed close to the ground level. The weather of the wind farm is favorable, with the absence of extreme weather condition which may damage the wind turbine. However, other issues which arise from the site include land lease from the owners, approval from all parties (landowners, government, council, and residents).

2.2 Initial Wind Resource and Wind Assessment

Weather data for the project sits is available for the years 1998–2011 through weather stations in Ajloun that were set at height of 3 m above the ground. In the proposed site, there is an adequate and regular supply of wind with annual mean wind speed of 19.9–23.1 km/h and mean temperatures between 21.5 and 11.2 °C.

The wind farm shall be designed for 100 MW. The available area is relatively small. Therefore, a higher-power-rated turbine is required. VESTAS V90 2.0 MW fulfills the requirements of the project. Figure 3 depicts the power curve of this wind turbine.

Wind speed is in a cubic relationship with output wind power, the wind turbine generates power only when wind speed is high, and it is between "cut-in wind speed (4 m/s) and cut-out wind speed (25 m/s)"; see Fig. 3.

For the following data analysis and the assessment of the wind energy potential, we base ourselves on the VESTAS V90 2.0 MW wind turbine. Wind speed is estimated at a height of 80 m (corresponding to the turbine hub height) where the turbine should be installed based on the local site characteristics.

The estimation of the wind speed bases on the logarithmic law. The wind data was recorded at around 3 m height, and the wind turbine's height is 80 m, so it is essential to extrapolate wind speed from 3 to 80 m height in order to acquire accurate power output.

Wind speed increases as the elevation from the ground increases. The proposed site of the wind farm consists of farmlands and few trees. The weather stations were constructed to be about 3 m high, using the logarithmic law to find the wind speed at the turbine height of 80 m (Manwell et al. 2009).

$$\frac{u(z)}{u(z_r)} = \frac{\ln\left(\frac{z}{z_o}\right)}{\ln\left(\frac{z_r}{z_o}\right)} \tag{1}$$

At the height of 80 m, the wind speed will be a multiplication of the wind speed at 3 m high by factor 1.39. It can be represented as [u (at 80 m) = 1.39 × u (at 3 m)], where u is representing the wind speed.

Using this result and the power curve of the wind turbine (Fig. 3), the electricity production at the site can be estimated based on the available wind data from 1998 to 2011. Figure 4 depicts this estimation graphically.

Based on Table 1, the expected average energy production per year and wind turbine is 8436.88 MWh or an average energy production of 0.96 MWh per hour.

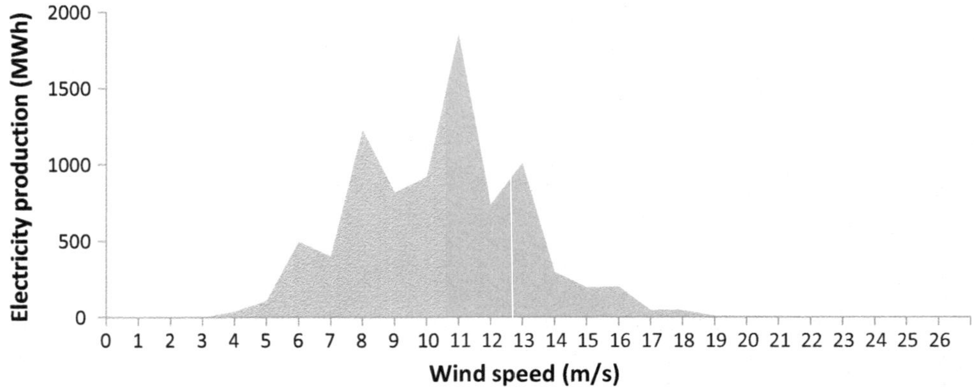

Fig. 4 Annual electricity productions MWh for each wind speed band

Table 1 Energy production estimation for a single 2 MW wind turbine

Wind speed band at turbine heights (m/s)	Number of hour per year for each wind speed band	Annual electricity production (MWh)
0	138.213	0.000
1	48.099	0.000
2	67.681	0.000
3	384.030	0.000
4	403.802	40.380
5	542.205	108.441
6	1382.129	497.567
7	728.897	400.894
8	1498.289	1228.597
9	694.297	819.270
10	612.548	918.821
11	976.236	1854.848
12	367.110	734.221
13	504.563	1009.125
14	148.669	297.338
15	99.049	198.099
16	101.521	203.042
17	24.335	48.669
18	23.954	47.909
19	4.943	9.886
20	3.422	6.844
21	3.992	7.985
22	0.570	1.141
23	1.711	3.422
24	0.190	0.380
25	0.000	0.000
26	0.000	0.000
Total	8760	8436.88

3 Results and Discussions

In order to examine the viability of building a 100 MW wind farm to be located in Ajloun, a place in the north of Jordan, there was an estimation of the power that can be generated from the wind farm, and then the infrastructure required to construct the turbines as well as the interconnections with the external transmission need to be introduced to the area. Likewise, some issues relating to the social and environmental impact of the wind farm need further investigations.

3.1 Power Generated

The following figure shows the data collected for power generation from wind turbine in the proposed location. Making an inference, from the result, the average energy production per year is seen to be 8436.88 MWh, which is equal to an average energy production of 0.96 MWh per hour. This is shown in Fig. 5.

3.2 Weibull Distributions

Statistical analysis is used to determine wind energy potential of a given site. If projection of measured data is required or only a small amount of data is available, statistical analysis has its advantages. The probability density function $p(U)$ can be used to describe the frequency of occurrence of wind speeds. The probability density function can be described as follows (Manwell et al. 2009).

$$p(U) = \left(\frac{k}{c}\right)\left(\frac{U}{c}\right)^{k-1} \exp\left[-\left(\frac{U}{c}\right)^{k}\right]$$

where

Long-term average wind speed	$\bar{U} = \frac{1}{N}\sum_{i=1}^{N} U_i = \frac{375{,}877}{46{,}080}$	=8.157 m/s	(2)
Standard deviation of the individual wind speed averages	$\sigma_U = \sqrt{\frac{1}{N-1}\sum_{i=1}^{N}(U_i - \bar{U})^2}$	=3.283	(3)
Shape factor, higher k is better for wind farm than lower value of k (which indicates a greater deviation away from mean wind speed which is untuneful energy because wind speed exceeds 25 m/s)	$K = \left(\frac{\sigma_U}{\bar{U}}\right)^{-1.086}$	=2.016	(4)
Scale factor	$c = \bar{U}\left(0.568 + \left(\frac{0.433}{k}\right)\right)^{\frac{-1}{k}}$	=9.2106 m/s	(5)

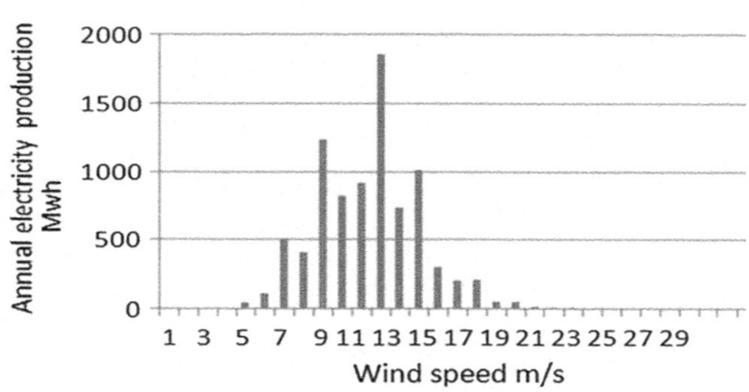

Fig. 5 Annual electricity productions MWh for each wind speed band

From Weibull distributions, the number of hours per year of wind speed is less than cut-in wind speed for the turbine, which is equal to 1821.159 h/year (20.789%) of the year the turbine was not used to generate electricity. Also, the number of hours per year of wind speed is higher than cut-out wind speed for the turbine which is 3.56 h per year (0.0407%) of the year the turbine was not used to generate electricity. This is a much desired value, and it is shown in Fig. 6.

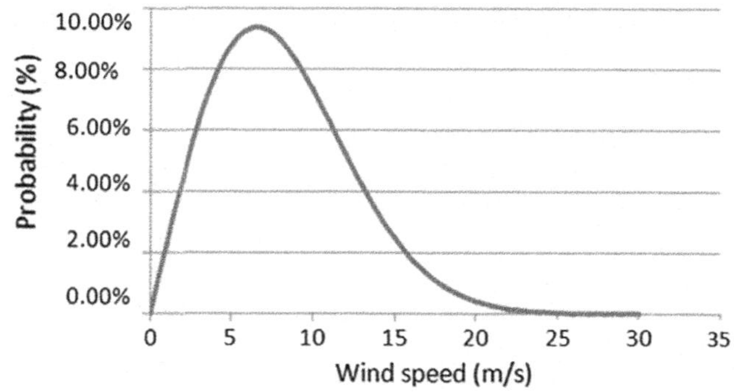

Fig. 6 Weibull distributions for the proposed location

3.3 Turbines' Layout

For the construction of a 100 MW wind plant, 50 units of the chosen 2 MW wind turbine are required. The layout of the array must take into account the site area and turbine characteristics to minimize array losses.

Array losses are bound by a maximum value of 10% if turbines are spaced 8–10 rotor diameters apart in the prevailing wind direction and 5 rotor diameters apart in the crosswind direction (Manwell et al. 2009). The rotor diameter of VESTAS V90 2.0 MW is 90 m (Vestas 2013). This results in:

Required distance in crosswind direction: 5 * 90 m = 450 m
Required distance in downwind direction: 10 * 90 m = 900 m.

The proposed location experiences mainly wind from southeast as shown in Fig. 7.

The project site has an area of 13,770,000 m^2. In order to guarantee the maximum array efficiency by respecting the minimum distances of 450 and 900 m between the turbines, the 50 turbines will be installed as depicted in Fig. 7 with 16 turbines on the first line, and 17 each for the second and third lines as shown in Fig. 8.

Wind power is a renewable energy source which has a promising future and can never be depleted. It has been used for a long-time past for generating electricity for local electricity supply and consumption. However, there are

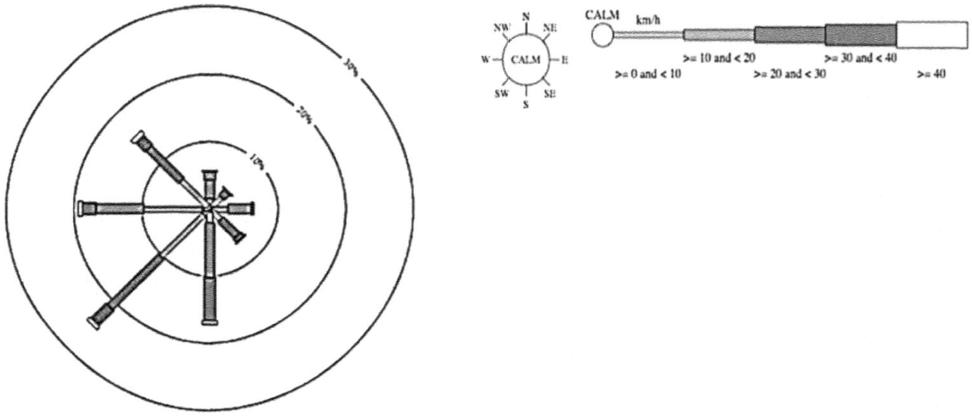

Fig. 7 Rose of wind direction versus wind speed in km/h

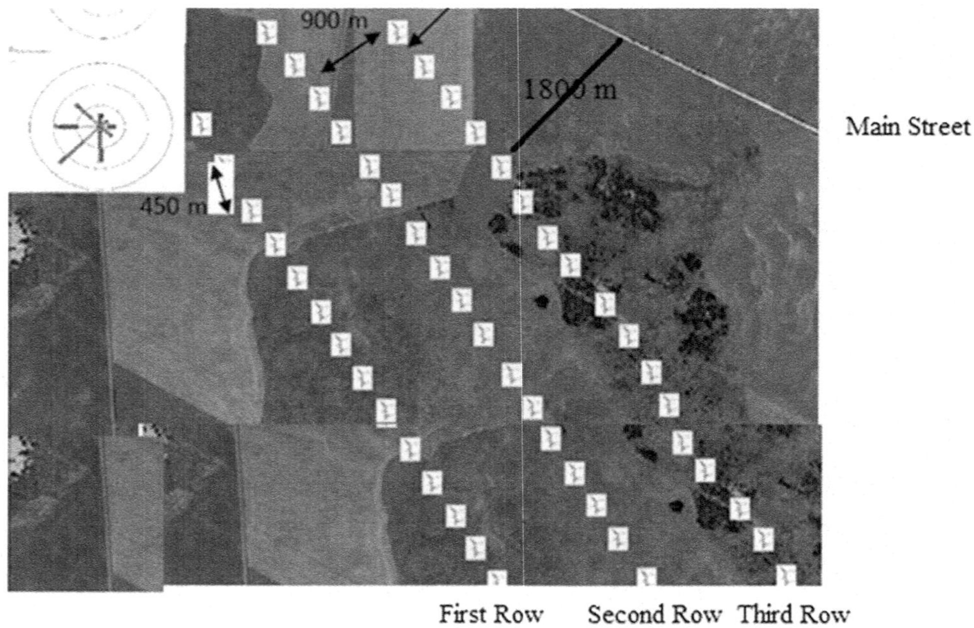

Fig. 8 Turbines' array for the proposed site

certain environmental issues which might arise from the construction of a wind farm, and they should be adequately considered.

3.4 Infrastructure Requirements

The selected location of the project is linked by a road extending from the highway. The site needs to have a distance more than 500 m from the road in order to minimize the noise effect and ensure the well-being of the people being around that place.

In order to meet the requirement for turbine erection, crane transportation and construction, the main access roads should be 10 m wide. Others roads which do not need to transport crane need to be around 6 m wide. For easy transportation, of the turbine blade, the horizontal and vertical alignments of the main access roads would be 45 m long, so civil work requires building the road for blade and crane transportation purpose. This is indicated in Fig. 9.

In order to connect the wind farm to the grid, a specific infrastructure is required. This comprises of a substation and converter system; the substation will be used in the conversion of the electricity from 33 to 220 kV. There will also be a control room and switch room. There will be two substations, one located at the northwest corner and the second at the southeast corner. This is shown in Fig. 10.

On transmission of energy, the power from the turbine will be transmitted to the substation via a 33 kV cable. Most of the cables will be buried along the road, while some of them need to be above the ground for specific conditions. Turbines in each row are connected due to their same output and then to the substations. The detailed interconnection is as represented in Fig. 11. The red lines show the interconnection between each turbine and connect to the substation. The external transmission connects the substations of the

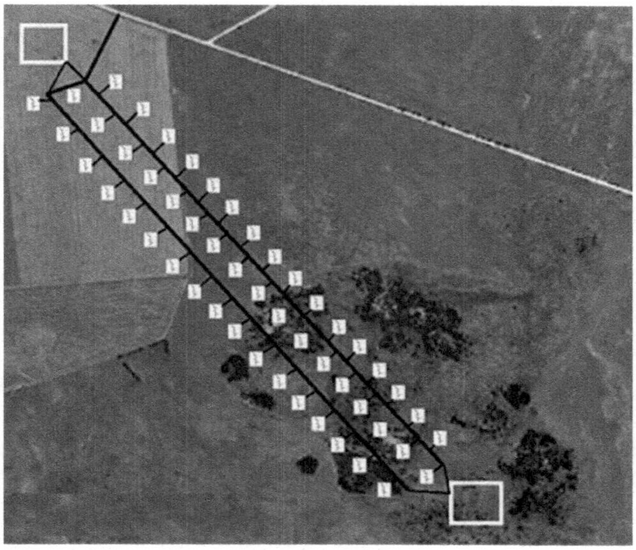

Fig. 9 Inner road layout of wind farm

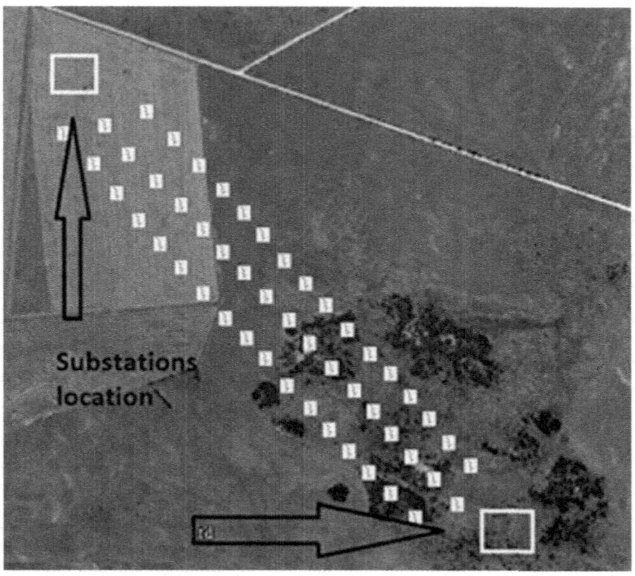

Fig. 10 Two substations' location for the proposed wind farm

proposed farmland to the national grid substation. It is possible to connect to the national grid directly.

A network of computer systems, signal cables from the turbine, transducers on the masts as well as electrical transducers will be installed and used for communication, control, and data collection.

3.5 Assessment of the Proposed Design

The wind farm consists of 50 wind turbines of type VESTAS V90 2.0 MW which are installed (see Sect. 4) at the required

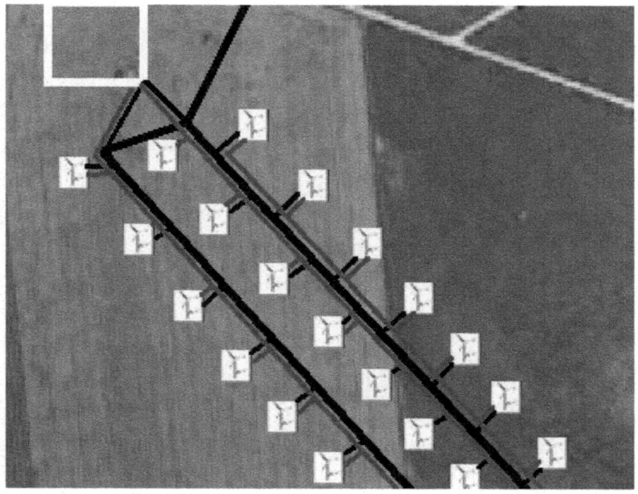

Fig. 11 Interconnection layout of turbines

distance to achieve an efficiency of above 90%. The efficiency includes array efficiency (mutual interference among turbines) and availability of programmed maintenance. Based on the assumed annual power generated by an isolated turbine of 8436.878 MWh (see Table 1), we can calculate the expected total energy of the whole wind farm as

$$90\% * (8436.878 \text{ MWh} * 50) = 379.65951 \text{ GWh}$$

3.6 Social and Environmental Issues Associated with Wind Power

Wind power is a renewable energy source known for its intermittent availability. It has been in use for decades for generating electricity for local energy supply and local consumption. However, there are certainly environmental issues which might arise from the construction of a wind farm, and they should be adequately considered, some of which includes as following.

3.6.1 Noise

Noise pollution has always been a big problem to the existence of wind farm. However, evidence has shown that certain measures can be put in place to bring it to the barest minimum. The noise produced by wind turbines has two distinct sources: the aerodynamic and mechanical ones. The first type is caused by interaction between rotor blades and relative wind, and the second type is due to mechanical parts of the system such as gears. The sound level depends on two main factors which include design and wind speed. Although there is no evidence to show that noise produced by wind turbines is harmful to human's health, we still need to ensure a reasonably low noise level.

It has been proven by modern wind turbine design that mechanical noise is insignificant and can be reduced by changing blade design and operation, so the dominant noise caused by wind turbines is aerodynamic (Kunz et al. 2007). This type of noise is ranging from infrasound to normal audible range and can be minimized by careful design of the blades by manufacturers, such as changing the blade pitch or the shape of the trailing edge. For example, the sound level for a 1 MW wind turbine at a distance of 300 m from it is 45 dBA.

The chosen VESTAS V90 2 MW turbine is operated with innovative technology and is built to reduce sound production and energy consumption effectively. When the electrical components and the moving parts are cooled down using some kind of heat exchangers, the turbine has low noise production level and could be operated in defined sound levels. It has been chosen to reduce the noise level of the planned farm. In addition, the area around the wind farm is only sparsely occupied.

3.6.2 Visual Impact

The health impact of visual burdens cannot be underestimated. Rotating wind turbine blades interrupt the sunlight producing unavoidable flicker bright enough to pass through closed eyelids, and moving shadows cast by the blades on windows can affect illumination inside the building. Though it is difficult to set standards to be accepted by everyone, it seems quite essential to be supported by local communities, especially in some areas where the landscape amenity is of high value (Bolton 2007). Some studies have shown that the wind farms are more acceptable visually to people who have been informed of the benefits derived from their use. It also has been indicated that farming and livestock are unaffected by existing wind farms, and nearly 99% of the land is available for farming or other uses.

The constantly changing light intensity triggered by moving blades is the cause of shadow flickers. It produces shadow which is cast on stationery objects like windows at the dwelling (Kunz et al. 2007). This shadow is also cast on land and houses when the turbines rotate having the sun behind it (Knopper and Ollson 2011). This may lead to dizziness for people living around there, and it also causes nausea when they look at the movement of the blades. However, this project is not affected by this as only a few people live close to the wind farm. Also, the impact of shadow flickers on roads should be considered, but because the location is 2 km away from the highway, it is of minor importance.

3.6.3 Wildlife

The great danger may be given by the transmission lines and wind turbines to wildlife especially birds considering their species, location, and time. The danger may come in three ways (Kunz et al. 2007); they are displacement of the birds, either temporarily or permanently, and they are sent away from the site. In another way, the birds may collide with rotating blades of the turbine, and this can cause serious injury or death of the birds. Also, noise from construction activities and rotation of turbines may have negative effect on the birds (European Wind Energy Association 2012). Sometimes, local birds are also affected when feeding, breeding, and laying eggs.

The birds can actually avoid collision with turbine blades, and they still continue to breed and feed when construction is going on. However, if the migration route of the birds passes through the project site, it may pose a serious danger to the birds. Nowadays, the slowly turning blades and lack of perching potential have minimized this (Devabhaktuni et al. 2010). Most birds travel when wind speeds are not too high where the turbines are off when wind speeds are low which minimize the time of potential impact between birds and the blades. Radar also can be used to detect when birds are approaching; turbines could be slowed or shut down when the radar determines birds are within a certain zone.

3.7 Economic Analysis

After assessing the technical feasibility as well as the socio-environmental issues related to the planned wind farm, an economic analysis is required before a final decision regarding the project can be taken.

Table 2 Electricity selling prices, profitability, and IRR

Electricity selling price ($/MWh)	Annual return ($)	Net profit/losses ($)	IRR (%)
20	7,593,190.20	−3,806,809.80	−3
25	9,491,487.75	−1,908,512.25	0
30	11,389,785.30	−10,214.70	3
35	13,288,082.85	1,888,082.85	6
40	15,186,380.40	3,786,380.40	9
45	17,084,677.95	5,684,677.95	11

The following are the projected cost of constructing the wind farm:

Installation cost 1.5 $/W (operation: 2.5%, other capital: 7%, site works: 24.5%, wind turbines: 66%).
Operation and maintenance costs—annual costs for wind turbines 2.5% of the original turbine cost ($3.75M/year).
Capital cost: 30% from investors ($45M), 70% bank loan ($105M), interest rate 7%, and inflation 3%.
Time of operation in years: 20 years.

The profitability of the wind farm in the proposed site depends on the electricity selling price as shown in Table 2.

Any average price less than $30.03/MWh will lead to losses from the capital cost while higher than $36.65/MWh (IRR = 7%) or more will be profitable based on the assumptions mentioned above based on the existing roads with no extra costs to build any roads.

4 Conclusions

Jordan is a country that is so much dependent on foreign energy sources, with 96% of an average of the country's energy need derived from oil and natural gas importation from middle-eastern countries. This has made the government of the country to make an investment plan of $15 billion for renewable and nuclear energy. Also in an attempt to curb these problems, a National Energy Strategy for 2007–2020 was formulated and targeted at boosting reliance on homemade energy sources from 4 to 40% at the end of the decade using shale oil and some sources of renewable energy which include the wind and solar energy.

The appraisal of the wind farm revealed that in about 79% of the whole year, this turbine generates electricity, and for 20.8%, it generates electricity on full rated power 2 MW. The wind farm as proposed to be located in Ajloun, north of Jordan, can be highly profitable with an annual expected electricity production of 379,660 MWh/year.

The estimated annual electricity production is around 380 GWh/year. Assuming an average selling electricity price of $30.027/MWh which is highly competitive compared with the price of other renewable energy sources, the wind farm is economical feasible. Geotechnical, environmental, and safety requirements were outlined which do not give any counter argument against proceeding with the project. The designs of the turbine array as well as the required roads on the farm were equally laid out. After the basic design was performed and the assessment successfully concluded, it is recommended to pursue the project. There have been some issues with geotechnical, environmental safety, and permitting review concerning proceeding with the project required further investigations.

Acknowledgements The authors acknowledge the financial support of the Deanship of Scientific Research at the German Jordanian University for the presented work in this paper.

Appendix

Technical specifications for VESTAS V90 2.0 MW

Power regulation	Pitch regulated with variable speed		**Main dimensions**	
Operating data	**IECIIA**	**IECIIIA**	**Blade**	
Rated power	1800 kW	2000 kW	Length	4.4 m
Cut-in wind speed	4 m/s		Max. chard	3.5 m
Rated wind speed	12 m/s		Weight	6700 kg
Cut-out wind speed	25 m/s			
Frequency	50 Hz/60 Hz		**Nacelle**	
Operating temperature	Standard range −20 to 40 °C		Height for transport	4 m
			Height installed (including cooler top™)	5.4 m
	Low-temperature option −30 to 40 °C		Length	10.4 m
			Width	3.4 m
Sound power			Weight	70 metric tonne
(10 m above ground, hub height 80 m air density 1225 kg/m^2)				
4 m/s	94.4 dB(A)		**Hub**	
5 m/s	99.4 dB(A)		Max. diameter	3.3 m
6 m/s	102.5 dB(A)		Max. width	4 m
7 m/s	103.6 dB(A)		Length	4.2 m
>8 m/s	104 dB(A)		Weight	18 metric tonne
Rotor			**Tower**	
Rotor diameter	90 m		**80 m** weight	148 metric tonne
Swept area	6362 m^2		**95 m** weight	206 metric tonne
Nominal revolutions	14.5 rpm		**105 m** weight	245 metric tonne
Operational interval	9.3–16.6 rpm		**125 m** weight	335 metric tonne
Air brake	Full balance feathering with three pitch cylinders			
Tower				
Type	Tubular steal tower		Power curve V90-1.8/2 MW	
Hub heights	80, 95, and 105 m (IECIIA)		Noise-reduced sound power modes are available	
	95, 105, and 125 m (DIBt I)			

References

Union of Concerned Scientists. (2017). Barriers to Renewable Energy Technologies. [online] Available at: http://www.ucsusa.org/clean_energy/smart-energy-solutions/increase-renewables/barriers-to-renewable-energy.html [Accessed 31 Mar. 2017].

IEA, key energy statics, Country specific indicator numbers 2016 available on; https://www.iea.org/publications/freepublications/publication/KeyWorld2016.pdf.

Central Electricity Generating Company, Wind Energy, available on; http://www.cegco.com.jo/?q=en (2015).

Kaldellis, J. K. (2008). The wind potential impact on the maximum wind energy penetration in autonomous electrical grids. Renewable Energy, 33(7), 1665–1677.

Jordan Vacation, Tours, Travel Packages & Experiences. Goway Travel. N.p., n.d. Web. 13 Apr. 2017. Ajloun north of Jordan.

Transmission lines and generators in Jordan." National Electric Power Company - NEPCO HomePage. N.p., n.d. Web. 13 Apr. 2017.

Manwell, J.G. McGowan, A.L. Rogers, Wind Energy Explained: Theory, Design and Application, John Wiley & Sons Ltd. (2009).

Vestas, 2 MW wind turbine, available on; https://www.vestas.com/en/products/turbines/v90%202_0_mw#!options-available (2013).

Bolton, R. H. "Evaluation of Environmental Shadow Flicker Analysis for "Dutch Hill Wind Power Project"." Environ. Compliance Alliance, New York (2007).

Kunz, Thomas H., et al. "Assessing impacts of wind-energy development on nocturnally active birds and bats: a guidance document." Journal of Wildlife Management 71.8 (2007): 2449–2486.

Knopper, Loren D., and Christopher A. Ollson. "Health effects and wind turbines: A review of the literature." Environmental Health 10.1 (2011): 78.

European Wind Energy Association. Wind energy-the facts: a guide to the technology, economics and future of wind power. Routledge, 2012.

Devabhaktuni, Vijay, et al. "Wind energy: Trends and enabling technologies." Proceedings of World Energy Congress. 2010.

Printed by Printforce, the Netherlands